Algorithms for Intelligent Systems

This book series publishes research on the analysis and development of algorithms for intelligent systems with their applications to various real world problems. It covers research related to autonomous agents, multi-agent systems, behavioral modeling, reinforcement learning, game theory, mechanism design, machine learning, meta-heuristic search, optimization, planning and scheduling, artificial neural networks, evolutionary computation, swarm intelligence and other algorithms for intelligent systems.

The book series includes recent advancements, modification and applications of the artificial neural networks, evolutionary computation, swarm intelligence, artificial immune systems, fuzzy system, autonomous and multi agent systems, machine learning and other intelligent systems related areas. The material will be beneficial for the graduate students, post-graduate students as well as the researchers who want a broader view of advances in algorithms for intelligent systems. The contents will also be useful to the researchers from other fields who have no knowledge of the power of intelligent systems, e.g. the researchers in the field of bioinformatics, biochemists, mechanical and chemical engineers, economists, musicians and medical practitioners.

The series publishes monographs, edited volumes, advanced textbooks and selected proceedings.

All books published in the series are submitted for consideration in Web of Science.

More information about this series at https://link.springer.com/bookseries/16171

Basant Agarwal · Azizur Rahman ·
Srikant Patnaik · Ramesh Chandra Poonia
Editors

Proceedings of International Conference on Intelligent Cyber-Physical Systems

ICPS 2021

Springer

Editors
Basant Agarwal
Department of Computer Science and Engineering
Indian Institute of Information Technology
Jaipur, Rajasthan, India

Srikant Patnaik
Department of Computer Science and Engineering
SOA University
Bhubaneswar, India

Azizur Rahman
School of Computing and Mathematics
Charles Sturt University
Wagga Wagga, NSW, Australia

Ramesh Chandra Poonia
Department of Computer Science
CHRIST (Deemed to be University)
Bangalore, Karnataka, India

ISSN 2524-7565 ISSN 2524-7573 (electronic)
Algorithms for Intelligent Systems
ISBN 978-981-16-7135-7 ISBN 978-981-16-7136-4 (eBook)
https://doi.org/10.1007/978-981-16-7136-4

This Springer imprint is published by the registered company Springer Nature Singapore Pte Ltd.
The registered company address is: 152 Beach Road, #21-01/04 Gateway East, Singapore 189721, Singapore

To all who have supported, contributed, and participated in the ICPS-2021.

Preface

This book constitutes the selected papers of the International Conference on Intelligent Cyber-Physical Systems (ICPS-2021), held on 24–26 June 2021, in Jaipur, Rajasthan, India.

The International Conference on Intelligent Cyber-Physical Systems (ICPS-2021) aims to showcase advanced technologies, techniques, innovations, and equipment in Cyber-Physical Systems. It provides a platform for researchers, scholars, experts, technicians, government officials, and industry personnel worldwide to discuss and share their valuable ideas and experiences. ICPS-2021 has acted as a significant international platform to share and demonstrate novel approaches, state of arts interdisciplinary research development, technology improvement, and considerable scientific debates among academic researchers, students, industry developers, and practitioners. It covered a wide range of essential topics with various plenary and invited keynote talks by internationally distinguished speakers.

ICPS-2021 was held in a virtual mode at the Indian Institute of Information Technology, Kota (IIIT, Kota) in the Rajasthan State of India. The main keynote lectures were delivered by Prof. Seeram Ramakrishna, Vice President, Research Strategy, NUS, Singapore, Prof. Azizur Rahman, Leader, Statistics and Data Mining Research Group, School of Computing and Mathematics, Charles Sturt University, Australia, and Professor Prashant Jamwal, Nazarbayev University.

The ICPS-2021 received 105 submissions from all over the world. At least three reviewers comprised of experts from the technical program committee and the external review panel have carefully reviewed each paper. After the rigorous review process, only 29 high-quality papers were accepted for presentation at the conference and publication in the book proceedings.

The editors would like to thank all the reviewers for their time and sincere efforts. We are also thankful to the management of the Indian Institute of Information Technology, Kota (IIIT, Kota), Rajasthan, India, for providing the best infrastructure and required logistics to organize the conference in virtual mode. We also wish to express our sincere gratitude to Profs. Azizur Rahman and Seeram Ramakrishna for accepting our invitation to give keynote addresses. Furthermore, a special appreciation from the IIIT, Kota, is due to Prof. Azizur Rahman, Charles Sturt University, Australia, for his outstanding international mentorship supports. Finally, we are also very thankful to Springer for supporting ICPS-2021.

We hope that this book would be handy for the researchers working in the relevant areas.

Jaipur, Rajasthan, India — Basant Agarwal
Wagga Wagga, NSW, Australia — Azizur Rahman
Bhubaneswar, India — Srikant Patnaik
Bangalore, India — Ramesh Chandra Poonia
August 2021

Contents

About the Editors

Dr. Basant Agarwal is working as an Assistant Professor at the Indian Institute of Information Technology Kota (IIIT-Kota), India, which is an Institute of National Importance. He holds a Ph.D. and M.Tech. from the Department of Computer Science and Engineering, Malaviya National Institute of Technology Jaipur, India. He has more than 9 years of experience in research and teaching. He has worked as a Postdoc Research Fellow at the Norwegian University of Science and Technology (NTNU), Norway, under the prestigious ERCIM (European Research Consortium for Informatics and Mathematics) fellowship in 2016. He has also worked as a Research Scientist at Temasek Laboratories, National University of Singapore (NUS), Singapore. His research interest is in Artificial Intelligence, Cyber-physical systems, Text mining, Natural Language Processing, Machine learning, Deep learning, Intelligent Systems, Expert Systems and related areas.

Prof. Azizur Rahman, Ph.D. is an applied statistician and data scientist with expertise in developing and applying novel methodologies, models and technologies. He is the Leader of the "Statistics and Data Mining Research Group" at Charles Sturt University, Australia. He can assist in understanding multi-disciplinary research issues within various fields, including understanding the individual activities that occur within very complex scientific, behavioural, socio-economic, and ecological systems. Prof. Rahman develops "alternative methods in microsimulation modelling technologies", which are handy tools for socio-economic policy analysis and evaluation. He has more than 120 publications, including a few books. His 2020 and 2016 books have contributed significantly to the fields of "data science and policy analysis" and "small area estimation and microsimulation modelling", respectively. The Australian Federal and State Governments fund his research, and he serves on a range of editorial boards, including the International Journal of Microsimulation (IJM) and Sustaining Regions. He received several awards, including the SOCM Research Excellence Award 2018 and the CSU-RED Achievement Award 2019.

Dr. Srikant Patnaik is presently working as Director of International Relations and Publication of SOA University. He is a full professor in the Department of Computer

Science and Engineering, SOA University, Bhubaneswar, India. He received his Ph.D. (Engineering) in Computational Intelligence from Jadavpur University, India, in 1999. He has supervised more than 25 Ph.D. Theses and 60 Master theses in the areas of Computational Intelligence, Machine Learning, Soft Computing Applications and Re-Engineering. Dr. Patnaik has published around 100 research papers in international journals and conference proceedings. He is the author of 2 textbooks and 52 edited volumes, and few invited book chapters published by leading international publishers like Springer-Verlag, Kluwer Academic, and other. Dr. Srikant Patnaik has been serving as the Editors-in-Chief of the International Journal of Information and Communication Technology and International Journal of Computational Vision and Robotics, published from Inderscience Publishing House, England; and International Journal of Computational Intelligence in Control, published by MUK Publication; the Editor of the Journal of Information and Communication Convergence Engineering; and Associate Editor of the Journal of Intelligent and Fuzzy Systems (JIFS), which are all Scopus Index. . He is also Editors-in-chief of Book Series on "Modeling and Optimization in Science and Technology" published from Springer, Germany.

Dr. Ramesh Chandra Poonia is an Associate Professor at the Department of Computer Science, CHRIST (Deemed to be University), Bangalore, India. He recently completed his Postdoctoral Fellowship from CPS Lab, Department of ICT and Natural Sciences, Norwegian University of Science and Technology, Ålesund, Norway. He received his Ph.D. degree in Computer Science from Banasthali University, Banasthali, India, in July 2013. He has published more than 60 research articles in refereed journals and international conferences and edited six books and five conference proceedings. His areas of interest are Sustainable Technologies, Cyber-Physical Systems and Intelligent Algorithms for Autonomous Systems.

Micro Phasor Measurement Unit (μPMU) in Smart Distribution Network: A Cyber Physical System

Santoshkumar Hampannavar, M. Swapna, B. Deepa, and Udaykumar Yaragatti

1 Introduction

The recent advancements in power system and restructuring policy allowed private players to participate in power generation due to which rapid penetration of distributed energy resource (DER) is observed in the distribution network (DN). Distributed generators (DG), electric vehicle (EV) and other interactive loads and some new features emerge in DN. Due to the intermittent nature of renewable sources like Photovoltaic (PV) and Wind, the presence of a battery energy storage system (BESS) becomes crucial. India holds 5th position in renewable energy installed capacity in the world and the government has taken aggressive steps to top the list. In line with this, the wind energy installed capacity is 38GW as of today and the National Institute of Wind Energy (NIWE) proposed that 302GW wind energy can be extracted from the windy states of India. Solar energy installed capacity is around 41GW and the state of Karnataka tops the list in India. Apart from solar and onshore wind, the Ministry of Power (MoP) and NIWE have ventured to take up offshore wind projects (ZERO to 5GW) in the states of Tamilnadu and Gujarat. It is clear from the developments that the Indian government is showing tremendous interest in promoting renewable energy. This gives rise to the rapid penetration of renewable sources in the distribution network which makes it difficult for the distribution system operator (DSO) to monitor and control. The intermittent nature of DGs need

S. Hampannavar (✉)
School of Electrical and Electronics Engineering, REVA University, Bengaluru, India

M. Swapna
Department of Electrical Engineering, National Institute of Technology , Silchar, India

B. Deepa
Department of Electrical and Electronics Engineering, RYM Engineering College, Bellary, India

U. Yaragatti
Director, Malaviya National Institute of Technology, Jaipur, India

B. Agarwal et al. (eds.), *Proceedings of International Conference on Intelligent Cyber-Physical Systems*, Algorithms for Intelligent Systems,
https://doi.org/10.1007/978-981-16-7136-4_1

large-scale battery storage systemand in this case EV batteries can serve the purpose. EVs act as a distributed source or a connected load. A group of EVs used for power transactions with the grid is known as Vehicle-to-Grid (V2G) [1–4] which supports shaving peak, filling valley, load levelling and provides support to the grid. A conceptual framework of EV aggregation as a Cyber Physical System (CPS) is shown in Fig. 1 [5–7].

Some DN features include intermittent and random power sources, changing operating modes of the power grid and islanding morphology which complicate the operation of DN. DNs are complex due to the presence of multiple nodes, short distances, amplitude and angle difference in between the nodes is smaller, dynamically changing and lack of standard documentation. Due to these complexities, dynamic monitoring system with high accuracy and precision for situational awareness is preferred. Most of the utilities worldwide use SCADA to monitor DN. SCADA suffers from the data rate which is around 2–4 samples/sec, whereas PMUs work at higher data rates with high-level accuracy and has exhibited excellent dynamic performance. For real-time monitoring and assessment, PMUs are used in WAMS [8]. EVs have a significant impact on the grid if there is no coordinated/uncoordinated charging and it might lead to grid instability. So, coordinated charging can curb this issue. There are four modes

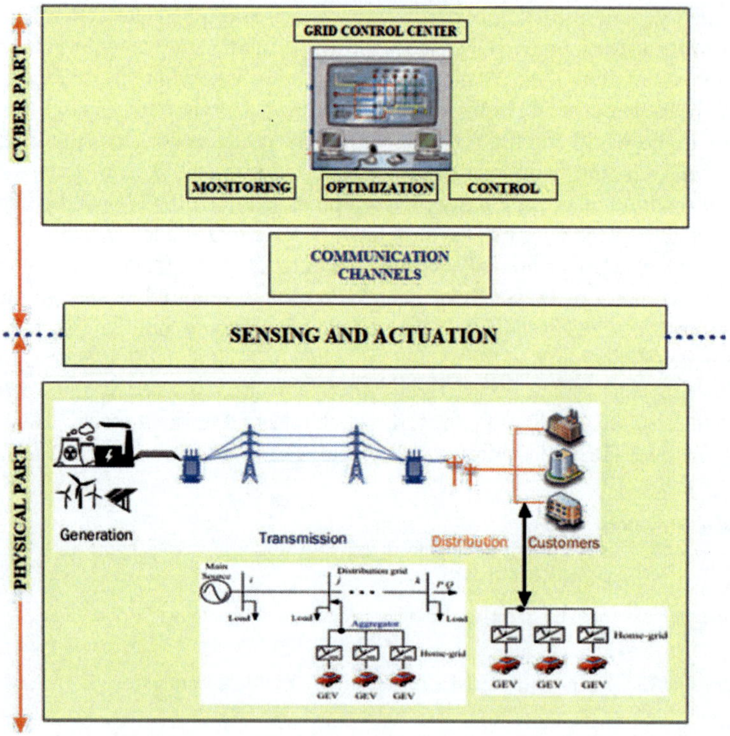

Fig. 1 EV Aggregation as a Cyber Physical System

of operation in each of the quadrants. In this work, we propose EVs to function in the fourth quadrant in which active power (P) is consumed and reactive power (Q) is injected into the grid. This strategy helps the grid to meet the time-varying intermittent load demand. In the existing literature, the application of multiagent-based μPMU communication is not reported and still unexplored for coordinated charging in the distribution network. Agent-based μPMU communication is proposed for coordinated EV charging in smart distribution network where EVs operate in the fourth P-Q quadrant.

2 Phasor Measurement Unit (PMU)

The data sensed and processed is time-stamped and sent to PDC located at the control center through the communication network as shown in Fig. 2 [3–5]. PMUs have an excellent dynamic performance which could be used to solve challenges in the distribution network. A low-cost micro PMUs (μPMU) are developed and are used in the distribution network for monitoring and surveillance. μPMU are very handy in addressing distribution network issues as compared to conventional PMUs used in transmission network [6–15]. The information of each PMU is being communicated to the PDC and Energy Management System, (EMS)/WAMS using communication links. Communication links, sensors and actuators are prone to and highly vulnerable to cyber physical attacks.

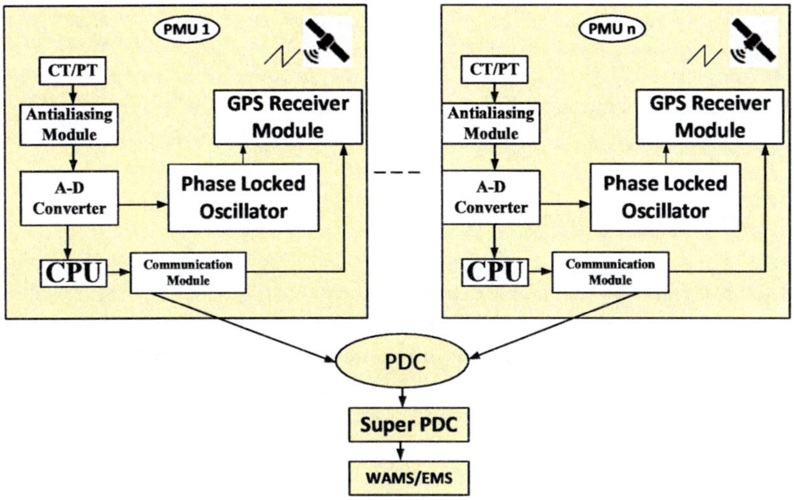

Fig. 2 Phasor Measurement Unit

3 EV Aggregation in Smart Distribution Grid

EVs are widely accepted and have great potential to address demand response due to their flexibility and sizeable power rating. V2G supports the grid requirements and will be an effective tool to manage the time-varying load demand [16]. Aggregator agents act as an interface between EV owners and the DSO and facilitate interested EV owners to participate in V2G operation by providing suitable resources such as parking lot, charging slot with free parking space.

The grid and load balance equation is given by

$$P_G + \sum_{n=1}^{N} P_{EV,i,k,disch\arg e} = P_L + \sum_{n=1}^{N} P_{EV,i,k,ch\arg e} \quad (1)$$

where P_G and P_L generation plant power and grid load.

$P_{EV,i,k,disch\arg e}$ and $P_{EV,i,k,ch\arg e}$ EV discharging and charging rate.

SOC of EV battery is given by

$$SOC_{i,e,t} = SOC_{i,e,t-1} + \eta_{i,e} \frac{P_{i,e,t}^{EV} \Delta k}{E_{i,e}^{\max}} \quad (2)$$

$$\begin{aligned} &SOC_{\min} \leq SOC_{EV,i,k} \leq SOC_{\max} \\ &SOC_{EV} \leq SOC_{EV,\min};\ ch\arg ing \\ &SOC_{EV} \geq SOC_{EV,\max};\ disch\arg ing \end{aligned} \quad (3)$$

where

η charging efficiency;$E^{\max}$ EV capacity; t time.

3.1 Distribution Grid Model

Distribution grid is a very complex structure due to the participation of many entities such as DGs, EVs and other loads. Flexible loads (FL) help the grid in flattening the load curve, and the objective of the grid is to maximize the FL penetration [17], and is given by

$$O = \sum_{i,k} P_{i,k}^{FL} \quad (4)$$

where

$$P_{i,k}^{FL} = \Re(V_{i,k} \overline{I}_{i,k}^{FL})$$

$$Q_{i,k}^{FL} = \Im(V_{i,k}\overline{I}_{i,k}^{FL}) \tag{5}$$

DSO sends P^{FL} and Q^{FL} to the aggregators at each node i.

Fairness index for the equal proportion of EV load penetration corresponding to the base loads at all nodes

$$F_k = \frac{P_{i,k}^{FL}}{P_{i,k}^{zl} + P_{i,k}^{ml} + P_{i,k}^{pl}} \tag{6}$$

The inequality constraints are

$$V_i^{\min} \leq |V_{i,k}| \leq V_i^{\max} \tag{7}$$

To operate EV in the fourth quadrant, the following limits are imposed on net reactive power dispatch

$$-Q_{i,k}^{\max} \leq Q_{i,k}^{FL} \leq 0 \tag{8}$$

3.2 EV Load Model

EVs arrival to the charging station is random and highly stochastic in nature in nature. Objective function for EVs charging total cost minimization

$$\psi_m = \sum_k \rho_k \sum_e P_{i,e,k}^{EV} \Delta k \tag{9}$$

where, ρ is energy price, e is EV number, EV is load and Δk is time interval.

EVs consuming power must follow:

$$P_{i,e,k}^{EV\ 2} + Q_{i,e,k}^{EV\ 2} \leq R_{i,e}^2 \tag{10}$$

where, R is rating of charging slot.

For EVs to operate in the fourth quadrant

$$P_{i,e,k}^{EV} \geq 0$$
$$Q_{i,e,k}^{EV} \leq 0$$

Grid constraints are incorporated by

$$\sum_{e} P_{i,e,t}^{EV} \leq P_{i,t}^{FL}$$
$$\sum_{e} Q_{i,e,t}^{EV} \geq Q_{i,t}^{FL} \tag{12}$$

3.3 Communication Framework for μPMU-Based Coordinated EV Charging

DSO communicates with μPMUs and exchanges the crucial information for EV integration to the grid such that the P is consumed and Q is injected to avoid adverse effects on the grid. Uncoordinating charging give rise to feeder losses, voltage deviation and overloading distribution transformers. Research studies demonstrate that 45% penetration of EVs leads to about 50% transformer overloading and 25% increase in losses in the distribution network. It is also reported in the literature that 40% uncoordinated EV penetration will result in the replacement of 50kVA transformers. EVs can supply/consume Q at any SOC. Figure 3 shows μPMU based coordinated EV charging.

Multi-agent system (MAS) communication reduces the computational burden and avoids overhead bits. A conceptual view of the proposed communication framework is shown in Fig. 4. The agent-based μPMUs are placed at each bus where EV aggregators are connected [18]. The static agents (SA) will communicate this information

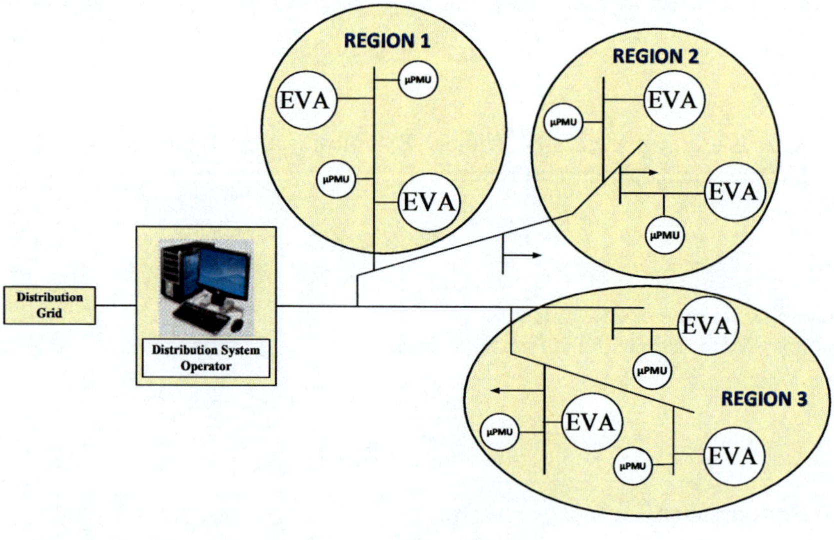

Fig. 3 Framework for μPMU based coordinated EV charging

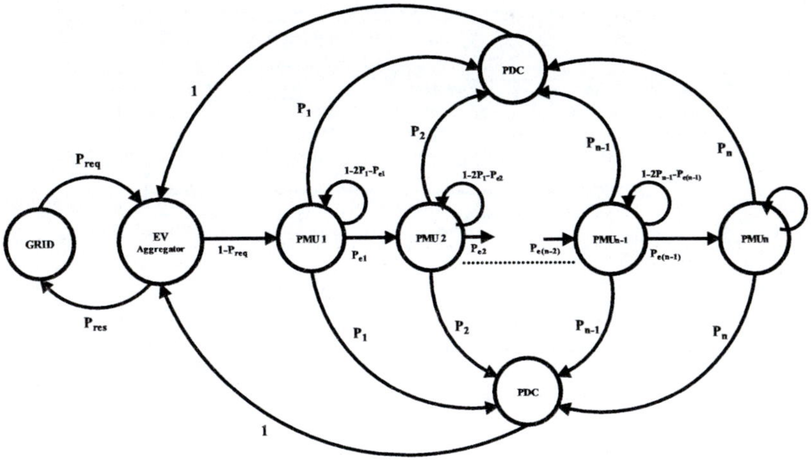

Fig. 4 Agent-based analytical model

to the mobile agent (MA) and send the same information to PDC. PDC will forward it to the DSO. DSO will send this information to the grid. EVAs are apprised of the P consumption and Q injection bounds. The agents proposed in this model are grid agent (GA), DSO Agent (DSOA), Electric Vehicle Aggregator agent (EVAA) [19–22].

Mobile Agent (MA): DSO periodically generates MA and is sent to every μPMU requesting for the power transaction with the grid. After nth PMU MA dies.

Static Agents (SA): Available at both PMU and PDC.

PMU Agent (PMUA): PMUA provides information about power exchange with the grid to PDC.

DSO Agent (DSOA): DSOA is a static agent (SA) which estimates all parameters associated with power and voltage. It will estimate the amount of power that can be with the grid and subsequently send this response to GA.

Grid Agent (GA): GA receives the information from AA and acts based on the load's requirement.

4 Results and Discussion

The mathematical models were developed in General Algebraic Modeling Language (GAMS) and solved using CIPLEX and KNITRO solvers. The case study was carried out on IEEE 13 bus test systems. In coordinated and controlled charging, DSO provides bounds (P and Q) to EVA at each node. Mobile C was interfaced with

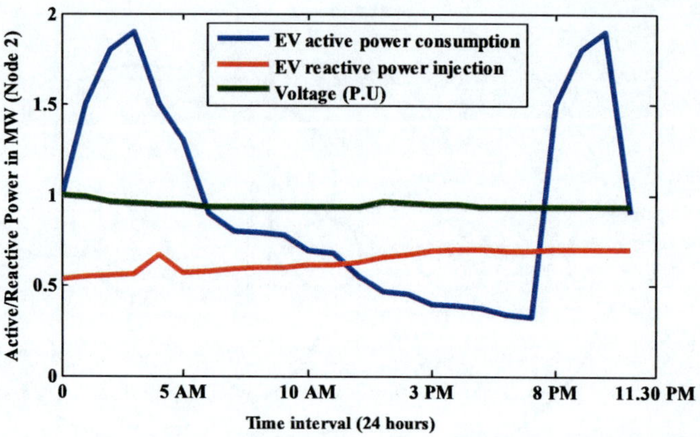

Fig. 5 Coordinated EV charging with EVA at bus 2

MATLAB for MAS communication. IEEE 13 bus test system was used for the simulation study.

EVs were made to operate in the fourth P-Q quadrant, where EVs consumed active power and supplied reactive power to the grid. For case 1, EVA was connected at bus 2, and using the coordinated EV charging based on the information received by DSO, it is observed from Fig. 10 that there is constant reactive power support to the grid which varies from 0.536 to 0.699 MVAr. Figure 5 also depicts that EVs are charged throughout the day without violating grid constraints using coordinated EV charging. EVs are charged during the night based on DSO information i.e they draw active power to the tune of 1.9 MW from the grid with a reactive power support of 1.65MVAr. Base case results demonstrate that active and reactive power losses are 0.8 MW and 0.6278 MVAr, respectively with voltage profiles maintained at bus 1 and bus 2 respectively.

In case 1, DG with a capacity of 6 MW was connected at bus 2 and it is observed that the power losses are reduced to 0.5287 MW and 0.4151 MVAr. There is improvement in voltage level also at bus 2, bus 3 and bus 4, respectively. In case 2, DG with a capacity of 6 MW and 3 MW are connected at bus 2 and bus 11, respectively. It is very interesting to note that the voltage profile improves at all the buses as per the limits specified by IEEE 1547 and there is a significant reduction in active and reactive power losses of 0.1827 MW and 0.1443 MVAr.

In case 2, EVAs are connected at bus 2 and bus 11 operating in the fourth P-Q quadrant. From Fig. 6, it is clear that the voltage profile is well within the prescribed limits and EVs consume active power and supply reactive power. At 2 PM, EVs draw active power of 1.5 MW and supply a reactive power of 0.572 MVAr. From 3 to 5 AM, the reactive power support of 0.672 MVAr and draws active power of 0.5 MW. From 7 AM to 3 PM, the average active power drawn is 0.41 MW and reactive power support is 0.61 MVAr. From 8 to 10.30 PM, EVs charging reaches a value

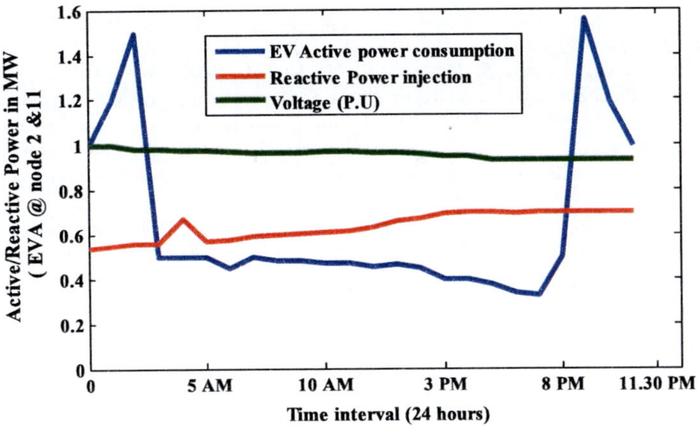

Fig. 6 Coordinated EV charging with EVA at bus 2 and 11

of 1.558 MW and supports reactive power of 0.699 MVAr. The voltage levels are maintained throughout and coordinated charging serves the purpose of promoting EVs connecting to the grid.

5 Conclusion

Mathematical models were developed and presented for EVs in the distribution grid and intermittent EV load. Multiagent-based micro PMU communication for coordinated EV charging in smart distribution network is presented. It is observed that when EVs are made to operate in the fourth P-Q quadrant, there is a continuous supply of reactive power which helps in maintaining the voltage levels and at the same time EV can consume active power as prescribed by DSO. It is found that the results obtained are significant in terms of reduced power losses and improved voltage profile. The work presented is a concept of Cyber Physical System and serves as a platform for the researchers venturing into intelligent transportation system (ITS) and EV integration to the grid.

References

1. Willett, K.: Jasna, T.: Vehicle-to-grid power fundamentals: Calculating capacity and net revenue. J. Power Sources **144**(1), 268–279 (2005)
2. Christophe, G., George, G.: A conceptual framework for the vehicle-to-grid (V2G) implementation. Elsevier Energy Policy **37**(11), 4379–4390 (2009)
3. Liu, C., Chau, K.: Gao, S.: Opportunities and challenges of vehicle-to-home, vehicle-to-vehicle, and vehicle-to-grid technologies. Proc. IEEE **101**(11), 2409–2427 (2013)

4. Wang, Z., Wang, S.: Grid power peak shaving and valley filling using vehicle-to-grid systems. IEEE Trans. Power Deliv **28**(3), 1822–1829 (2013)
5. Hampannavar, S., Chavhan, S., Mansani, S., Yaragatti, U.: Electric vehicle traffic pattern analysis and prediction in aggregation Regions/Parking lot zones to support V2G operation in smart grid: a cyber-physical system entity. Int. J. Emerg. Electr. Power Syst. **21**(1), 20190176 (2020)
6. Yinfeng, W., Kamwa, I., Chen, F.: An adaptive filters based PMU algorithm for both steady-state and dynamic conditions in distribution networks. Int. J. Electric. Power Energy Syst. **117** (2020)
7. Yilmaz, M.: Krein, P.: Review of the impact of vehicle-to-grid technologies on distribution systems and utility interfaces. IEEE Trans. Power Electron. **28**(12), 5673–5689 (2013)
8. Phadke, A.G.: Synchronized phasor measurements in power systems. IEEE Comput. Appl. Power **6**(2), 10–15 (1993)
9. IEEE Standard for Synchrophasor Measurements for Power Systems, IEEE, C37.118.1-2011 (2011)
10. IEEE Standard for Synchrophasor Data Transfer for Power Systems, IEEE, C37.118.2-2011
11. Hampannavar, S., Teja, C.B., Swapna, M., Uday Kumar, R.Y.: Performance improvement of M-class Phasor Measurement Unit (PMU) using hamming and blackman windows. In: 2020 IEEE International Conference on Power Electronics, Smart Grid and Renewable Energy (PESGRE2020), Cochin, India, 2020, pp. 1–5. https://doi.org/10.1109/PESGRE45664.2020.9070382
12. Borghetti, C., Solari, A.: Synchronized phasors monitoring during the islanding maneuver of an active distribution network. IEEE Trans. Smart Grid **2**(1), 82–91 (2011)
13. Krumpholz,, G.R., Clements, K.A., Davis, P.W.: Power system observability: a practical algorithm using network topology. IEEE Trans. Power Appar. Syst. **PAS-99**(4), 1534–1542 (1980). (July/Aug 1980)
14. Voltage Characteristics of Electricity Supplied by Public Distribution Systems, BSI Standards, EN 50160 (2010)
15. Ali, I., Aftab, A., Suhail, K.: Performance comparison of IEC 61850–90–5 and IEEE C37.118.2 based wide area PMU communication networks. J. Mod. Power Syst. Clean Energy **4**(3), 487–495 (2016)
16. Pei, Y., Xiao, W., Han, X.: PMU placement protection against coordinated false data injection attacks in smart grid. IEEE Trans. Ind. Appl. **56**(4), 4381–4393. (2020). https://doi.org/10.1109/TIA.2020.2979793. (July-Aug. 2020)
17. Hampannavar, S., Chavhan. S., Udaykumar, Y., Naik, A.: Gridable electric vehicle (GEV) aggregation in distribution network to support grid requirements: a communication approach. Int. J. Emerging Electric Power Syst. **18**(3), 153–163 (2017)
18. Wang, J., Paudyal, S., Khan, I.: Distribution grid voltage support with four quadrant control of electric vehicle chargers. IEEE Power Energy Soc. Gen. Meet. (PESGM) **2019**, 1–5 (2019). https://doi.org/10.1109/PESGM40551.2019.8973479
19. Hampannavar, S., Yaragatti, U., Chavhan, S.: A stochastic model based on markov chain to support vehicle-to-grid (V2G) operation in smart distribution network. Int. J. Emerg. Electr. Power Syst. **20**(3), 20180347 (2020). https://doi.org/10.1515/ijeeps-2018-0347
20. Kamboj, S., Kempton, W., Decker, K.: Deploying power grid-integrated electric vehicles as a multi-agent system. In: Proceedings of the Tenth International Joint Conference on Autonomous Agents and Multiagent Systems (AAMAS), pp.1–6, New York (2011)
21. Gregor, R., Hahnel, J., Simon, F.: Multi-agent systems asset for smart grid application. IEEE ComSIS **10**(4), 1799–1822 (2013)
22. Sharma, A., Dipti, S., Tivedi, A.: A decentralised multiagent system approach for service restoration using DG islanding. IEEE Trans. Smart Grid **6**(6), 2784–2793 (2015)

Emerging Role of Intelligent Cyber-Physical Systems in Monitoring Stability of Engineered and Natural Slopes

Abhipsa Kar and Manas Ranjan Das

1 Introduction

Failure of natural and engineered slopes is one of the significant geological phenomena which occurs due to extreme events of climate change and variations in topography. It causes several ranges of ground movements that need proper planning in order to avoid its risk on lives and economy by use of appropriate stabilizing methods [1, 2]. Slopes may be artificial, i.e., man-made, as in earth dams, embankments for highways and railroads, etc. or natural as in coastal and river cliffs and hillside and valleys, etc. Instabilities of man-made and natural slopes are the major factors responsible for huge fatalities, injuries and destruction of properties every year [3, 4]. Researchers have carried out multiple studies on slope stability for the reduction of losses caused due to sliding of slopes and minimizing the cost incurred due to disaster prevention and mitigation [5]. Majority of the problems in the stability of slopes are statically indeterminate; hence, some simple assumptions are made for determining a unique factor of safety. Uncertainties in geotechnical parameters (e.g., shear strength parameters, i.e., cohesion and angle of internal friction, pore water pressure, unit weight, etc.) are unaddressed in the deterministic approach [6].

Another major matter of concern is the requirement for a better understanding of instability phenomena. Due to the involvement of uncertain factors, like those of a geological-geotechnical order, human activities and climate, etc., these processes are quite complex [7–9]. Therefore, real-time monitoring and forecasting of slope must be incorporated, in order to quantify the variations in the model input parameters and study the influence of these variations on the risk factor.

A. Kar (✉) · M. R. Das
Siksha 'O' Anusandhan (Deemed to be University), Jagamara, Khandagiri, Bhubaneswar, India
e-mail: abhipsakar@soa.ac.in

M. R. Das
e-mail: manasdas@soa.ac.in

B. Agarwal et al. (eds.), *Proceedings of International Conference on Intelligent Cyber-Physical Systems*, Algorithms for Intelligent Systems,
https://doi.org/10.1007/978-981-16-7136-4_2

There has been a huge development in various fields of computer science and information technology. Mainframe computers appeared around 1960s–1970s. Internet and desktop computers were created during 1980–1990's for personal as well as commercial purposes [10]. Around the twentieth Century, pervasive computation to perform calculations at any place or time appeared. These events are the influencing factors in the development of information society. Serious attention is being paid by various researchers towards the evolution of a novel engineering system known as the cyber-physical systems (CPS) [11]. These are multi-disciplinary systems that combine computation, communication and control technologies to conduct feedback control on widespread embedded computing systems. The transformation of the existing network systems integrated with traditional embedded systems has given rise to CPS. CPS has the ability to perceive real-time and dynamic information by maintaining safety and reliability through collaboration with physical systems represented by the embedded system [12, 13]. CPS has a wide variety of applications, such as aerospace and aircraft control, digital medical instruments, industrial control, distributed energy systems, structural health monitoring, water resource management, etc. [14]. The application of CPS in the fields of slope failure prediction and disaster mitigation is quite limited. Hence, this study introduces the evolution of Intelligent Cyber-Physical Systems from IoT and CPS. Then it proposes an ICPS-based slope monitoring system which will help in real-time monitoring of slope. The last part is the prospect of applications of ICPS in slope stability monitoring of natural and engineered slopes.

2 Major Slope Failures in India

Slope failures are natural or man-made hazards that lead to severe destruction of lives and economy. Some of the major slope failures of India along with their impacts are described in this section.

A landslide occurred on 7th August 2020, near the TNEB colony, Nilgiri district, Tamil Nadu, after a spell of heavy rainfall of about 34.6 cm resulting in damage to four houses. The major cause of the damage was debris flow. However, many small-scale landslips were also recorded in the area. The slide (Lat: 11°19′14.94"N and Lon: 76°37′31.34"E) is a complex slide, which initiated as a cut slope failure which subsequently resulted in earth flow. Ground cracks ranging from 5 to 10 cm were observed in the close vicinity of the slide, which also indicates chances of future reactivation, if untreated. Ground subsidence of about 1 m was observed in the slide zone.

The landslide in Talacauvery had occurred on 06th August 2020, at 02:30 am, on a day of heavy rainfall, near Talacauvery Temple on Brahmagiri hills near Bhagamandala in Kodagu district, Karnataka, along the approach road to the temple. Talacauvery, the place from where the Cauvery river originates is an important pilgrimage and tourist centre. The area is situated 43.5 km east of Madikeri town. It is a reactivated slide and it was first initiated during the year 2007. The present slide got reactivated on 06th August 2020, due to intense rainfall covering part of the old slide zone (2007). It can be inferred that the Talacauvery landslide underwent four episodes of events, i.e., in the years 2007, 2018, 2019 and 2020.

A landslide occurred at Pettimudi, near Munnar, Idukki district, Kerala, on 06th August 2020 around 22:30 h, due to heavy and incessant rainfall. As a result, four housing lines of tea garden workers of Kannan Devan Hill Plantations (KDHP) situated at the foot slope area of Rajamala Hill ranges, got buried under debris (more than 80 people lived). 66 casualties were reported. The landslide (Lat: 10°10′18.10"N and Lon: 77° 0′40.70"E) initiated as shallow planar at rock overburden contact and the distressed material was directed through the topographic hollow scouring entire material from both the flanks and flow path. It was inferred that the failure must have occurred/initiated at the head of a steep gulley/streamlet.

On 9th May 2019, a landslide occurred in the riverbank of the NE flowing Rengma River. The continuous rainfall because of Cyclone Fani triggered the slide and erosion on the banks led to the collapse of the road.

A rock slide took place in the Amagarh area (200–250 m towards the south east of Jaipur–Delhi NH-8) at around 15:30 h due to intensive rainfall resulting in one casualty, injuring six residents and damage of numerous houses situated in the foot of Amagarh hill. The cause of the rockslide was reduction in shear strength due to rainwater percolation along the bedding plane [15].

An unprecedented heavy rainfall with an increased rainfall intensity of 32% as compared to the average annual rainfall of past 20 years led to multiple landslides in Kodagu district. The landslides were catastrophic leading to loss of life and property and damaged communication and infrastructure service lines. The communication line along Shiradi and Sampaje Road, which connects Mangalore with Bangalore, was disconnected. The landslides were triggered by heavy rainfall which resulted in natural slope modification, blockages in natural drainage and flash flood due to blockages of streams.

The Western Ghats of Kerala have become a highly hazardous zone as they are prone to serious mass movements like landslides, debris flow, slump and rockfall [16]. The factors responsible for the triggering mechanisms are site specific. A slide occurred along a steep overhanging road cut cliff in Ninumullipara. The valley side had a very steep gradient which was subjected to perennial water seepage.

A devastating debris flow occurred in Pasukkadavu hill on August 04, 2004, causing ten casualties and damaging three houses. The highest and lowest elevation of the hill slope is 775 m and 300 m, respectively, from mean sea level. A typical slump failure occurred in the lateritic profile at Vadavathoor on July 10, 1995, leading to damage of agricultural lands and partial destruction of a flour mill.

For the adequate representation of the slope movements, it is necessary to include geometrical and structural characterization of the slope, identification of the evidence of ground movements, monitoring the key drivers of slope failure, physical and hydro-mechanical characterization and analysis of these data for identifying the probable mechanism of the ground distortions [17]. Proper forecasting of slope failure is a complex process that needs an accurate and real-time monitoring system for generating early warnings for disaster mitigation and management. Therefore, an

Intelligent Cyber-Physical System has been proposed in the present study which may prove effective in minimizing the huge losses caused due to these catastrophes.

3 Evolution of ICPS

3.1 CPS

Recently, Cyber-Physical System (CPS) have become an emerging topic in the field of information technology. Cyber-Physical Systems (CPS) are an integration of computation, networking and physical processes. It consists of various units like sensors, actuators, control units and communication networks [12]. Systems existing in the physical world can have interaction with each other through the exchange of data. The combination of computation and communication with the physical world in Cyber-Physical System has multiple advantages: (i) System become more efficient and safer. (ii) The cost of set up and operation of these systems is minimized; and individual machines form complex systems by working together which will provide new capabilities. CPSs enable the mapping of real-time spatio-temporal data. Furthermore, it also aids in the creation of new products and services by understanding the relations between unrelated events [18]. The network administrators find difficulty in transferring data and managing network devices as CPS applications possess several standard protocols [19].

The social and economic relevance of these systems is significantly high, and considerable investments are being made globally for the development of this technology. New analysis and design approaches are provided through the integration of physical processes with software, internet and networking. Nowadays, CPS is widely used in diverse areas such as defence, industrial automation, transportation, energy, healthcare and biomedical, critical infrastructure, agriculture, etc. Researchers from various parts of the world have started paying high attention to CPS.

3.2 CPS Versus IOT

In majority of the projects and researches, the dissimilarity between "Internet of Things" and "Cyber-Physical Systems (CPS)" is not distinct. Hence, it becomes tough to find a source that can draw a clear-cut distinction. An IoT system can considerably evolve to CPS by networking of identified objects which can control a certain scenario in a coordinated manner.

It consists of a mechanism in which various physical entities are controlled by collaborating computational elements. This occurs when the mechanical and electrical systems are networked using software components. Knowledge and information are shared from processes for independently controlling logistics and production systems.

IoT refers to the integration of web and internet into the physical world through extensive deployment of devices which are distributed spatially along with embedded identification, sensing and/or actuation capabilities. The aim of IoT and CPS is to define the relationship between the cyber and physical world through information sensing and sharing [20]. They can be distinguished from each other in the following ways: IoT is an open platform in which emphasis is given to networking and interlinking the things present in the physical world, whereas in CPS, emphasis is given to the information interchange and feedback, where feedback and control of the physical world is operated by the system by sensing the physical world, forming a closed-loop system. Furthermore, CPS systems target the control of a combination of physical and organizational processes, and hence specifically ensure sound human–machine interaction, which is not addressed in case of IoT.

3.3 ICPS

Presently available network embedded systems like Internet of Things generally depend upon sophisticated sensors for gathering data from an environment, communication processes for exchange of information and application design procedures based on data collected by the sensors. However, sometimes the acquisition of data is not completely accurate because of calibration issues, reduction in efficiencies in the compensation mechanisms, variations in the environmental conditions in which the embedded systems operate. Errors in the collected data streams might seriously affect the Quality of service and performance of the application/service. Moreover, the importance of design of security and privacy policies of CPS is underscored by privacy leaks and cyberattacks which affect the IoT and distributed systems.

Nowadays, the focus is given to researching *Intelligent Cyber-Physical Systems* (ICPSs) to address the issues mentioned above. ICPSs are new-generation cyber-physical systems equipped with intelligent capabilities. The major important characteristic of ICPSs is their potentiality to interact with the environment and adapt to changing conditions through distributed intelligent operations at a single unit, groups of units (cluster) and network of units (network) [21]. After the acquisition of data from the environment, it is analyzed and interpreted through the help of machine learning and intelligent solutions, which activates proper control mechanisms to ensure the Quality of service and performance of the application. Furthermore, privacy and security intelligent mechanisms must be considered in such systems to preserve the privacy by avoiding privacy leaks and enhancing cybersecurity through approaches based on machine learning.

4 ICPS-Based Stability Monitoring System of Slopes

4.1 Need

Huge economic losses and casualties are caused by natural and engineered slope failures every year. This demands a requirement of an early warning system to mitigate economic losses and casualties. Multiple researches in the field of development of early warning systems for areas vulnerable to natural slope failures have been performed but works related to real time prediction of occurrence of natural slope failures by realtime monitoring of slope movements are still insufficient. A steady year-wise increase in natural slope failures caused by extreme events of climate change strongly warrants a great need for the development of real-time monitoring technology for natural slope failures that can reduce resulting economic losses and casualties. Many researches have been conducted on early warning systems developed on real-time rainfall data. Existing methods like Frequency Domain Reflectometry, Time Domain Reflectometry, Ground Penetrating Radar come under this category need high frequency to operate causing environmental hazards. These facts give rise to the importance of CPS based slope stability monitoring system to accurately interpret and understand data in order to minimize/prevent slope failure by implementing sound slope management strategies [22]. The need of integrating intelligent mechanisms with CPSs is emphasized due to an increased demand for autonomy and the requirement of reduction of time both in decision-making and the transmission bandwidth. This new generation advanced system is expected to be effective in predicting failure, adapting to changes and autonomous behaviour directly in CPSs units.

4.2 Architecture of the Proposed System

In the present study, a new generation of Intelligent Cyber-Physical System (ICPS) has been proposed to monitor the stability of man-made and engineered slopes. The general architecture of the proposed ICPS consists of three layers, namely: Intelligent Embedded Sensors, Intelligent Coordinators and Intelligent Server. This ICPS interacts with the physical world comprising geotechnical database for slope stability. The ICPS represents a novel and robust mechanism for supporting the stability monitoring of slopes and thus preventing catastrophes.

4.2.1 Intelligent Embedded Sensors (IES)

Slope movement is a very complex and time-dependant phenomenon. Hence, for identification and characterization of the depth of movements occurring in the slope, Intelligent Embedded Sensors will play a significant role. Nature of geologic material, hydrological, physical and geotechnical properties are some of the major controlling factors of slope behaviour. Therefore, for the establishment of these properties, IES serves an important purpose. The hardware of the proposed IES consists of various sensor nodes like accelerometers (for measuring ground movements), rain gauge sensor (for measuring pore water pressure), temperature sensor, humidity sensor, moisture sensor (for measuring soil water content) and tiltmeter (for measuring slope distortions). Hardware part of these units is similar to the ones proposed by Cogliati et al. [21] both in terms of sensing activity and communication with the Intelligent Coordinator. Intelligent mechanisms are organized in the embedded software. Intelligent techniques of the IESs depends upon various software modules such as:

(i) Model Learning: The main aim of this module is to learn a predictive model of the geotechnical data collected from the Geotechnical Data Base.
(ii) Residual Generation: This computes and monitors the deviations in the acquired and predicted data over time inspecting for changes.
(iii) Change Detection: Change detection tests for variations in ground movements, temperature, humidity, water content of the slope are carried out by this module to analyse the residuals. Change detection library stores the change detection tests.
(iv) Adaptation: This module is activated when a change is detected so that the sensors get adapted to the fresh working environment in which the sensor operates. The time instant at which the change started is detected by the sensor. Then the previously acquired knowledge is discarded and the model learning module is automatically activated to re-learn a new model based on changing conditions. The algorithms of this module are stored in the adaptation library.

4.2.2 Intelligent Coordinators (IC)

Hardware parts of these units are similar to the ones proposed by Cogliati et al. [21]. These units depend upon off-the-shelf PC embedded platforms like Raspberry and Ordoid. Software-wise these units assist and manage the intelligent processing of the associated group of Intelligent Embedded Sensors. An Intelligent Coordinator consists of the software services like MQTT BROCKER, MAIN_COORDINATOR, DEPENDENCY GRAPH, DB_NOSQL and APACHE_SERVER. Coordinator software library stores these services. MQTT broker is responsible for the transfer of information from IES to IS and vice versa. After a change is detected by any one of the IES, the IC identifies and isolates the reason for the change and takes decisions about the reaction process.

4.2.3 Intelligent Server (IS)

Intelligent Server makes all the data procured and handled by the Intelligent Embedded Sensors accessible to the final user for stability monitoring of slopes. This unit provides the mechanism to store, process and visualize the information related to both IES and IC. The characterization of the event and magnitude of a slope failure is processed by a trained classifier. This will help in decision-making and reaction process of the end-user. The change detection data are stored inside IS.

4.3 Working of the Proposed System

The detailed architecture of the proposed Intelligent Cyber-Physical System (ICPS) for stability monitoring of slopes is depicted in Fig. 1. Organization of the proposed ICPS consists of a three-layer architecture as described earlier. Physical World of the proposed system consists of a Geotechnical Data Base comprising geotechnical information influencing the stability of engineered and natural slopes such as information on cracks, deflections, tilt, shear strength parameters, surface deformations, underground displacements, water level, matric suction, etc. [23]. ICPS interacts with this physical world through intelligent sensors and actuators (if required). The intelligent embedded sensors are application specific for the purpose of stability monitoring of slopes. The intelligent mechanisms of the IES rely on various software modules as described in the previous section. Each module performs a specific job assigned to it. Intelligent Embedded Sensors are equipped with WiFi X-NUCLEO-IDW01M1 for communication with Intelligent Coordinator. IC aims to assist and manage the

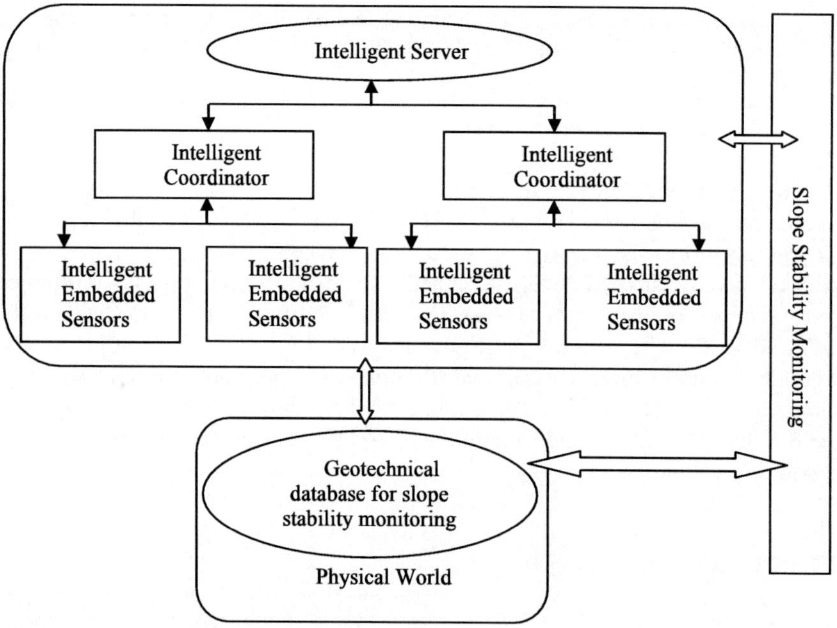

Fig. 1 Proposed ICPS-based stability monitoring system for slopes

intelligent processing of the group of associated sensors. The objective of IS is to ensure the availability and processing of all the acquired information to the final user for stability monitoring of slopes and their management. IS is responsible for the provision of storage mechanisms. It also processes and visualizes the information related to both IES and IC. The proposed Intelligent Cyber-Physical Systems can prove to be effective in real- time stability monitoring of natural and engineered slopes in a large-scale database.

5 Application Scenarios

The main objective of an effective slope monitoring programme is maintaining safe operational practices and providing awareness of instability in advance. Recently, on 7th February 2021, an avalanche struck the Chamoli district of Uttarakhand, India, after a portion of the Nanda Devi glacier broke off. The sudden flooding of the Rishi Ganga, Dhauli Ganga and Alaknanda rivers led to large-scale devastation along with the damage of two important power projects (NTPC's Tapoban-Vishnugad hydel project and Rishi Ganga hydel project). It created a huge loss of life and economy. The serious loss in life and economy could have been avoided by proper real-time slope monitoring and necessary warning. The proposed system is believed to be effective in mitigating these types of problems. The Hirakud Dam is located at latitude 21.31°N and longitude 82.52°E across Mahanadi river, 15 km upstream of Sambalpur, Odisha. It is the longest earthen dam in India. The dam is the first multi-purpose mega-dam project of India. The dam is facing massive siltation due to the heavy erosion in the catchment area. The failure of the dam will create havoc and would add high risk to human life and property. Therefore, the proposed system will play a vital role in predicting potential flood damage and to prepare an emergency action plan (EAP) in advance. It is necessary to carry out such type of monitoring not only for newly planned dam projects, but also for the existing ones.

A landslide that occurred on 7th August 2020, had killed at least 63 people in the state of Kerala, southern India. The slope failure occurred around 25 km away from Munnar. Multiple houses were devasted and numerous people were seriously affected. Heavy rainfalls that struck the region days before triggered the landslide. The ground was saturated due to intense precipitation which leads to an increase in built-up of pore water pressure resulting in the reduction of effective stress and shear strength of the ground. Real time monitoring of the slope conditions would have helped in the prevention of the huge loss caused due to this disaster. Based on the ICPS technology, stability monitoring of slopes becomes an important trend of future development. It will help in detecting the deformations which lead to failure in the pre-failure stage by determining the threshold values/ boundary conditions of the triggering variables. It can send early warnings before and after failure by generating alerts through the help of various alert generating units.

An effective solution could be provided by the proposed ICPS for establishing a robust and reliable system for real-time monitoring and an early warning system for slope failure of engineered and natural slopes. The proposed system consists of a three-layered architecture, namely: Intelligent Embedded Sensors (IES), Intelligent Coordinators (IC) and Intelligent Server (IS). IES are responsible for real-time monitoring of slope movements through various sensor nodes, viz., accelerometers (for measuring ground movements), rain gauge sensor (for measuring pore water pressure), temperature sensor, humidity sensor, moisture sensor (for measuring soil water content) and tiltmeter (for measuring slope distortions). The intelligent techniques of IES rely on model learning, residual generation, change detection and adaptation (if a change is detected) as mentioned in the earlier section. The intelligent mechanism of associated IES is supported and managed by IC. IC identifies and isolates the reason for change detected by IES which helps in the decision- making and reaction process. The data acquired by IES are then made available to the end users through IS. It provides the techniques to store, process and visualize the information related to both IES and IC. IS is also responsible for generating alerts beforehand to the end users through an alert generating unit when there is a possibility of failure.

6 Conclusion

Several factors influence the stability of natural and engineered slopes or tailing dam slopes. Failure of a slope not only leads to loss of property and lives, but also causes damage to the environment. In this scenario, there is a bare necessity for an effective stability monitoring system for different types of slopes. Moreover, there is a need for the prediction of the timeline of occurrence of slope failures which is possible with the help of monitoring of parameters of slope stability. In this context, an Intelligent Cyber-Physical System has been proposed. Its architecture and working specifically to serve the purpose of slope stability monitoring have been described in detail. Application scenarios of such a system in terms of assessment of the vulnerability of slopes to failure have been discussed in the present study.

References

1. Zaki, A., Chai, H, K., Razak, H, A., Shiotani, T.: Monitoring and evaluating the stability of soil slopes: a review on various available methods and feasibility of acoustic emission technique. C.R. Geosci. **346**, 223–232 (2014)
2. Chandarana, P.U., Momayez, M., Taylor, K.: Monitoring and predicting slope instability: a review of current practices from a mining perspective. IJRET: Int. J. Res. Eng. Technol. **05**(11) (2016). eISSN: 2319-1163 | pISSN: 2321-7308
3. Jayanthu, S., Karthik, G.: Recent innovations in wireless systems for slope stability monitoring in opencast mines–an appraisal for indeginisation. RPIMI-Raipur-Feb 19–20 (2016)
4. Moulata, M.E., Debauche, O., Mahmoudi, S., Brahim, L.A., Manneback, P., Lebeau, F.: Monitoring system using internet of things for potential landslides. In: The 15th International Conference on Mobile Systems and Pervasive Computing (MobiSPC 2018) (2018). https://doi.org/10.1016/j.procs.2018.07.140

5. Hou, X.: Geotechnical engineering slope monitoring based on internet of things. IJOE **14**(6) (2018)
6. Kanan, M., Kumar, A.: Slope stability analysis and design their control measures. Int. J. Eng. Res. Technol. (IJERT) (2019). ISSN: 2278–0181, www.ijert.org. (CONFCALL-2019 Conference Proceedings (2019))
7. Lee, S., Choi, J., Kim, Y.: A probabilistic smart system to monitor unsaturated slope instability induced by rainfall infiltration. In: First International Symposium on Geotechnical Safety & Risk (ISGSR) 2007 Shanghai Tongji University, China (2007)
8. Pei, H., Zhang, S., Borana, L., Zhao, Y., Yin, J.: Slope stability analysis based on real-time displacement measurements. Measurement **131**(2019), 686–693 (2018)
9. Lienhart, W.: Case studies of high-sensitivity monitoring of natural and engineered slopes. J. Rock Mech. Geotech. Eng. **7**, 379e384 (2015)
10. Liu, Y., Peng, Y., Wang, B., Yao, S., Liu, Z.: Review on cyber-physical systems. IEEE/CAA J Autom Sin **4**(1) (2017)
11. Bhrugubanda, M.: A review on applications of cyber physical systems. IJISET-Int. J. Innov. Sci. Eng. Technol. **2**(6) (2015). ISSN 2348–7968
12. Wang, P., Xiang, Y., Zhang, SH.: Cyber-physical system components composition analysis and formal verification based on service-oriented architecture.In: 2012 Ninth IEEE International Conference on e-Business Engineering (2012). https://doi.org/10.1109/ICEBE.2012.60
13. Lin, C.Y., Zeadally, S., Chen, T., Chang, C.: Enabling cyber physical systems with wireless sensor networking technologies. Int. J. Distrib. Sens. Netw. **2012**, Article ID 489794, 21 (2012). https://doi.org/10.1155/2012/489794. (Hindawi Publishing Corporation)
14. Bai, Z.Y., Huang, X.Y.: Design and Implementation of a Cyber Physical System for Building Smart Living Spaces. Int. J. Distrib. Sens. Netw. **2012**, Article ID 764186, 9 (2012). https://doi.org/10.1155/2012/764186. (Hindawi Publishing Corporation)
15. Landslide recent incidents, Geological Survey of India, www.gis.gov.in
16. Sreekumar, S., Aslam, A.: Spatio-temporal distribution of slope failures in the Western Ghats of Kerala, India. In: WIT Transactions on Information and Communication Technologies, vol. 43, ©2010 WIT Press (2010). https://doi.org/10.2495/RISK100361, www.witpress.com, ISSN 1743-3517
17. Asch, T.V., Malet, J.P., Beek, L.V., Amitrano, D.: Techniques, advances, problems and issues in numerical modelling of landslide hazard. Bulletin de la Société Géologique de France, Société géologique de France, **178**(2), 65–88 (2007). (ffhal-00172644 (2007))
18. Radanliev, P., Roure, D.D., Kleek, M.V., Santos, O., Ani, U.: Artificial intelligence in cyber physical systems. AI & SOCIETY (2020)
19. Changanti, R., Gupta, D., Vemprala, N.: Intelligent network layer for cyber-physical systems security (2021). arXiv:2102.00647v1 [cs.CR]
20. Greer, C., Burns, M., Wollman, D., Griffor, E.: Cyber-Physical Systems and Internet of Things. NIST Special Publication 1900-202 (2019)
21. Cogliati, D., Falchetto, M., Pau, D., Roveri, M., Viscardi, G.: Intelligent cyber-physical systems for industry 4.0. In: 2018 First International Conference on Artificial Intelligence for Industries, IEEE Computer Soceity (2018). https://doi.org/10.1109/ai4i.2018.00013
22. Park, S., Lim, H., Tamang, B., Jin, J., Lee, S., Chang, S., Kim, Y.: A study on the slope failure monitoring of a model slope by the application of a displacement sensor. Hindawi J. Sens. **2019**, Article ID 7570517, 9 (2019)
23. Mametja, T.D., Zvarivadza, T.: Slope stability enhancement through slope monitoring data interpretation. In: 51st US Rock Mechanics/Geomechanics Symposium held in San Francisco, California, USA, pp. 25–28 June (2017)

Battery Management System in Smart City: An Application of Cyber-Physical System

B. S. Sagar, Hampannavar Santoshkumar, and B. P. Divakar

1 Introduction

Smart cities are being developed across the world to provide better dwelling facilities to people, which ultimately leads to a secure, cleaner and harmless eco-system. The purpose of smart cities is to improve the quality of the life of the people by harnessing technology that leads to smarter outcomes. A smart city uses various sensors to collect the data, which in turn is utilized in monitoring and enabling for efficient utilization of the available resources. With respect to transportation, the data is analyzed to monitor and manage the traffic and also enable electric vehicles to charge efficiently without overloading the grid. Information and Communication Technologies (ICT)-an intelligent framework, are developed and deployed for sustainable development practices in building smart cities. Special attention is given to meet the commuting requirements of people that provide innovative services leading to safer, coordinated and eco-friendly modes of transportation. The concept of Vehicle-to-Grid (V2G) is about grouping Electric Vehicles (EVs) to the grid for power exchange. This helps in meeting various challenges in power systems like peak shaving during peak charging hours, valley filling and load shifting [1–4]. The conceptual framework of the aggregator regions with parking zones enabled with charging facilities is shown in Fig. 1.

It shows smart city (smart grid with Intelligent Transport System (ITS)), aggregator regions and charging stations. Multiple aggregators can be connected to the

B. S. Sagar (✉) · H. Santoshkumar · B. P. Divakar
School of Electrical and Electronics Engineering, REVA University, Bangalore, India
e-mail: sagar.bs@reva.edu.in

H. Santoshkumar
e-mail: santoshkumar.sh@ieee.org

B. P. Divakar
e-mail: divakar@reva.edu.in

B. Agarwal et al. (eds.), *Proceedings of International Conference on Intelligent Cyber-Physical Systems*, Algorithms for Intelligent Systems,
https://doi.org/10.1007/978-981-16-7136-4_3

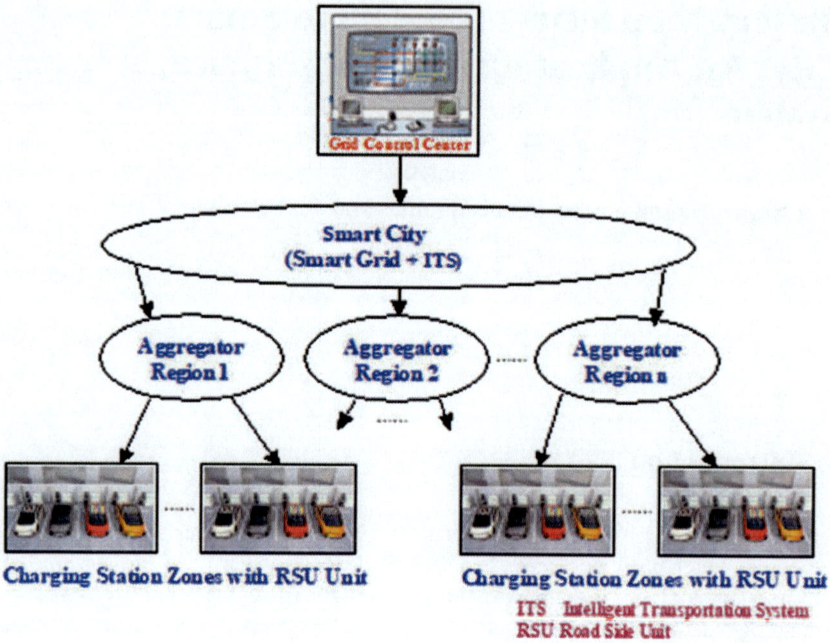

Fig. 1 Conceptual framework of the aggregator regions with parking zones enabled with charging facility

grid and it is assumed that they are connected to a bus in the distribution feeder. Each aggregator region is assumed to have a certain number of parking slots enabled with power exchange slots to have seamless connectivity to the grid. Further, the generation, distribution and aggregator network from the physical part is shown in Fig. 2, which shows the framework of aggregator connected to the distribution grid through home or any other interface. To this physical system, various sensors are connected to obtain voltage, current and power, which captures real time data of the system. This can then be utilized to process and provide mechanisms for an efficient transportation system. This layer of a system consisting of sensors, monitoring, computing, controlling and communication with various interfaces is termed as a cyber system. Thus, a cyber- physical transportation system is needed for creating an eco-friendly and more coordinated transportation system.

EV aggregator architecture is shown in Fig. 3 and it portraits that there are multiple communication lines that exist in a smart system. The communication lines are linked to energy market transaction, communication to EV fleet with flexible and inflexible loads where communication can either be unidirectional or bidirectional. All these details are even updated to the manufacturing end and even to the consumer end.

The need for a smart grid is essential in order to coordinate and manage various plants that are involved in power exchange. Smart grid infrastructure can be depicted as shown in Fig. 4.

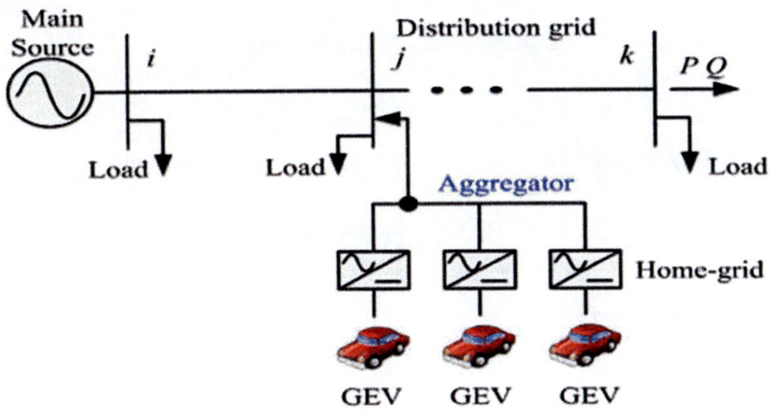

Fig. 2 Generation, distribution and aggregator network

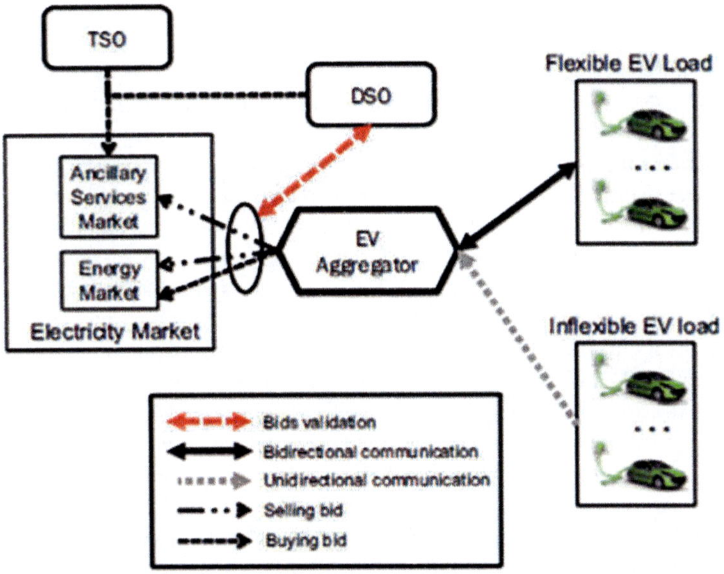

Fig. 3 EV aggregator architecture

A small security breach in the cyber layer of CPS impacts the physical layer. A malicious user can take control of the computing or communication components and cause damage to the property. Cyber security is essential for various blocks of smart grid operational systems including control systems, SCADA systems, smart meters and substations. A robust integrated cyber security protection is a must in the smart grid [5].

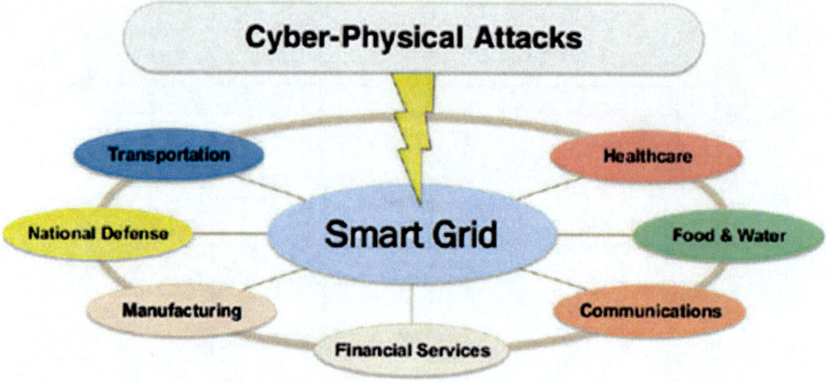

Fig. 4 Smart grid infrastructure

2 Recent Works

EVs to be connected to the grid as per IEEE 1547 2003 standard are employed in V2G integrators [6]. Demonstration of plug-in electric vehicle is provided in [7] along with an optimization behaviour and battery ageing methods analysis. V2G conceptual framework is considered along with the role of aggregators and independent service operators (ISO) [8]. V2G impact on the grid is studied in [9, 10] presents a bidirectional DC-DC converter with input/output voltage ranges for V2G energy transfer capability. A secure and privacy-oriented protocol for V2G communication networks is presented in [11]. A decentralized multi-agent system is proposed for restoring service in case of islanding segment using distributed generation considering larger EV fleet [12]. Literature survey provides information on V2G need and its importance, V2G impact on the grid, charging infrastructures and EV integration to the grid.

The requirement of battery ageing calculation and corresponding replacement of the batteries at right time intervals and the concept of charge equalization is to be addressed so as to ensure the longevity of the EV battery life. These aspects are covered under Battery Management System (BMS) interface unit. The investigation study on various methods presented in research papers [13–16] needs extra circuitry for cell equalization purpose with more time depending upon the equalization algorithm.

3 Battery Management System (BMS)

Battery Management System (BMS) is an integral part of the physical part of cyber-physical system as shown in Fig. 5. Various sensors used in BMS and communicating the monitoring of sensed data and communication system can be made use of the

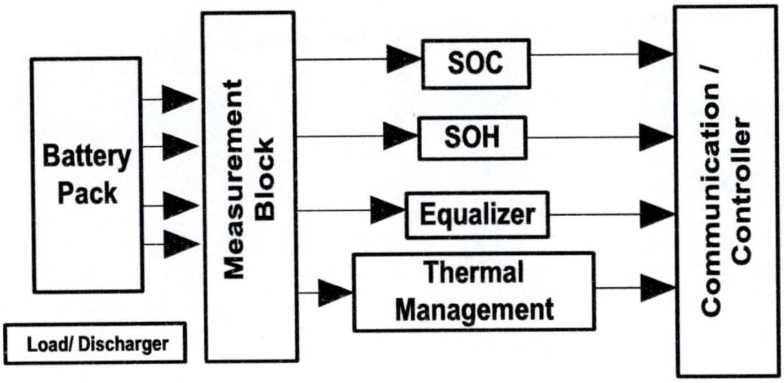

Fig. 5 Basic battery management system architecture

same infrastructure as that in CPS. Thus, with a single infrastructure, integration of BMS to CPS is made easier.

The major functions of BMS are to protect batteries from over-charging/under-charging when the EVs are connected to the grid. Cell balancing can also be employed to maintain uniform cell voltage. Communication among sub-systems and across the CPS system is another aspect where BMS plays a critical role. Data acquisition unit can be employed to collect the data and can even be stored for further analysis. Centralized, semi-centralized or localized BMS can be employed in CPS for actuating the battery pack of the EV fleet.

4 Methodology

4.1 Battery Ageing

The health of batteries employed in EV will be affected and degraded over a period of time due to factors like temperature variations, operating at a high and low state of charge, high current, usage cycles and extreme climate and charging type. As per the battery degradation prediction by Geotab, an EV analyst [17], assess how the batteries are able to perform over the years that are subjected to real time conditions. Analysis on six thousand plus fleet of consumer EV of more than 20 distinct vehicle models, representing close to 2 million days of data gives an insight on how the batteries are holding up over a period of time. The degradation curves displayed in Fig. 6 are the average trend line from the data analyzed. Understanding the degradation of EV batteries will give a fair idea cost of ownership and expected residual value, especially considering the huge cost of battery pack replacement at the end of the specified cycle. As expected, the older the vehicles more the degradation.

According to the suggestions of EV manufacturers, battery replacement is to be carried out once in 8 years. However, due to failure in some of the batteries in

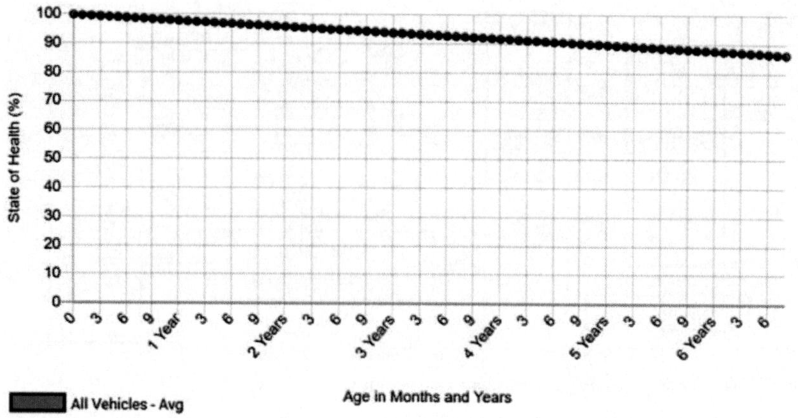

Fig. 6 EV battery degradation analysis

the bank, the batteries need to be replaced in less than 8 years. Batteries start their life with 100% state of health (SOH) and over time they deteriorate. For example, a 60 kWh battery that has 90% SOH would effectively act like a 54 kWh battery. Accounting the calendar ageing of the battery, the overall estimates of life of the battery can be considered as follows: 2% of batteries to be replaced at the end of the first year, 4% after completion of the second year, 6% at the end of the third year, 8% after the fourth year, 10% at the end of 5th year, 15% after 6th year, 25% after 7th year and remaining 30% at the end of 8 year cycle. Considering the above set of battery replacement, the battery can be 0,1,2,3,4,5,6 or 7 years old. If a battery is '*k*' year old then it can become either 0 year old or '*k + 1*' year old. If the battery is 7 years old, then it will be replaced next year and it will become 0 year old. Thus, it is evident that the battery age is dependent on its age in the previous year, where Markov chain-based model technique can be aptly employed to represent the replacement age of the batteries.

Let 'k' denote the state of the Markov model. The transition diagram of the battery age model is shown in Fig. 7. In practical applications like considering EVs, the interest is focussed on the probability distribution of p_n with the knowledge of the initial state of the system to be modelled. The initial state of the system is defined by an initial probability row vector as given in Eq. 1.

$$P(0) = \left[p_1(0)\ p_2(0)\ \ldots\ p_m(0) \right] \tag{1}$$

where $p_i(0)$ is the probability that the system is initially in state i. In the case that the initial state of the system is known, then

$$p_i(0) = 1.0\ and\ p_k(0) = 0\ for\ k \neq i \tag{2}$$

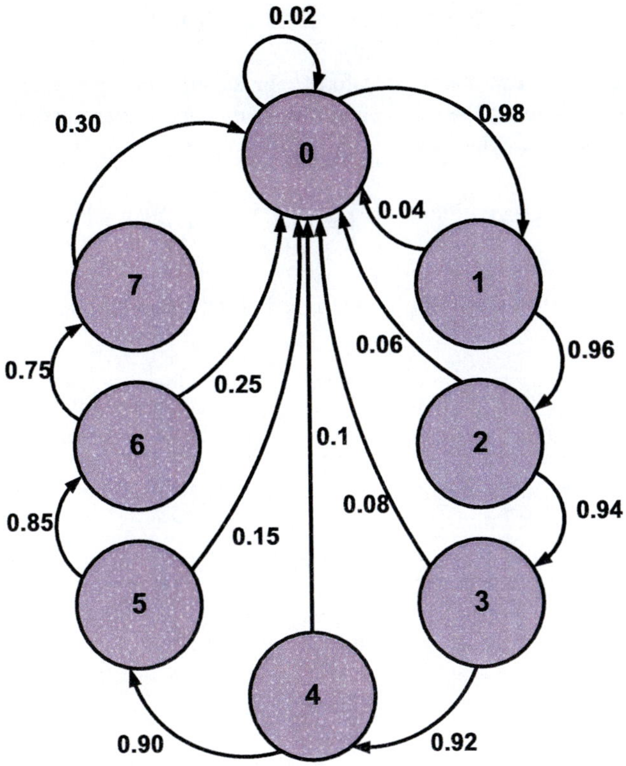

Fig. 7 Transition state diagram for battery ageing model

After the first transition, the probability that the system is in the state j can be obtained by the theorem of probability as in Eq. 3.

$$p_j(1) = P(X_1 = j) = \sum_1 P(X_1 = j) \mid X_0 = i)\, P(X_0 = i) \tag{3}$$

Thus

$$p_j(1) = \sum_1 p_i(0)\, p_{i,j} \tag{4}$$

For a matrix representation, the single-stage state probability becomes

$$P(1) = P(0)\, P \tag{5}$$

Consecutive stage transitions can be obtained by applying the theorem of total probability represented by

$$P(n) = P(n-1)\,P = P(0)\,P^n \quad (6)$$

Markov chain is a two-state closed-form expression, which can be obtained for the vector m(n). By denoting the states of the system as 0 and 1, the transition probability matrix (TPM) can be written as

$$m = \begin{bmatrix} m_{0.0}\ m_{0.1} \\ m_{1.0}\ m_{1.1} \end{bmatrix} = \begin{bmatrix} 1 - m_{0.1} & m_{0.1} \\ m_{1.0} & 1 - m_{1.0} \end{bmatrix}$$

The n step transition probability matrix m^n is the mth power of m which can be obtained by

$$m^n = \frac{1}{m_{0.1} + m_{1.0}} \begin{bmatrix} m_{1.0}\ m_{0.1} \\ m_{1.0}\ m_{0.1} \end{bmatrix} + \frac{(1 - m_{0.1} - m_{1.0})^n}{m_{0.1} + m_{1.0}} \begin{bmatrix} m_{0.1} & -m_{0.1} \\ -m_{1.0} & m_{1.0} \end{bmatrix} \quad (8)$$

Considering the initial distribution as

$$m(0) = [\beta,\ 1 - \beta] \quad (9)$$

Then as per the application of the theorem of total probability to the second and consecutive stage transitions, we obtain

$$m(n) = m(n-1)\,m = m(0)\,m^n \quad (10)$$

The state probabilities after n stages of transition can be determined alternatively by applying the theorem of total probability. This can be represented by

$$\begin{aligned} m_{ij}(n) = m\,(X_n = j \mid X_0 = i) &= \sum_l m(X_n = j \mid X_0 = i,\ X_s = l) \\ &\quad m(X_s = l \mid X_0 = i) \\ &= \sum_l m_{l,j}\,(n - s)\, m_{i,l}\,(s), \quad 0 < s < n \end{aligned} \quad (11)$$

Equation (11) is known as the Chapman-Kolmogorov equation.

The steady-state probabilities exist in the cases where the number of states in the state space is finite. If these probabilities are finite, then the Markov chain is said to be ergodic. The values of p_j must also be unique, and therefore, totally independent of the initial condition.

At steady-state condition, we get

$$P(n) = P(n-1) = P^* \quad (12)$$

P^* is the row vector of steady-state probabilities. Also, from (1), we have

$$P(n) = P(n-1) = P \tag{13}$$

And hence

$$P^* = P^* P$$

The above equation possesses one degree of freedom. This can be solved for vector P^* by

$$\sum_j p_j^* = 1 \tag{15}$$

4.2 Battery Balancing

In applications such as EV, hybrid EV, plug in EV require high capacity battery banks which are arranged in series and parallel combination to meet the load requirement. These banks are subjected to harsh operating conditions which result in an imbalance in charge levels. This imbalanced condition results in reduction in the efficiency and life span of the battery bank. To overcome this challenge, cell balancing is vital to ensure batteries maintain a uniform charge level, such that the battery life is extended and made easier for connecting such EV aggregators to the grid.

Battery balancing techniques can be broadly classified into two groups: passive technique and active technique. The passive balancing technique is also known as resistor balancing, where overcharged cell is discharged by using a resistor (R_1–R_n) to equalize the low charged cells in a battery pack by operating switches S_1–S_n, respectively. This is shown in Fig. 8 [18].

In the active balancing mechanism, the charge is transferred from higher energy cells to lower energy cells for equalization. A relatively simple control strategy is used to transfer the energy between the cells with a difference in energy. Passive elements such as capacitors and inductors are used to store and transfer the energy between the cells. A single capacitor switched cell balancing scheme is shown in Fig. 9.

It is found that a Single-switched capacitor cell balancing needs $n + 5$ switches to balance n cells, which leads to switching loss. On similar lines, single-switched inductor cell balancing technique is developed to equalize the cells. The scheme of single-switched inductor cell balancing is shown in Fig. 10.

Battery balancing method can be employed in three step process. The steps are constant current charging equalizing and trickle charging. Firstly, the battery pack is subjected to constant current charging till it reaches a set value around 70–80% of its charge level. Secondly, rearranging of the battery pack and reading of individual battery voltage. Depending on the voltage levels of various batteries in the pack, the batteries are subjected for constant voltage charging scheme, where the batteries

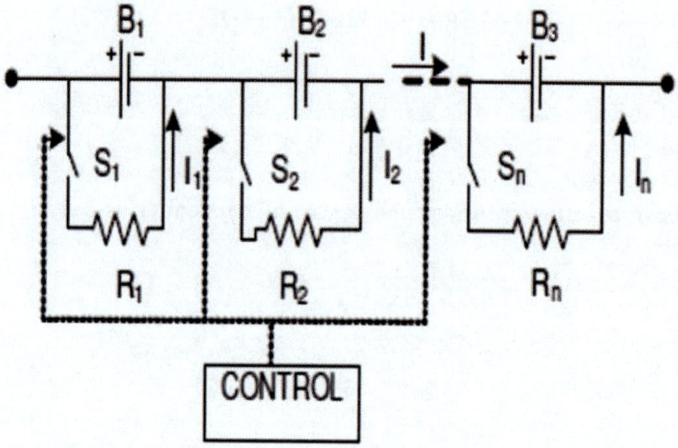

Fig. 8 Passive balancing mechanism

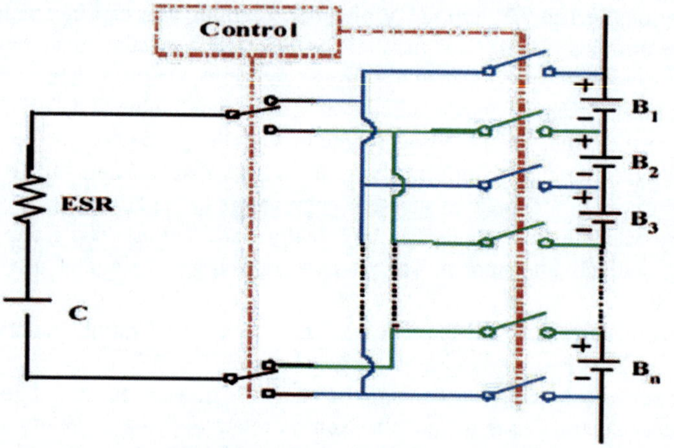

Fig. 9 Single-switched capacitor cell balancing

will be subjected to charge for certain set time period. This process repeats until all the batteries in the pack reaches a uniform level. In this step, battery might be overcharged in few cases so as to avoid sulphation in case of lead acid batteries. The third step is trickle charging method where a fully charged battery is subjected to charging at a rate equal to its self-discharge rate. This maintains the battery at its completely charged level, especially when the battery is not loaded.

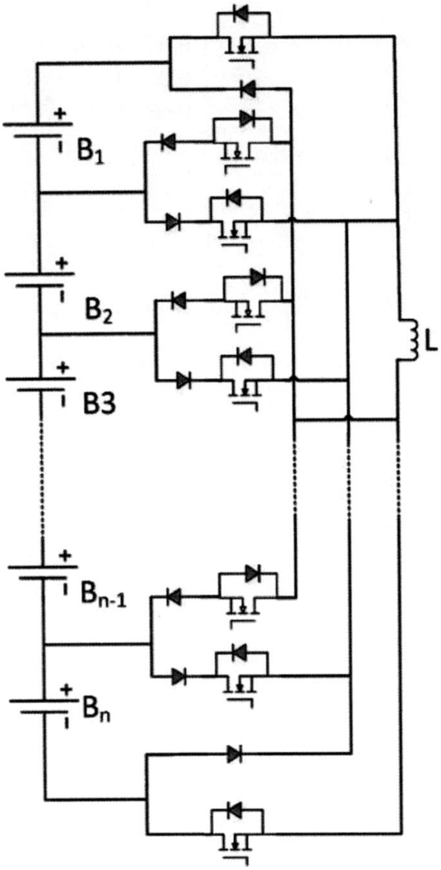

Fig. 10 Single-switched inductor cell balancing technique

5 Results and Discussion

5.1 Ageing Model

The transition probabilities is obtained as follows. If the battery is considered to be replaced last year, that it, is 0 year old, the probability that it will be replaced this year is 0.02 and the probability that it become be 1 year old is 0.98. This states that p0.0 = 0.02 and p0.1 = 0.98. If the battery is already a year old, then the probability that it will become 0 year old because of replacement is 0.04 and probability that it become 2 years old is 0.96. This states that p1.0 = 0.04 and p1.2 = 0.96. Thus, the probability matrix can be given as below

$$p = \begin{bmatrix} 0.02 & 0.98 & 0 & 0 & 0 & 0 & 0 & 0 \\ 0.04 & 0 & 0.96 & 0 & 0 & 0 & 0 & 0 \\ 0.06 & 0 & 0 & 0.94 & 0 & 0 & 0 & 0 \\ 0.08 & 0 & 0 & 0 & 0.92 & 0 & 0 & 0 \\ 0.1 & 0 & 0 & 0 & 0 & 0.90 & 0 & 0 \\ 0.15 & 0 & 0 & 0 & 0 & 0 & 0.85 & 0 \\ 0.25 & 0 & 0 & 0 & 0 & 0 & 0 & 0.75 \\ 1.0 & 0 & 0 & 0 & 0 & 0 & 0 & 0 \end{bmatrix} \tag{16}$$

For a two state Markov chain, the steady state probabilities can be obtained by

$$m_0 = \left[\frac{m_{1.0}}{m_{0.1} + m_{1.0}}\right] \; and \; m_1 = \left[\frac{m_{0.1}}{m_{0.1} + m_{1.0}}\right] \tag{17}$$

In the EV application as being discussed, the steady state probabilities are obtained by solving Eqs. (3) and (4). The matrix can be obtained by solving

$$\begin{bmatrix} p_0^* & p_1^* & p_2^* & p_3^* & p_4^* & p_5^* & p_6^* & p_7^* \end{bmatrix} \begin{bmatrix} 1 & -0.98 & 0 & 0 & 0 & 0 & 0 & 0 \\ 1 & 1 & -0.96 & 0 & 0 & 0 & 0 & 0 \\ 1 & 0 & 1 & -0.94 & 0 & 0 & 0 & 0 \\ 1 & 0 & 0 & 1 & -0.92 & 0 & 0 & 0 \\ 1 & 0 & 0 & 0 & 1 & -0.90 & 0 & 0 \\ 1 & 0 & 0 & 0 & 0 & 1 & -0.85 & 0 \\ 1 & 0 & 0 & 0 & 0 & 0 & 1 & -0.75 \\ 1 & 0 & 0 & 0 & 0 & 0 & 0 & 1 \end{bmatrix} = \begin{bmatrix} 1 \\ 0 \\ 0 \\ 0 \\ 0 \\ 0 \\ 0 \\ 0 \end{bmatrix} \tag{18}$$

The solution for this set of linear equation is obtained as below

$$\begin{bmatrix} p_0^* & p_1^* & p_2^* & p_3^* & p_4^* & p_5^* & p_6^* & p_7^* \end{bmatrix} = \begin{bmatrix} 0.153 & 0.152 & 0.146 & 0.137 & 0.126 & 0.113 & 0.096 & 0.072 \end{bmatrix} \tag{19}$$

This shows that, in any given year, 15.3% of the batteries have just been replaced. Also, only 15.2% of the batteries are 1 year old and so on. The fact that the battery will not fail in within the scheduled replacement cycle as per the manufacturers is only 7.2%. Thus, it is clear from the above result that, replacing battery pack after

8 year cycle is not reasonable as most of the batteries in the pack are to be replaced before the specified time duration. In fact, the average age of the batteries computed form the result is equal to 3 years.

5.2 *Battery Balancing*

A 12 V, 42 AH valve regulated lead acid (VRLA) batteries are employed in the battery stack at the laboratory to test the balancing algorithm. As evident, the battery terminal voltage shows variation in the voltage levels as in Fig. 11. Employing battery balancing algorithm reduces this variation and ensures uniformity with the voltage levels as shown in Fig. 12. Comparison of battery voltage levels with/without balancing as in Fig. 13 and it clearly reveals the requirement of the battery balancer

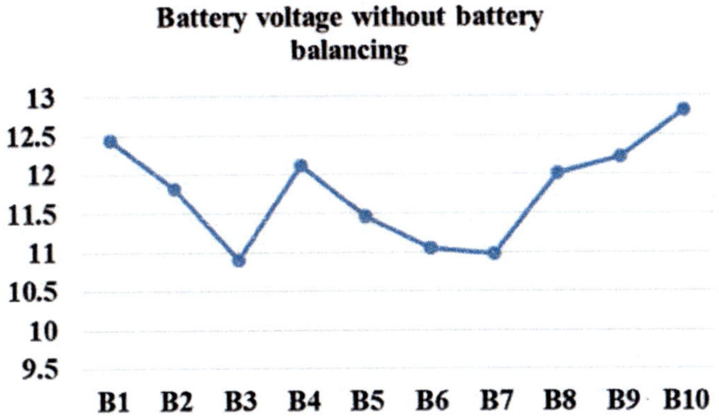

Fig. 11 Battery voltage without battery balancing

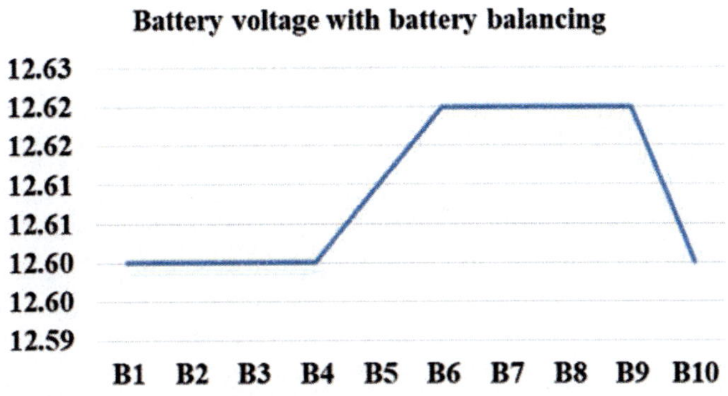

Fig. 12 Battery voltage with battery balancing

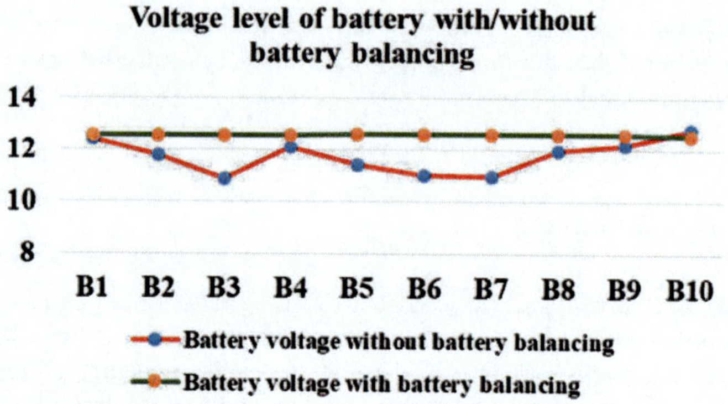

Fig. 13 Comparison of battery voltage with and without balancing

to ensure proper functioning of battery pack, to enhance the battery life for seamless connectivity to the grid.

6 Conclusion

In this paper, firstly battery management system (BMS) in smart city is explained and presented. Secondly, stochastic model of battery ageing is developed which helps in systematic replacement of batteries in a pack. Finally, experimental set up is developed for battery/cell balancing technique and it is found that there is significant improvement in battery voltage before and after battery balancing. The balancing enhances the life span of the battery so as to ensure proper functioning of EV fleet over longer periods.

References

1. Indian Smart Grid Forum (ISGF). Electric Vehicle as a Sustainable Solution for Pollution Free Air, (ISGF New Delhi) 120.
2. Chunhua, L., Chau, K., Diyun, W., Gao, S.: Opportunities and challenges of vehicle-to-home, vehicle-to-vehicle, and vehicle-to-grid technologies. Proc. IEEE **101**, 2409–2427 (2013)
3. Wang, Z., Wang, S.: Grid power peak shaving and valley filling using vehicle-to-grid sytems. IEEE Trans. Power Deliv. **28**, 1822–1829 (2013)
4. Diyun, W., Chunhua, L., Shuang, G.: Coordinated control on a vehicle–to-grid system. In: Proceedings of the International Conference on Electrical Machines and Systems, Beijing, China, pp. 1–7 (2011). (Smith, T.F., Waterman, M.S.: Identification of common molecular subsequences. J. Mol. Biol. 147, 195–197 (1981))
5. Haibo, H., Yan, J.: Cyber-physical attacks and defences in the smart grid: a survey. IET Cyber-Phys. Syst. Theory Appl. **1**, 13–27 (2016)
6. Yilmaz, M., Kurien, P.T.: Review of the impact of vehicle-to-grid technologies on distribution systems and utility interfaces. IEEE Trans. Power Electron. **28**, 5673–5689 (2013)

7. Dallinger, D., Link, J.: Smart grid agent: plug-in electric vehicle. IEEE Trans. Sustain Energy **5**, 710–717 (2014)
8. Christophe, G.D.: A conceptual framework for the vehicle-to-grid (V2G) implementation. Energy Policy **37**, 4379–4390 (2009)
9. Robert, C.G., Lingfeng, W., Mansoor, A.: The impact of plug-in hybrid electric vehicles on distribution networks: a review and outlook. Renew. Sustain. Energy Rev. **15**, 544–553 (2011)
10. Khan, M., Husain, I., Sozer, Y.: A bi-directional DC-DC converter with overlapping input and output voltage ranges and vehicle to grid energy transfer capability. In: Proceedings of the IEEE International Electric Vehicle Conference (IEVC), pp. 1–6 (2012)
11. Huei, T.: A secure and privacy-preserving communication protocol for V2G networks. In: IEEE Wireless Communications and Networking Conference: Mobile and Wireless Networks, pp. 2706–2711 (2012)
12. Sharma, A., Dipti, S., Tivedi, A.: A decenetralised multiagent sytem approach for service restoration using DG Islanding. IEEE Trans. Smart Grid. **6**, 2784–2793 (2015)
13. Lu, S., Corzine, K.A., Ferdowsi, M.: High Efficiency Energy Storage System Design for Hybrid Electric Vehicle with Motor Drive Integration, 1-4244-0365-0/06/$20.00 ©2006 IEEE
14. Tanaka, Y., Tsuruta, Y., Nozaki, T., Kawamura, A.: Proposal of Ultra High Efficient Energy Conversion System (HEECS) for Electric Vehicle Power Train, 978-1-4799-3633-5/15/$31.00©2015 IEEE
15. Chang, W.-Y.: The State of Charge Estimating Methods for Battery: A Review, Hindwai Publishing Corporation, ISRN Applied Mathematics, vol. 2013, Article ID 953792, p. 7
16. Reynaud, J.F., Carrejo, C.E., Gantet, O., Aloïsi, P., Estibals, B., Alonso C., Chang, W.-Y.: Active balancing circuit for advanced lithium-ion batteries used in photovoltaic application. In: International Conference on Renewable Energies and Power Quality (ICREPQ'11) Las Palmas de Gran Canaria (Spain), 13th to 15th April, 201, RE&PQJ, vol. 1, no. 9 (May 2011)
17. Geotab: https://www.geotab.com/blog/ev-battery-health/
18. Daowd, M., Omar, N., Van Den Bossche, P., Van Mierlo, J.: A review of passive and active battery balancing based on MATLAB/Simulink. Int. Rev. Electric. Eng. (I.R.E.E.) (2011)

Security of Cyber-Physical Systems Through the Lenses of the Dark Web

Ashwini Dalvi, Samata Salve, Gauri Zape, Faruk Kazi, and S. G. Bhirud

1 Introduction

Cyber-physical systems (CPSs) manifest primarily in critical infrastructures like a power grid, transportation, water network, etc., with Industrial Control System (ICS) as their integral part. Security for ICS is a daunting task because of the sheer difference between the shelf life of installed ICS system and the advancement of Information and Communication Technology (ICT).

The researchers have offered statistics on the number of incidents reported by US ICS-CERT during 2012–2016 [1], where they asserted that cyber attacks on critical infrastructures have multi-folded implications, including an economic loss to life-threatening conditions. Therefore, there is a growing need towards finding vulnerabilities, assessing possible attack vectors related to the cyber-physical systems in critical infrastructure. The approaches, like the Red Team exercise, are backed by the research community in tightening the infrastructure security from the past several years [2].

The mitigation solutions for secure CPSs are discussed in the literature for proactive, as well as reactive, security measures along with the forensic investigation. The present discussion aims to open the possibility of collecting intelligence from the dark web for investigating the cyber attacks on CPSs.

The term darknet originated in 1970 when part of the ARPANET network was isolated. This network was designed to receive the messages but not to acknowledge the received messages. In 1971, Students of MIT (Massachusetts Institute of Technology) and Stanford University exerted drug transactions through ARPANET.

Further, in pioneer work [3], the authors predicted that the Dark Net would be a future challenge. The authors in their work advocated that the peer-to-peer nature of

A. Dalvi (✉) · S. Salve · G. Zape · F. Kazi · S. G. Bhirud
Veermata Jijabai Technological Institute, Mumbai, India
e-mail: aadalvi_p19@ce.vjti.ac.in

B. Agarwal et al. (eds.), *Proceedings of International Conference on Intelligent Cyber-Physical Systems*, Algorithms for Intelligent Systems,
https://doi.org/10.1007/978-981-16-7136-4_4

the darknet would make copyright infringement of digital content unavoidable and become a deterrent in executing effective Digital Rights management.

The growth and scope of the dark web forced related stakeholders, including law enforcement agency (LEA) and the research community, to consider the illicit interactions on the dark web platform as a rich source of data, and hence various approaches are employed to access the dark web content. Since the crawling of onion sites is mostly an inexpensive task and doesn't require much manual intervention, a significant amount of unlabelled data could be collected. Hence, specific semi-supervised approaches were explored to learn from both the labelled examples and the unlabelled examples.

Researchers discussed the threat intelligence analysis framework that helps law enforcement agencies analyze crimes and criminals with the relevant information from the dark web [4]. The framework implemented to carry out this analysis is known as the Dark Web Threat Intelligence Analysis (DWTIA) Platform. The DWTIA framework implemented as the traditional network investigation method based on IP address found it challenging to trace the cybercriminal's identity, but provided access to a large volume of information, combining the surface web and dark web together. It did so by providing or using the OnionScan Dark web crawler.

Also, commercial industries offering paid services to monitor specific organization-related data on the dark web are on the rise [5–8]. One of the standard features of dark web monitoring services is searching the dark web and alerting the organization about the spread of data breaches or potential threats by curating intelligence collected from the dark web.

The presented work offers the novel objective of collecting data from the dark web to convert it into investigating leads. The outcome of the objective is presented with results obtained for the Florida water supply cyber attack.

The following sections of work are arranged as follows: section two discusses the literature review on addressing cyber attack challenges in CPS, typically in the water sector. The next part of the literature involves work on how the dark web is studied as a source of threat intelligence. Section three includes methodology followed by results and discussion.

2 Background Work

2.1 Securing Cyber-Physical Systems

The research on the cybersecurity aspect of CPSs is evolving. The work published in 2020 documented the vulnerabilities, threats, and attacks associated with CPSs [9]. The work also offered a review on measures on protection, limitation of proposed mechanisms, and future directions.

The authors mentioned the cyber attack on the Florida water supply; therefore, a brief review on cyber attacks on water management-related infrastructure is mentioned.

Researchers documented the fifteen cyber attacks on water supply infrastructure from the past few years [10]. The attack methodologies and learnt lessons from attacks were mentioned, along with mentioned trends of cyber attacks in the form of ransomware, crypto jacking, insider threats, etc., [11].

The CPSs security is discussed by referring to the scope of quantum computing, brain-like structure approach. The researchers recommend exploring the possibility of quantum cryptography to protect CPSs [12]. As with recent advancements of technologies, the CPSs will be dealing with emerging technologies: industry 4.0, Fog computing, etc., on various levels. Therefore, the trade-off between security and privacy handling through peripheral technologies and CPSs needs to be addressed diligently. The researchers proposed "Brain-Like Distributed Control Security" for the protection of CPSs [13] which was a self-autonomous protection mechanism to identify a flag raised by the intrusion detection mechanism in CPSs infrastructure [14].

The approaches mentioned in the preceding paragraph are the gist of the work research community to secure CPSs. Still, there are incidents of cyber attacks on CPSs involved in critical infrastructure. Therefore, it is always better to keep an open mind to achieve secure CPSs infrastructure.

2.2 *Dark 'Web as an Investigative Mechanism*

The requirement of a data-driven mechanism and challenges associated with it to protect cyber-physical system is discussed in [15]. Thus, the authors of the presented work mentioned the need to extend the scope of dark web data consideration in the design of data-driven protection and mitigation mechanism.

The researcher represented an automatic crawling infrastructure termed as ZeroCrawler and a prototype called AMCL (Automated Multi-Categorization Labeling) over ZeroCrawler to identify the illicit web pages based on identified hidden themes in the ZeroNet [16]. ZeroNet is an emerging platform facilitating services to host illegal data. The online hacker forums for identifying the emerging threats in terms of popular trends and tool functionality were dissected with Diachronic Graph Embedding Framework (D-GEF) in [17]. Investigative research on image data from the dark web marketplace resulted in top vendor names, top markets, and top hash analysis results [18]. Further, a thorough evaluation of the emerging ransomware-as-a-service (RaaS) economy in the dark web, where cybercriminals or expert malicious users demand ransom or payment in return to release the infected digital assets, is presented in [19].

Thus, the above research attempts to highlight how the research community investigates the dark web data for threat intelligence or proactive measures.

Commercial solutions like DarkOwl Vision offer a proactive strategy of entity searching, monitoring, and tracking to identify and assess the threats present in the marketplace or environment to provide additional security measures if required for prevention and cybersecurity defences [8].

In summation, it is visible that CPSs driven infrastructure required considerate proactive and reactive security solutions, and dark web data is receiving interest from researchers for drawing meaningful insights. In the present work, the authors attempted to improvise an approach to gathering data from the dark web to collect all possible information related to respective cyber attacks.

3 Methodology

3.1 Crawling Mechanism

The in-house dark web crawler mechanism is depicted in Fig. 1. The dark web crawling can be initiated by providing a seed URL or Keyword(s) along with the depth of crawling. If the provided URL is already crawled and the results exist in the database, the results are extracted from the database and displayed on the web page. But, if the URL/ keyword has not been already crawled, it checks if it is the keyword or seed URL that has been provided. If it is a keyword, it uses Tordex to retrieve the results and store them in a database, whereas if a seed URL has been provided, it stores that URL in the database and takes this as the base URL.

The Tor (The Onion Router) browser is launched automatically to create a channel to crawl the dark web. The crawler checks if the number of pages visited is less than the depth provided. It then uses the Tor browser for IP rotation and pseudo-user agent generation to avoid tracking by the websites and then visits the page on the dark web while extracting the links from that page and storing them in the database. It proceeds further by crawling the stored links from the database and repeats the above process until the number of pages visited is greater than the depth provided.

Once the crawler runs until the pre-decided depth level, it proceeds to get extracted links from the database one by one to retrieve information such as title, page content, parent link, image URLs, link status, and stores them as a document in the database. Finally, these links are displayed on the crawler web page with a further option of iterative crawling. If it is not, the crawler terminates the crawling, but in case of a yes, the crawler scans whether the number of keywords crawled is less than five. If yes, it retrieves the five most occurring keywords during the previous crawl and provides the user with an option of choosing one. In case none is selected within a certain time, the crawler auto-selects the first keyword to initiate the next crawling process, and then the entire process repeats. When the number of crawls is greater than five, then it terminates the crawling. The results of the iterative crawls can be observed individually as well as a group on the crawler's dashboard. Overall, this is the crawling process for the dark web.

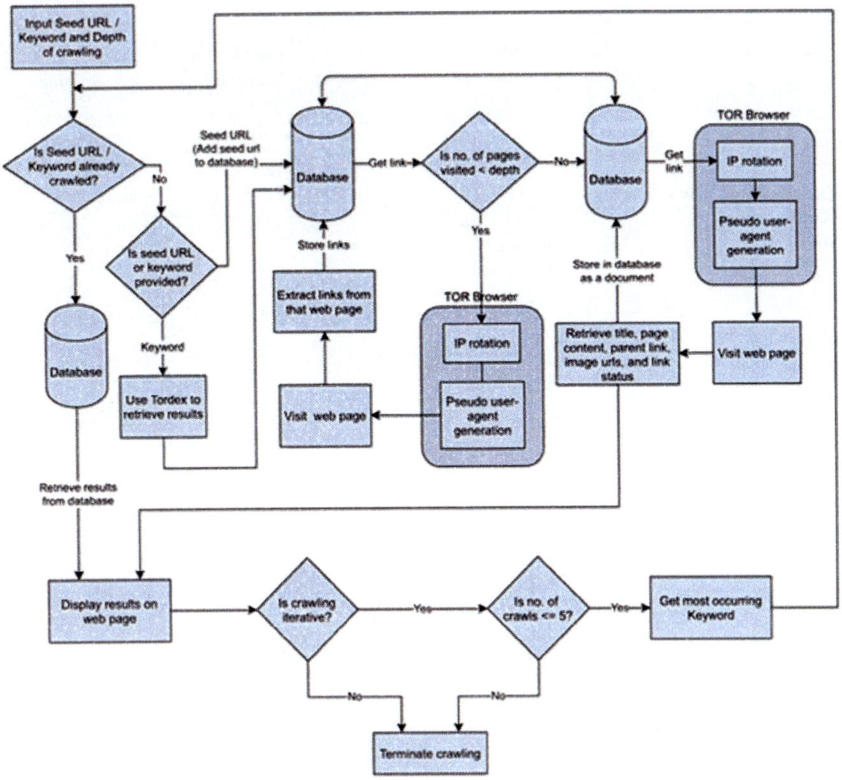

Fig. 1 Proposed dark web crawling mechanism

3.2 Discussion on Florida Water Supply Attack

The recent incident of cyber attack on the Florida water supply was reported on February 2021. In the particular attack, the attacker attempted to change the value of Sodium Hydroxide from 100 parts per million to 11,100 parts per million in the Oldsmar water treatment facility. Such increase value of Sodium Hydroxide in water is dangerous for human intake. Fortunately, due to alert human vigilance, the attack was averted.

The particular incidence motivated authors to investigate the Florida water supply cyber attack traces on the dark web. The authors ran the in-house dark web crawler with the keyword "Florida water supply," "cyber attack on water treatment plants" and "increased sodium hydroxide." The keyword "Increased sodium hydroxide" wasn't giving significant results, so the keyword was changed to "sodium hydroxide."

Table 1 presents link statistics covering the number of links crawled for entered keywords. Further, the links are compared among themselves to confirm how many links are common. For example, for the keyword Florida water supply cyber attack

Table 1 Link statistics with respect to the searched keywords

Keyword	Number of links crawled	Number of similar links w.r.t other related keywords	Number of active links during crawling	Number of inactive links during crawling	Percentage of active links
Florida water supply cyber attack	311	69 (w.r.t cyber attack on water treatment plants)	281	30	90.35%
		1 (w.r.t sodium hydroxide			
Cyber attack on water treatment plants	191	69 (w.r.t Florida water supply cyber attack)	165	26	86.39%
		2 (w.r.t sodium hydroxide)			
Sodium hydroxide	225	1 (w.r.t Florida water supply cyber attack)	183	42	81.33%
		2 (w.r.t cyber attack on water treatment plants)			

the total links collected are 311, and 69 links are common with the keyword "cyber attack on water treatment plants", and one link is common with respect to the keyword "sodium hydroxide".

The nature of the dark web is volatile. The hidden services that were active once could be inactive in the next instance. Therefore, the other two columns mention the number of active and inactive links during crawling observed over a period of one week, while the last column gives the value of active links in terms of percentage.

To comprehend the collected information without manual intervention, the Word Cloud visualization technique is employed. The dark web is famous for illicit activities; therefore, the proposed mechanism is designed so that without opening the links collected from the dark web, the essence of information is displayed with the help of Word Cloud. The Word Cloud is a popular visualization technique to represent words that frequently occur in the targeted text. The size of the word is an indicator of how many times the word appears in the text.

Fig. 2 Word cloud for "Florida Water Supply Cyber attack" (311 links)

Figure 2 presents the most occurring words in 311 links related to Florida water supply cyber attack. As depicted, the visibility of the words like "Buy," "Cash," "Money," "order," "get" infers that Florida water supply cyber attack is frequently discussed on the dark web marketplace. The word "PM" is also prominent on links crawled for the keyword "cyber attack on water treatment plants" as shown in Fig. 3.

The visual clues could be picked from Word Cloud to investigate ripples on the dark web regarding the cyber attacks on the surface web. On inspection of Word cloud generated for the keyword "sodium hydroxide" (225 links), depicted in Fig. 3, the word "Praveen" appears frequently but with low frequency on respective dark web pages (Fig. 4).

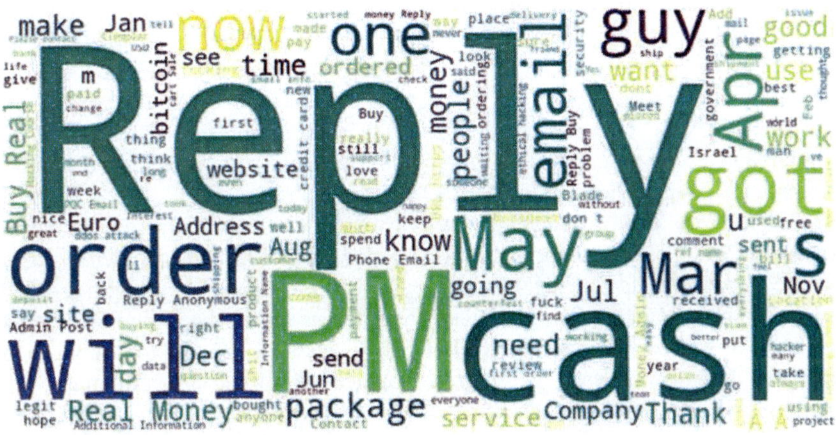

Fig. 3 Word cloud for "cyber attack on water treatment plants" (191 links)

Fig. 4 Word cloud for "sodium hydroxide" (225 links)

Interestingly, the word Praveen appears in the search related to another keyword as well, "web wolf". The keyword 'web wolf 'is picked from the Word cloud of the keyword "covid-19 offers". Thus, it is apparent from the result that the Florida cyber attack is most likely discussed in the dark web forum/marketplace. Once it is confirmed that the cyber attack is discussed on hidden service, the particular links are monitored and inspected further.

3.3 Results of Some Other Malware-Related Keywords

3.3.1 DoppelPaymer in Critical Infrastructure

DoppelPaymer is a ransomware that has been evolving since 2017 and is considered similar to another ransomware, BitPaymer. In a blog post in June 2020, it was reported that DoppelPaymer ransomware groups had successfully breached the network of Digital Management Inc. (DMI), a Maryland-based company providing managed IT and cybersecurity services with NASA as one of their clients [20].

The group is alleged to operate via hacking forums where they release the compromised data while blackmailing them. In the dark web, word cloud as shown in Fig. 5 has the word "reply" occurring most times indicating that most of the pages crawled were part of a forum. The word "Drake" also occurs several times and could be a reference to a threat group called "Gold Drake" which is rumoured to be consist of operators from "Gold Heron", a group of financially motivated cybercriminals [21].

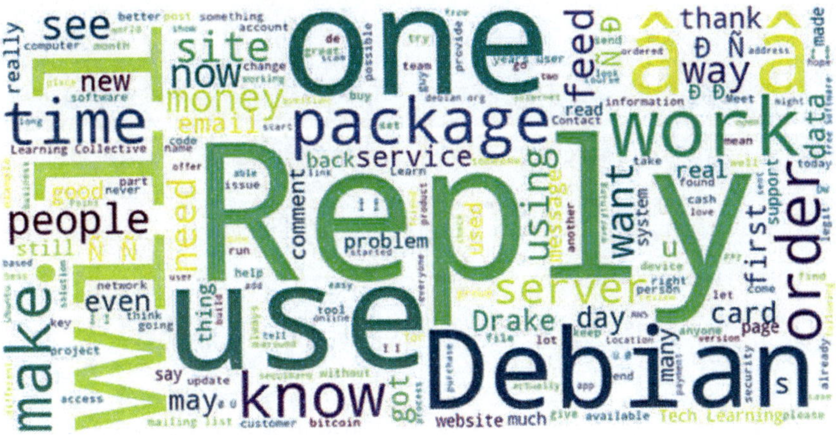

Fig. 5 Word cloud for "DopplerPaymer" (305 links)

3.3.2 New Groups of Cyber Criminals

Dragos, a private security consulting group, analyzed the trends of the past fifteen years and came across four new hacking groups, namely: Stibnite, Talonite, Kamacite, and Vanadinite [22]. According to them, these groups targeted Operational Technology (OT) and Industrial Control Systems (ICS), with each of them having different target specifications.

In Table 2, the rows indicate the searched keyword with the number of links extracted mentioned in brackets, while the columns indicate the number of links that are common between the keyword in the row and the keyword in the column. The number in the bracket indicates the total percentage of the similar links in these keywords from all the links extracted from the keyword mentioned in the row. The high percentage indicates that the four groups work around the same marketplaces in the dark web, similar to their clients.

Table 2 Number and percentage of links common between each of the keywords

Keyword	Stibnite	Talonite	Kamacite	Vanadinite
Stibnite (172)	–	159 (92.44%)	159 (92.44%)	172 (100%)
Talonite (161)	159 (98.75%)	–	161 (100%)	161 (100%)
Kamacite (199)	159 (79.89%)	161 (80.90%)	–	161 (80.90%)
Vanadinite (195)	172 (88.20%)	161 (82.56%)	161 (82.56%)	–

Fig. 6 Word clouds of the four words: stibnite, Talonite, Kamacite, Vanadinite. These are four new hacker groups active on the dark web

The word cloud compilation of the four words in Fig. 6 shows that the word "hacker", "need", and "hire" occur quite often indicating that the groups are active on the dark web and have an array of clients that desire their services.

4 Conclusion and Future Scope

The cyber attacks on CPSs, in critical infrastructure, are severe and challenging to mitigate. The security analysts have attempted proactive and reactive security measures to secure cyber-physical infrastructure, but the attacks continue. The anatomy of cyber attacks shows that attackers get creative while crafting the cyber attacks on infrastructure. Similarly, the defence and investigation mechanism need to be customized to curb the potential cyberattack. The proposed approach has the potential to reach the attacker or get zero-day exploits if utilized constructively. The stakeholders of the proposed works are Law Enforcement Agencies and the security research community. The case study of Florida is intriguing enough to realize that engaging in dark web data result in a stream of threat intelligence.

In further development, the authors aim to map the timeline of link finding over the stipulated timeframes like immediately after mentioned attacks, consecutive one week after the attack. The objective of plotting the timeline is to comprehend the cyber attacks are discussed concerning attack methodology, exchange of exploits among dark web forums, and dark web marketplace. Also, the webpage classification will be employed to confirm the type of webpage, i.e., whether the web page is a forum, dark web marketplace, blogs, etc. Once the web page type is confirmed, then a customized investigation approach will be applied. For example, if a web page belongs to a forum, the forum thread will be investigated.

Thus, the proposed work discussed one module of planned automated and intelligent mechanism to investigate the landscape of the dark web.

References

1. Noguchi, M., Ueda, H.: An analysis of the actual status of recent cyberattacks on critical infrastructures. NEC Techn. J. Spec. Issue Cybersec. **12**(2), 19–24 (2019)
2. Brown, G., Carlyle, M., Salmeron, J., Wood, K.: Defending critical infrastructure. Interfaces **36**(6), 536–544 (2006)
3. Biddle, P., England, P., Peinado, M., Willman, B.: The darknet and the future of content distribution. In: ACM Workshop on digital rights management, vol. 6, p. 54. (2002)
4. Zhang, X., Chow, K.P.: A framework for dark Web threat intelligence analysis. In: Cyber Warfare and Terrorism: Concepts, Methodologies, Tools, and Applications, IGI Global, pp. 266–276. (2020)
5. Zerofox Dark Web Monitoring https://www.zerofox.com/products/dark-web-monitoring. Accessed 20 April 2021
6. Network Box Dark web monitoring https://www.network-box.com/nb5-darkWeb_monitoring Accessed 20 April 2021
7. Acid Cyberintelligence https://www.acid-tech.co/?page_id=66 Accessed 20 April 2021
8. Dark owl monitoring https://www.darkowl.com/ Accessed 20 April 2021
9. Yaacoub, J.P.A., Ola, S., Hassan, N.N., Nesrine, K., Ali, C., Mohamad, M.: Cyber-physical systems security: Limitations, issues and future trends. Microprocess. Microsyst. 77:103201 (2020)
10. Hassanzadeh, A., Rasekh, A., Galelli, S., Aghashahi, M., Taormina, R., Ostfeld, A., Banks, M.K.: A review of cybersecurity incidents in the water sector. J. Environ. Eng. **146**(5), 03120003 (2020)
11. Tuptuk, N., Hazell, P., Watson, J., Hailes, S.: A systematic review of the state of cyber-security in water systems. Water **13**(1), 81 (2021)
12. Tosh, D., Galindo, O., Kreinovich, V., Kosheleva, O.: Towards security of cyber-physical systems using quantum computing algorithms. In: IEEE 15th International Conference of System of Systems Engineering (SoSE) (2020)
13. Yang, H., Zhan, K., Kadoch, M., Liang, Y., Cheriet, M.: BLCS: brain-like distributed control security in cyber physical systems. IEEE Netw. **34**(3), 8–15 (2020)
14. Kholidy, H.A.: Autonomous mitigation of cyber risks in the cyber-physical systems. Futur. Gener. Comput. Syst. **115**, 171–187 (2021)
15. Jiang, Y., Yin, S., Kaynak, O.: Data-driven monitoring and safety control of industrial cyber-physical systems: basics and beyond. IEEE Access **6**, 47374–47384 (2018)
16. Ding, J., Guo, X., Chen, Z.: Big data analyses of zeronet sites for exploring the new generation darkweb. In: 3rd International Conference on Software Engineering and Information Management p. 46–52 (2020)
17. Samtani, S., Zhu, H., Chen, H.: Proactively identifying emerging hacker threats from the dark web: a diachronic graph embedding framework (D-GEF). ACM Trans. Priv. Sec. **23**(4), 1–33 (2020)
18. Jeziorowski, S., Ismail, M., Siraj, A.: Towards image-based dark vendor profiling (2020)
19. Meland, P.H., Bayoumy, Y.F.F., Sindre, G.: The ransomware-as-a-service economy within the darknet. Comp. Secur. **92**, 101762 (2020)
20. The Business of Federal Technology https://fcw.com/articles/2020/06/04/johnson-dmi-nasa-ransomware-attack.aspx Accessed 11 April 2021

21. SecureWorks Threat Profiles https://www.secureworks.com/research/threat-profiles/gold-heron%20%20 Accessed 11 April 2021
22. Dragos ICS cybersecurity year in review. Available via https://hub.dragos.com/hubfs/Year-in-Review/Dragos_2020_ICS_Cybersecurity_Year_In_Review.pdf (2020) Accessed 11 April 2021

Sustainable and Secure IoT-Chain Architecture Integrated with Blockchain Technology

Sana Zeba and Mohammad Amjad

1 Introduction

With the time being a progression of the Internet of Things (IoT) computing and advancement network technology, the integration of blockchain with IoT makes large-scale self-directed IoT ecosystems. In general, the IoT Network contains heterogeneous devices that exchange massive amounts of critical data, as well as delicate information of IoT applications. IoT technology is increasing fast worldwide in various sectors as well as the number of interconnected devices increased in a variety of IoT applications. International Telecommunication Union (ITU) has defined the Internet of Things term as "The Internet of Things will connect the world's physical object in both intelligent and sensory way." Gloukhovtsev [1] proposed various areas of IoT applications such as Smart Supply Chain, Smart Transportation, Smart Agriculture, Smart Governance, Smart Healthcare System, Smart Grid, Smart Parking Management system, Smart Home, and Smart Cities, etc.

The IoT enables the connection of humans, devices, places, and products, etc., to offer chances to create the value of time, money, and easiness of services on a human daily basis. Security is a primary concern subject in the IoT that has overdue its large-scale deployment. IoT. Outdated IoT applications used the client–server model approach, which is not suitable most of the time. Many times, the central authority-based IoT system has created the critical condition for taking any instant decisions.

Recently, the worldwide, emerging blockchain technology has come to succeed over the third party's authorization. Blockchain depends on the distributed ledger technology (DLT), which is a platform where one virtuously has built trust over

S. Zeba (✉) · M. Amjad
Department of Computer Engineering, Jamia Millia Islamia University, New Delhi 110025, India

M. Amjad
e-mail: mamjad@jmi.ac.in

B. Agarwal et al. (eds.), *Proceedings of International Conference on Intelligent Cyber-Physical Systems*, Algorithms for Intelligent Systems,
https://doi.org/10.1007/978-981-16-7136-4_5

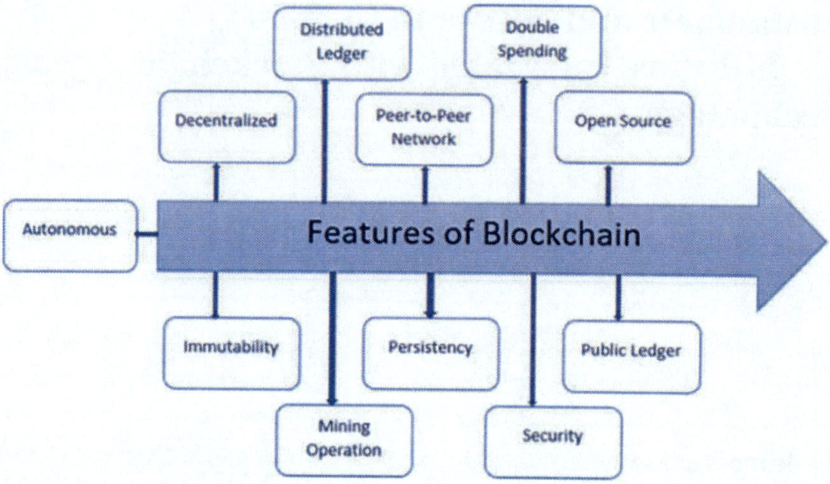

Fig. 1 Key features of blockchain

the IoT system such as it's not trust on a single central authority to commence the exchange or communication of the data in the network.

Blockchain technology has the succeeding key features that are shown in Fig. 1.

- Decentralized: Blockchain makes the IoT systems decentralized, which means there is no need for a third central authority for the validation of any transaction.
- Persistency: Through the Persistency feature users are determined to guarantee or accomplish something.
- Anonymity: With address, any user can interact directly with the blockchain.
- Immutability: The distributed ledger of blockchain is immutable, which means the majority of participant nodes verified the data alteration or modification in the IoT Network.
- Security: Cryptographic technologies like a public key, private key, and hashing concept of blockchain have increased the security parameters automatically.
- Double Spending: Blockchain technology can resolve the double-spending issues of the network.

Blockchain distributed ledger technology has combined with IoT network to make easy machine-to-machine transactions and a set of transactions which are recorded in a database.

Furthermore, this paper is arranged as follows. Section II discussed the previous work related to both Blockchain and IoT; section III has presented the background study of IoT and Blockchain technology in detail. Section IV briefly describes the motivations for Blockchain in IoT, and section V has discussed the proposed Paradigm for Integrated IoT-Blockchain Architecture. Finally, in the last, section VI provides the future direction of Blockchain architecture in IoT technologies, and section VII précis the conclusion of this paper.

2 Related Work

This section shortly presents a summary of previous works related to the security challenges of the IoT application Blockchain Technology solution in IoT and Blockchain framework. This literature review has discussed below all related previous work.

Mohanty et al. [2] have proposed a model to achieve the requirements of IoT as a Lightweight integrated Blockchain (ELB). This model is deployed in an intelligent home system as a case study to verify its applicability in numerous IoT circumstances. The proposed model contains two main levels, namely smart home and overlay where resources can merge to a shared BC which verifies devoted security. The offered ELIB model functions in three levels, namely certificate less cryptography (CC), lightweight consensus algorithm, and Distributed Throughput Management (DTM) scheme. In the future, it improved the energy consumption and deployed the proposed system in divers' applications.

Lao et al. [3] have discussed an outline of Blockchain–IoT architecture and also examining correspondence protocols and the structures. They discussed several consensus protocols for Blockchain IoT and comparison of several consensus algorithms of Blockchain. This paper also analyzes the model for Peer-to-Peer traffic model. They have provided a traffic model for Blockchain based on its system for traffic distribution.

Pavithran et al. [4] have analyzed the current literature on blockchain in IoT. In the particular literature review, the author has identified five vital components along with their design attentions and encounters that should be measured while creating blockchain architecture for IoT systems. Define the research gaps for creating a safe blockchain framework for IoT systems. However, few of them have used distributed storage, which does not have any protection on confidentiality of data.

Salim et al. [5] have offered a study of inclusive method which includes overall DDoS attack inspirations and specific explanations why attackers prefer the IoT devices to launch DDoS attacks. The author enumerated different apparatuses that are offered for attacking its devices to form a botnet and further apparatuses are discussed which allow using it bots to launch DDoS attacks and offered a complete and systematic organization of different DDoS attacks that take place on the cloud.

Sultana et al. [6] have proposed an IoT network for controlling access of systems where the objectives of that IoT system are to accomplish authorization, authentication, and truthfulness for data sharing in IoT networks. Various smart contracts like Judge Contract (JC), Access Control Contract (ACC), and Register Contract (RC) are used to justify objectives. For access control ACC, RC was used for authentication and JC was used for judging misbehavior of the system user and fixed the penalty for that user or subject. The result of the proposed system is effective in terms of cost.

Khalid et al. [7] have developed access control and novel authentication mechanism for lightweight decentralized IoT which is appropriate to a large number of scenarios. The fog computing technology and the thought of the public blockchain are used in the mechanism. The planned mechanism is based on blockchain technology to advantage from its cryptographic assets and distributed nature and trusts fog computing technology to address the latency issues. The proposed mechanism can be practically applied to several IoT network scenarios. There is a requirement of huge amounts of energy consumption by PoW in the mechanism and gives future focus to reduce energy consumption by using trust value in the protocol.

Ghadekar et al. [8] have proposed lightweight safe architecture for IoT by using Ethereum Blockchain. Because of the Blockchain decentralized nature, a single point of authentication issues resolves in the proposed model. Immutable feature of Blockchain is used to store the whitelist of devices in the proposed model. The proposed model exploits the Proof of Work Consensus model. A Smart Home case study has been employed for broader IoT applications. The two factors measured are intrusion detection and temperature. To scale up in the future at a better level, aim to implement the model using Proof of Stack protocol which provides more trust, less time, and more efficiency.

Lao et al. [9] have proposed a framework for gathering the forensics evidence information and that system collects, processes, and analyzes evidence. This framework used a permission blockchain to enhance authenticity, integrity, and non-repudiation for collected forensics evidence. One conceivable future effort is to develop the framework for IoT that comprises a diverse group of devices, to assess the framework consistency and benchmark the performance.

Kostal et al. [10] have proposed a new IoT network for management and monitoring architecture based on Blockchain technology. Administrators of IoT network control and record modification or device configuration indirectly with the help of blockchain. A private blockchain is used for this architecture design. There is a limitation of periodically checking the configuration of updates needed to implement sophisticated systems and, in the future, it might be useful to know more information like utilization of the CPU, battery level, and RAM requirements in the network operations.

Varghese and Jose [11] have implemented an IoT system with help of blockchain and make it into a decentralized network that offers transparency, authentication, integrity, and synchronization of network. For all the measures like synchronization, authentication, and integrity used consensus and distributed concept of blockchain. To overcome the limitations and issues in the IoT system, the author proposed an IoT system using Blockchain.

Fan et al. [12] have proposed a scheme to solve the issue of time announcement in the IoT system with blockchain. The Distributed ledger brings ease to verify and synchronize time in the IoT network and avoid a single point of failure. Future work also explains to focus on how to enhance the accuracy of time and reduce the offset as much as possible. Besides, the consensus mechanisms are also a direction of improvement. Time synchronization schemes can be more effective and secure by the consensus mechanism, which balances efficiency and security.

Dabbagh and Sookhak [13] have analyzed the bibliometric conference papers of Blockchains and related articles and reviewed papers that have been indexed in WoS from 2013 to 2018. They have analyzed those collected papers against five research questions. The results exposed valuable insights, including yearly publications and citation trends, etc., in influential papers, favorite publication venues, and most supportive funding bodies.

Mohammed et al. [14] have used Software Defined Networking (SDN) in proposed security decentralized architecture with blockchain. Also used fog, mobile edge computing in the projected architecture to detect different attacks effectively in IoT networks. SDN is continuously monitoring and examining the traffic data in the whole IoT network. Blockchain has made decentralized networks and removed the single point of failure problem with it. Mobile Edge and fog computing have been supported to detect attacks at fog nodes and then mitigated at the edge node.

Suchaad et al. [15] have applied blockchain mechanisms to the education or discipline of children's home IoT systems. A token-based mechanism has been proposed for a parental control home system in which parents can encourage good behaviors and discourage misbehavior of children with the help of token schemes. For this home IoT system, different devices like TV and computers are connected to the blockchain and accessible only when they have a token. Proof of concept is used for this system, and children are rewarded with playcoin when they do good sometimes.

Song et al. [16] have provided an IoT framework for securing sensor data. The goal of the system is to provide flexibility with blockchain design for a high throughput system. Application-specific strategies must cautiously consider the trade-off between security, privacy, and latency against the value of the task at hand. Docker Swarm is used for container orchestration in this framework and extends further study to the use of Kubernetes rather than Docker Swarm. There is a limitation of a set of peers in a permission network and improves the high throughput after completing the proof of concept (POC) stage.

3 Background Study

3.1 The Internet of Things (IoT)

In the recent era, the wireless Internet of Things (IoT) becomes a rich courtesy field in the smart environment. It rapidly increased the bigger number of intelligent devices in the system and it effectively emerged devices and physical environments through the Internet and created a smart system. Firstly, the Internet of Things network was invented in 1998 and developed in 1999 by Kevin Asthon. The market of the Internet of things increased from $547 in 2018 to $841 million in 2020 according to the report of Gartner and in 2018 worldwide, an estimated 23.14 billion devices were installed. The Internet of Things (IoT) is defined as the network of interconnected smart physical and sensor devices through a wireless or wired connection with a processor unit, actuators, and software to assemble data for control or exchange the IoT environment remotely.

The security requirement is the main anxiety in the Internet of Things (IoT) due to its open nature. Different Security parameters which are needed in the network are discussed below

- Integrity: The process to prevent the unauthorized nodes from modifying the information's.
- Privacy: To secure the personal or private data of users.
- Confidentiality: Confidentiality means prevent the leakage of user's information among unauthorized nodes.
- Availability: Availability means that information's are always available when it's needed.
- Authenticity: Sensitive IoT systems or data should not be reachable by any unauthorized end users.
- Authorization: Authorization is the process of permitting to do anything in the network (Fig. 2).

Layered Architecture of IoT: According to the operations of various IoT devices, the IoT architecture has been divided into different layers. There are many versions of IoT Layer architecture such as 3 Layer, 4 Layer, and 5 Layer IoT Architecture. However, discussed 4 Layer IoT Architecture in Fig. 3 with each layer.

1. Perception Layer: The perception Layer is the uppermost layer in IoT in which data or information arrive firstly. This layer used the sensors for data sensing that's why it's called as also as the "Sensors Layer." For this layer different devices such as RFID, sensors, and GPS are used.

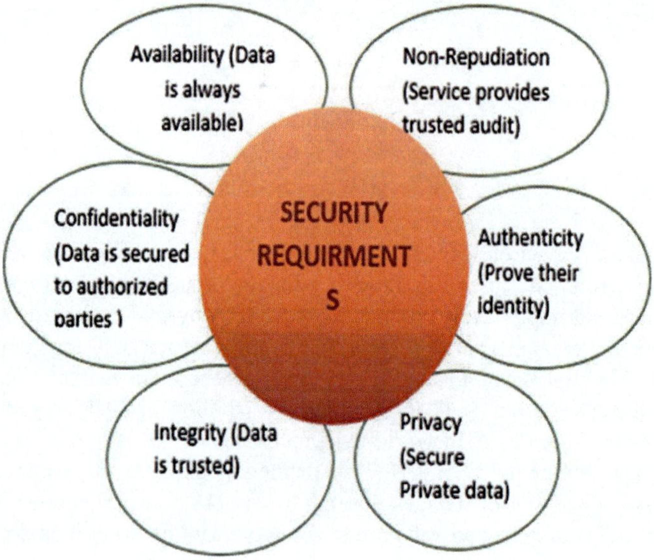

Fig. 2 Security requirement in IoT system

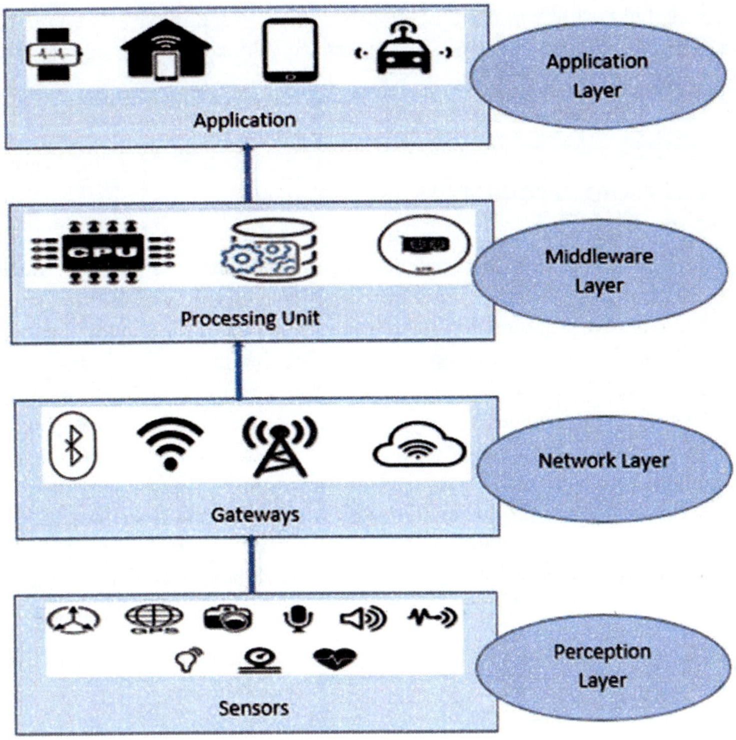

Fig. 3 Layered IoT system architecture

2. Network Layer: The Network Layer has performed the function of routing and transferring across the network. This operation has been performed with various technologies like 5G, 3G, Bluetooth, 4G, Zigbee, Wi-Fi, etc., at the network layer.
3. Middleware Layer: The process of manipulation in data has been done in this layer of the architecture. Its acts as the boundary between network and application layers. It is also called "Support layer" of the IoT architecture.
4. Application Layer: It is the most important layer of IoT architecture. This layer is responsible for accessing and ensures the integrity, privacy, authenticity of data, and confidentiality as well.

Recently, researchers moved toward Blockchain Technology for securing the IoT applications from security threats that occurred in the Internet of Things. In 2008, Blockchain Technology comes by Satoshi Nakamoto that is based on distributed ledger technology (DLT) concept, immutable, and transparency in transactions. In the traditional IoT system, many securities and privacy shortages occurred because of its centralized nature where all things and devices are controlled centrally. Blockchain has given the concept of distributed ledger, Peer-to-Peer network, and authentication mechanisms that ensure authenticity, security, privacy, and integrity in a distributed manner. Therefore, there is a huge requirement for blockchain to integrate with IoT systems in the future.

3.2 Blockchain Technology

Blockchain is defined as a distributed database that stores a nonstop increasing list of records of the network, called blocks. Blockchain has given the concept of decentralized network nature that handles the nodes or user's authentication and accessing without any central authority. Here, "chain" gives the feasibility to store and connection of all nodes like computers, devices, or users in the network. Moreover, Blockchain is defining as a group of devices, physical things, nodes, connected because of its peer-to-peer network and not depending on any central server. A blockchain holds two types of fundamentals such as

Transactions: The actions designed by participant's nodes of the system. Blocks: Blocks are a record of legal transactions in the correct sequence (Fig. 4).

The construction of blockchain is showed by a sequence of blocks with digital transactions in a precise order. Two strong data structures of blockchain are

- Pointers: It holds variables that reserve information about the position of the extra variable.

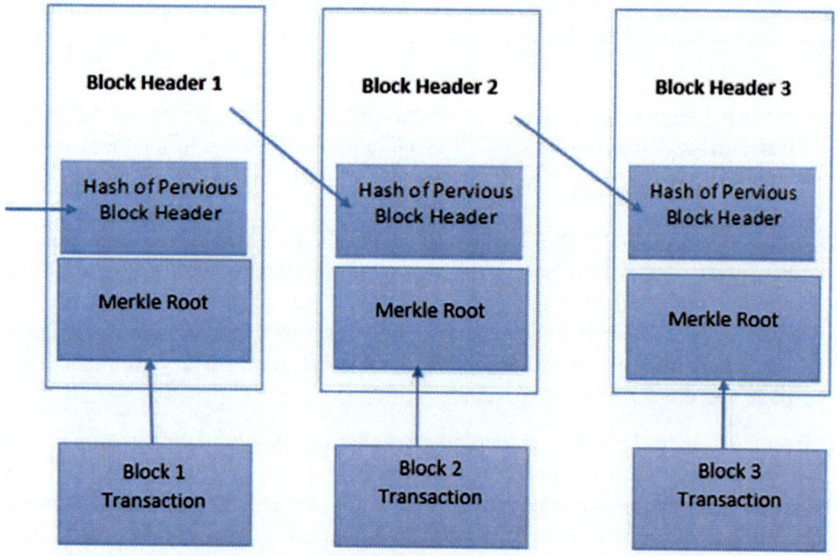

Fig. 4 List of blocks in blockchain sequence

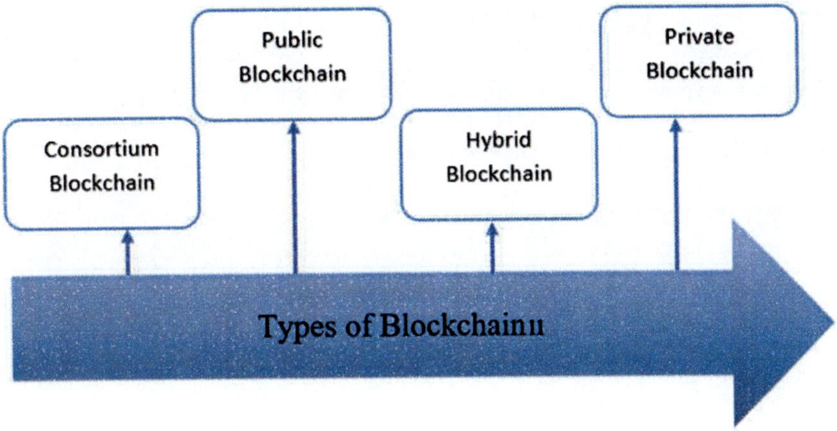

Fig. 5 Different types of blockchain

- Linked lists: Sequence of blocks where each block has clear data and links to the previous block with the pointer define through linked list.

Types of Blockchain: Blockchain is characterized based on the access power of the network. There are four kinds of the network which are

1. Public Blockchain: There is no restriction to access the public blockchain network. Every blockchain type used some consensus mechanism such as Proof of Work (PoW), and Delegated Proof of Stake (DPoS) which are used in the public blockchain. Public Blockchain examples are Monero, Dash, Litecoin, Bitcoin, Ethereum, etc.
2. Private Blockchain: There is a restriction to access the private network. Without network administration, permission nodes cannot join the network. The consensus mechanism of this blockchain is RAFT, Practical Byzantine fault tolerance (PBFT). The Private Blockchain example is Multichain, MONAX.
3. Consortium or Federated Blockchain: It is a semi-decentralized blockchain network that is controlled and restricted by more than one administrative or organization. The Consortium blockchain example is Corda, B3i (insurance), EWF (energy), R3(Bank), etc.
4. Hybrid Blockchain: This is the mixture of both private and public blockchain. This blockchain network has used the features or characteristics of both private and public blockchain (Fig. 5 and Table 1).

4 Motivations

Day by day researchers is working hard to plan a new framework to recover further meaningful information. Every day, smart objects or devices increased in the IoT

Table 1 Comparison of different blockchain

Public blockchain	Private blockchain	Consortium blockchain
Many, unknown participants	Known participants from one organization	Known participants from multiple organizations
Write by all participants	Write permissions centralized	Writes require consensus of several participants
Read by all participants	Reads may be public or restricted	Reads may be public or restricted
Consensus by proof of work	Multiple algorithms for consensus	Multiple algorithms for consensus
Many, unknown participants	Known participants from one organization	Known participants from multiple organizations

system. To regain meaningful data from the massive IoT generated data, efficient frameworks are required which can be able to analyze and control the data security of IoT networks. Due to the open nature of the architecture of IoT, these architectures are vulnerable to security threats and different attacks. Therefore, innovative techniques are required for controlling and analyzing real time highly scalable data generated by IoT systems. Security threats are the main concerns for its system. Some reasons discussed like

- It is not a secure environment because of the Internet networking of objects in the IoT.
- There is much possibility of executing any malware or any threats activities.
- In its system, things are communicated with each other in the network that's why there is an opportunity of hindering the privacy and integrity of data.
- Used Blockchain technology in IoT network so that integrity and authenticity of any nodes would not affect and information would not capture or modify by any compromised node and handle the single point accessing the problem as well as security concerns.

5 Paradigm for Integrated IoT-Blockchain Architecture

In the IoT, internet connectivity allows the physical objects to collect the data and executed any operation accordingly. The IoT architecture used different protocols for the different layers. Security threats are core concerns for the IoT system which is handled through blockchain. According to the literature review, some research gaps have left behind while processing IoT networks.

- Until now, there is no well-organized approach for processing and handling e-data generated by IoT devices connected through the internet
- Until now, there has been no detailed discussion on the integration of IoT with blockchain to eliminate the existing problems in the IoT network.

Blockchain Technology has peer-to-peer and distributed ledger concepts. This distributed ledger has three concepts like block, chain, and transaction. The Block is the storage part that contains the transactions, hash values, and records, etc. The Chain is the linking part of Blockchain which is used for connecting the blocks and creating chains. Any valuable information which is circulated in the network is called the Transactions. The Internet of Things is the internetworking of items like smart vehicles, smartphones, or any smart devices which are embedded with sensors, actuators, software, electronics, and internet connectivity.

Generally, the Blockchain concept is defined with the five terms that are used to handle the IoT threats:

- Peer-to-Peer Network: This concept removes central dependencies of all participant nodes from any central party in the network.
- Consensus Mechanism: It is defined as the legal agreement among all nodes of an IoT network.
- Distributed and Open Ledger: Each participant node of the IoT network is validated independently and represented transparently.
- Synchronization of ledger copies: Replicate the ledger among all nodes of the network and synchronize as well.
- Mining Operation: It is the process of adding the transactions in the public ledger of a previous transaction.

The best solution to solve the IoT security and threats problems is to establish the IoT network in a distributed or decentralized way. In this paradigm, IoT layered architecture is integrated with the blockchain layer to give more security in the network. The flow of data in integrated IoT systems is different from centralized IoT systems. In the Blockchain integrated IoT systems, the data processing is the sequence form sensors-cryptographic-network-processing unit-distributed blockchain-analytics-user.

The working of the Integrated IoT-Chain paradigm is similar to IoT layer architecture except for the summation of one extra Collateral Layer. The IoT network makes potentially durably in terms of security, privacy, integrity, and authenticity of data and nodes as well in case of the addition of Collateral layer. The data created from the sensor of the perception layer is gathered and then forward to the collateral layer. In the collateral layer, numerous cryptographic hash functions, public and private keys for encryption and decryption of data have been defined.

The public key of IoT nodes or devices can be kept in an open blockchain platform like Ethereum while the private key is kept in the IoT devices or nodes themselves. If any participant device of the IoT senses it to be doubtful of taking only private key for all transactions, it may select a new key for each transaction for the same IoT network. Many times this process becomes chaotic to the allocation of public keys to each node in the network. The proposed blockchain integrated IoT layered architecture is shown in Fig. 6.

After performing the key exchange operation at the Collateral layer, the security of the transaction increased. All transactions with public and private keys move to the next network layer for further processing. This integrated IoT-Chain layered

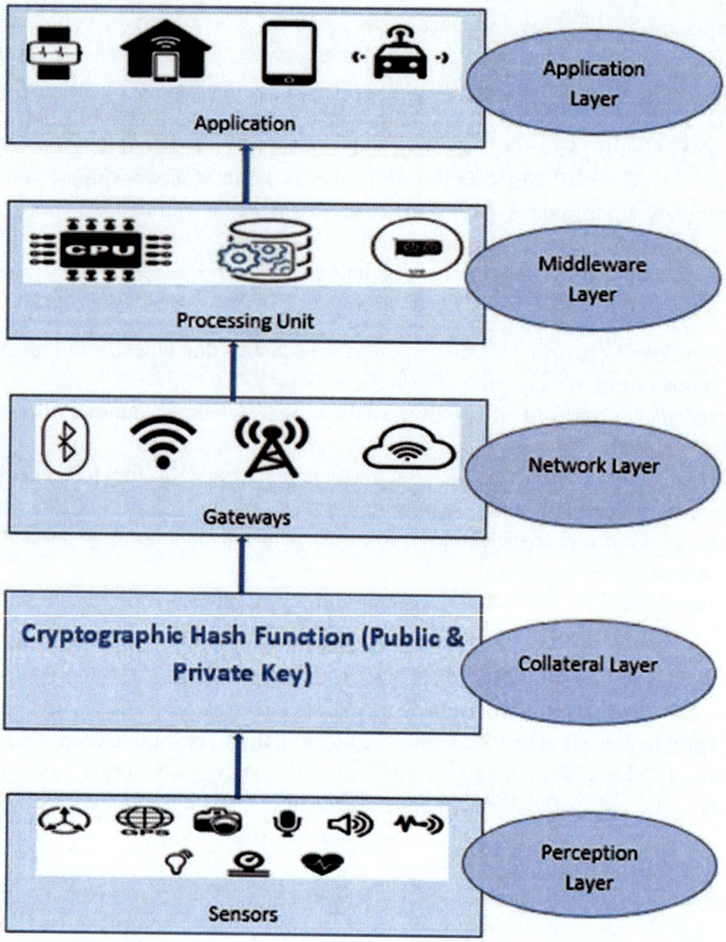

Fig. 6 Proposed blockchain integrated IoT (IoT-Chain) layered architecture

architecture increased the security one level up by integrating the blockchain-related layer in the traditional IoT layer architecture.

6 Future Scope

However, this paper has discussed Security IoT layered architecture integrated with blockchain as IoT-Chain. In the future, require more studies on cryptographic hashing security algorithms that are much extra proficient at operating on IoT system resource-constrained devices (Light Weight Crypto Algorithm) to reduce the complexity and chaos of the network that is generated. Also explained different technologies like DAG and IOTA architectures with IoT architecture as alternatives of Blockchain in the future.

7 Conclusion

Rapid and emerging expansion of IoT applications increased the attention toward IoT architecture, security, attacks, and threats of the Internet of Things (IoT) system. However, it has discussed IoT layered architecture and blockchain technology in the details. This paper proposed integrated IoT and Blockchain paradigm explained as the IoT-Chain paradigm based on layered architecture, to prevent any security threats and give more secure IoT architecture development. Hopefully, this paper would give knowledge to understanding the requirement of integration of IoT with blockchain and paradigm of Integrated IoT–Blockchain layered architecture.

References

1. Gloukhovtsev, M.: Iot Security: Challenges, Solutions & Future Prospects. (2018)
2. Mohanty, S.N., Ramya, K.C., Rani, S.S., Gupta, D., Shankar, K., Lakshmanaprabu, S.K., Khanna, A.: An efficient lightweight integrated blockchain (ELIB) model for IoT security and privacy. Futur. Gener. Comput. Syst. **102**, 1027–1037 (2020). https://doi.org/10.1016/j.future.2019.09.050
3. Lao, L., Li, Z., Hou, S., Xiao, B.I.N., Hong, T., Polytechnic, K.: A survey of IoT applications in blockchain systems: Architecture, consensus, and traffic modeling. ACM Comput. Surv. **53**(1) (2020)
4. Pavithran, D., Shaalan, K., Al-Karaki, J.N., Gawanmeh, A.: Towards building a blockchain framework for IoT. Clust. Comput. **23**(3), 2089–2103 (2020). https://doi.org/10.1007/s10586-020-03059-5
5. Salim, M.M., Rathore, S., Park, J.H.: Distributed denial of service attacks and its defenses in IoT: a survey. J. Supercomput. **76**(7), 5320–5363 (2020). https://doi.org/10.1007/s11227-019-02945-z
6. Sultana, T., Almogren, A., Akbar, M., Zuair, M., Ullah, I., Javaid, N.: Data sharing system integrating access control mechanism using blockchain-based smart contracts for IoT devices. Appl. Sci. **10**(2), (2020). https://doi.org/10.3390/app10020488
7. Khalid, U., Asim, M., Baker, T., Hung, P.C.K., Tariq, M.A., Rafferty, L.: A decentralized lightweight blockchain-based authentication mechanism for IoT systems. Clust. Comput. **23**(3), 2067–2087 (2020). https://doi.org/10.1007/s10586-020-03058-6
8. Ghadekar, P., Doke, N., Kaneri, S., Jha, V.: Secure access control to IoT devices using blockchain. Int. J. Rec. Technol. Eng. **8**(2), 3064–3070 (2019). https://doi.org/10.35940/ijrteF2273.078219
9. Lao, L., Li, Z., Hou, S., Xiao, B., Guo, S., Yang, Y. A survey of IoT applications in blockchain systems: Architecture, consensus, and traffic modeling. ACM Comput. Surv. **53**(1) (2020). https://doi.org/10.1145/3372136
10. Kostal, K., Helebrandt, P., Ries, M.: Management and monitoring of IoT devices using blockchain. Sensors **19**(4), 856 (2019). https://doi.org/10.3390/s19040856

11. Varghese, C., Jose, J.: IoT device management using blockchain. Int. J. Sci. Eng. Technol. Res. **8**(3), 79–84 (2019)
12. Fan, K., Wang, S., Ren, Y., Yang, K., Yan, Z., Li, H., Yang, Y.: Blockchain-based secure time protection scheme in IoT. IEEE Internet Things J. **6**(3), 4671–4679 (2019). https://doi.org/10.1109/JIOT.2018.2874222
13. Dabbagh, M., Sookhak, M.: The evolution of blockchain: a bibliometric study. IEEE Access **7**, 19212–19221 (2019). https://doi.org/10.1109/ACCESS.2019.2895646
14. Mohammed, M., Shailendra, S., Jong, R., Park, H.: Distributed denial of service attacks and its defenses in IoT : a survey. J. Supercomput. (2019). https://doi.org/10.1007/s11227-019-02945-z
15. Suchaad, S.A.L., Mashiko, K., Ismail, N.B., Zainal Abidin, M.H.: Blockchain use in home automation for children incentives in parental control. ACM Int. Conf. Proc. Ser. 50–53 (2018). https://doi.org/10.1145/3278312.3278326
16. Song, J.C., Demir, M.A., Prevost, J.J., Rad, P.: Blockchain design for trusted decentralized IoT networks. 2018. In: 13th System of Systems Engineering Conference SoSE, pp. 169–174. (2018). https://doi.org/10.1109/SYSOSE.2018.8428720

LIMBO: Telecom Signal Strength Coverage in Different Regions

R. Yuvaraj, N. R. N. Sivasurya, M. S. Vijayprasanth, and M. Buvana

1 Introduction

An Application-based solution may be developed for Detecting Poor Telecom Connectivity (Cellular) regions using user devise signal strength along with geo-coordinates of the user to a central server. Government authorities can use the information to assess the poor coverage regions and take necessary steps to address issues of poor coverage. Signal strength is represented in -dBm format (0-—100). This is the power ratio in decibels (dB) of the measured power referenced to one milliwatt. That means the closer the value is to 0, the stronger the signal. Users will be able to see the complete set of network parameters of their cell phones in the app like RSSI, ASU level, RSSNR for the respective type of network and operator. They will be able to submit comprehensive signal strength reports to the central database. They can view the coverage of the type of network for the given operator based on similar reports submitted by other users in the vicinity (within a radius of 500 m from their location). Users will be provided a rich map-based UI to graphically view coverage stats. Government authorities and officials from different operators will be provided a separate portal where they will be provided access to the signal data along in a map-based UI so that they can have insights on improving connectivity in different regions [1, 2].

The Purpose of the project is to detect poor telecom connectivity regions by aggregating signal strength data from different regions.

With the raising amount of cellphone users and telecom network subscriptions across India it is difficult to detect poor connectivity areas across different telecom

R. Yuvaraj · N. R. N. Sivasurya · M. S. Vijayprasanth · M. Buvana (✉)
Department of Computer Science and Engineering, PSNA College of Engineering and Technology, Dindigul, India
e-mail: buvana@psnacet.edu.in

B. Agarwal et al. (eds.), *Proceedings of International Conference on Intelligent Cyber-Physical Systems*, Algorithms for Intelligent Systems,
https://doi.org/10.1007/978-981-16-7136-4_6

circles as there are no solutions to aggregate location-based connectivity data from location across India.

The main objective is to find the poor network coverage [3–5] areas of the cellular networks provided by them to the users. By showing the data of the signal strength of the network the user can identify which network has better performance in particular area, in which user can report the particular network to increase their services in those particular areas. Most of the existing systems only show the connectivity strength of your mobile device. Some systems do aggregate data from different locations, but they do only for a specific network operator and do not provide technical details of the connectivity strength except the quality of signal strength at the area. There exists only one system that aggregates data from different locations and network operators. But they do not guarantee user privacy with their location data and other sensitive information.

The reason for need for the proposed system is that we need a system that guarantees user privacy at the same time is robust with data so that the government and network operators can improve connectivity in regions where there is poor connectivity.

2 Proposed System

The proposed system LIMBO is represented in Figure 1. The system guarantees complete user privacy with sensitive data like location and other cellular parameters [6–8]. It provides a simple dashboard that displays data that is fairly technical. The system is designed in such a way that it collects and stores data of signal strength for different network operators in different locations and telecom circle. It uses geofencing to query and analyze signal data which are in the specified region or radius.

The Firebase Realtime Database is a cloud-hosted database. Data is stored as JSON and synchronized in realtime to every connected client. When you build cross-platform apps with our iOS, Android, and JavaScript SDKs, all of your clients share one Real-time Database instance and automatically receive updates with the newest data. Every record has a time-based chronological unique timestamp as a key. This helps to uniquely identify each record and acts as a primary key.

The system uses the RSSI value that is retrieved from the android system API as the main indicator for signal strength. The location is positioned using the GPS service which connects to one of the satellites of the Global Positioning Service and sets the location of the device. The user can view the signal strength parameters and other network details from the dashboard for his/her reference. The user can submit the signal strength value at any particular instant of time by clicking the submit report button. The data is uploaded to the backend NoSQL database hosted in firebase. Each record has the location data of the position of the signal strength measured. Each record has a unique chronological key as its identifier so uniquely identify each record. In the maps activity, the user can view the signal strength values of his/her

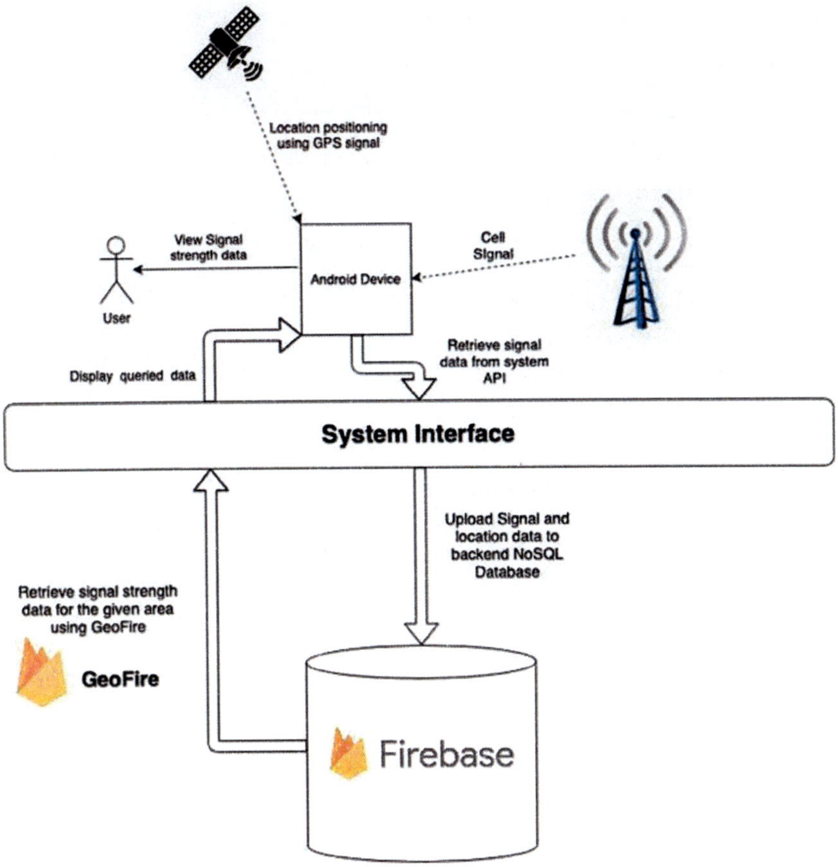

Fig. 1 LIMBO system architecture diagram

primary network, plotted within the 500 m radius of the user. This is queried using the GeoFire library.

3 Database Design

In this system Data is stored in JSON format. The primary node is the reports node. It has all the signal reports identified with a chronologically unique key. The second node is the data node within the geo node. This node has records with the chronologically unique key as the foreign key. It has location data structured in a specific structure in order to allow the GeoFire library to perform location-based queries. The keys retrieved from here is used to fetch reports within the 500 m radius of the user

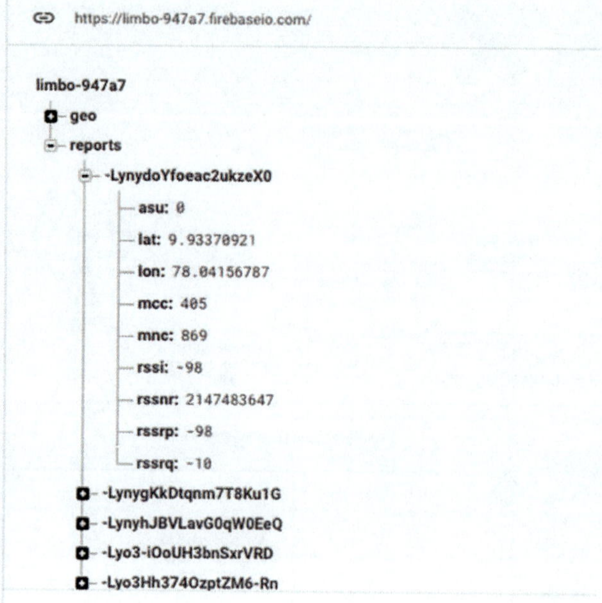

Fig. 2 Signal data records

for the primary network she/he uses. Figs. 2 and 3 represent the database design of the system (Table 1).

4 Implementation and Testing

The Proposed LIMBO system have the different modules; the following section describes in detail about the modules present in this system. The process flow of permission module, database module, Application page of db module, Process flow of upload module, Map module is described in Fig. 4, 5, 6, 7, and 8, respectively.

4.1 Permission Management Module

Since android Marshmallow (6.0) (API Level 23) a new permission model was implemented in which critical and sensitive app permissions were assigned in the runtime by the user. Since the app accesses sensitive location and telephony data from the phone permissions to access these data must be handled in the runtime, without which the app won't function. On the other hand the user will be able to allow permissions in the settings menu.

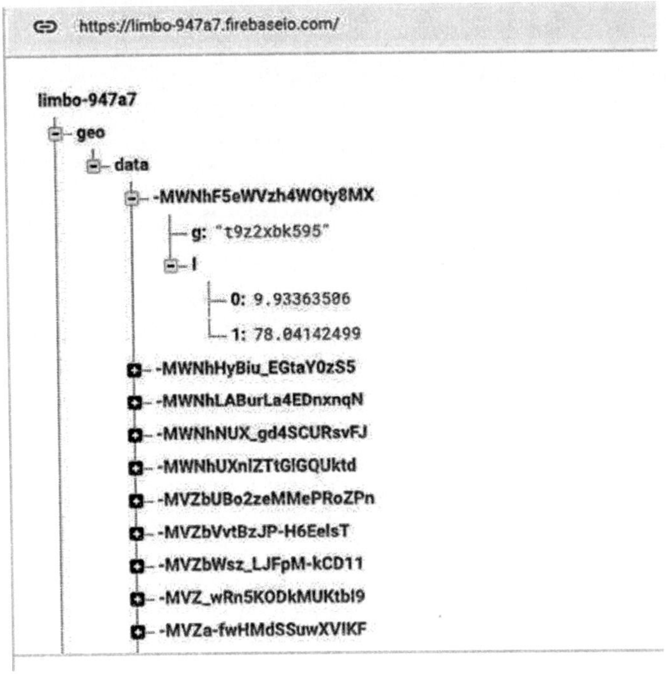

Fig. 3 Geofire records

Table 1 Geofire records

S. No	Fields	Datatype
1	**Unique key**	**String**
2	Latitude	Double
3	Longitude	Double

First the user has to check whether the OS is in the necessary version (6.0) by using the API level. If it is older than that the system tries to access the location and telephony data. If it is allowed then the data is displayed or else it prompts the user to grant permissions.

If the API level is 23 or higher, the system checks the availability of the permissions. If it is available, the system accesses the data or else the system asks the permissions in the runtime.

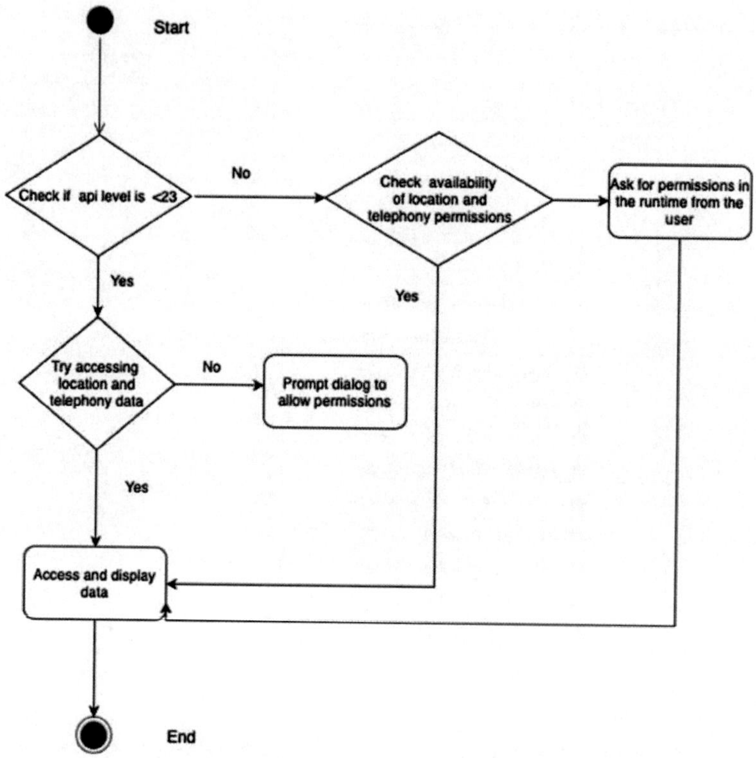

Fig. 4 Permission management module

4.2 Dashboard Module

Home page is a basic dashboard in which the signal strength data for the two SIMs will be displayed in decibel milliwatts along with other data like the operator name and its MCC/MNC number, service state, network type, and other parameters. The module detects the changes in the signal strength changes in both the networks registered in the mobile and displays the value of signal strength in decibel milliwatts.

The system accesses the SIM card details from the system API. Then, it listens for changes in signal strength changes in the network. If there is a change it updates the value in the dashboard. It continuously detects the changes in signal strength and updates it in the dashboard until it is manually terminated by the user.

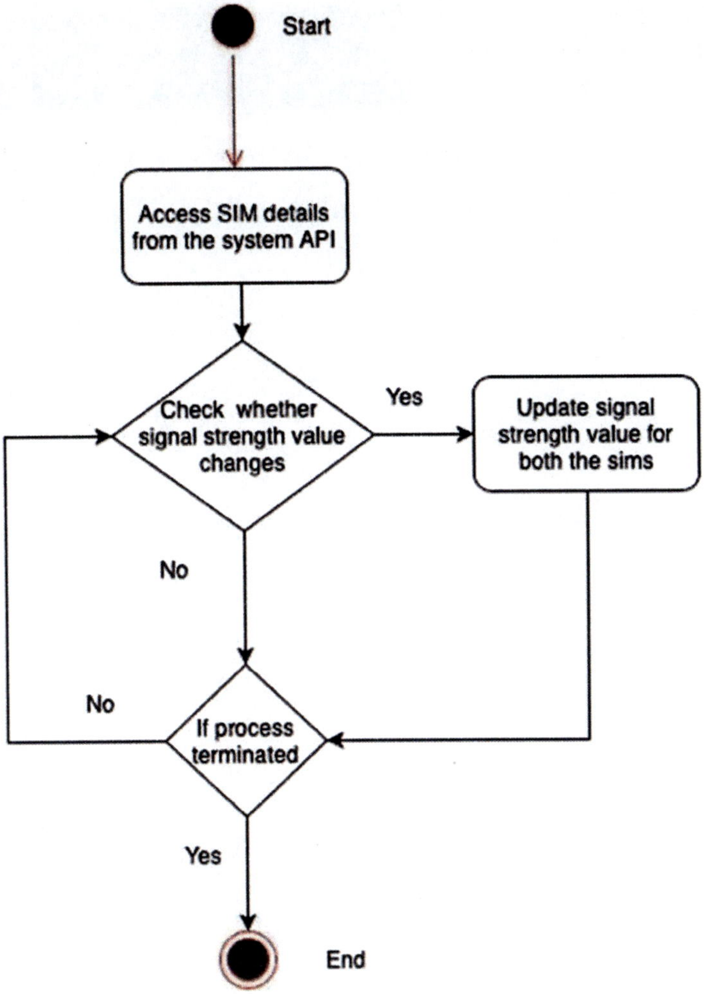

Fig. 5 Process flow of dashboard module

4.3 Database Upload Module

After the user views the signal strength values from the dashboard, the user can submit a record of the signal strength parameters along with other metadata to the backend NoSQL database. Each record written to the database has a unique chronological key value based on timestamp as its identifier. The key is used as a foreign key in a separate nosql tree and location data alone is appended there in order for the Geofire library to work.

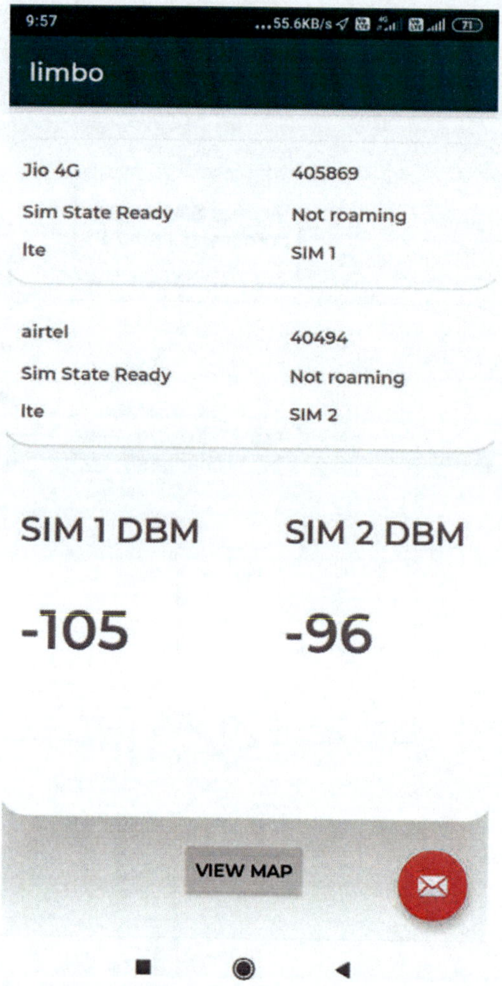

Fig. 6 Application dashboard module

The module checks whether the location has been set by the GPS. If not it waits for the GPS receiver to set the location. If the location is set, it retrieves the latitude and longitude. It retrieves the signal strength parameters, network details and uploads it to the backend using a chronologically unique key as an identifier.

4.4 Map Module

This page fetched signal strength values which are reported within 500 m radius of your current location, and displays the points as markers in the map inside the map.

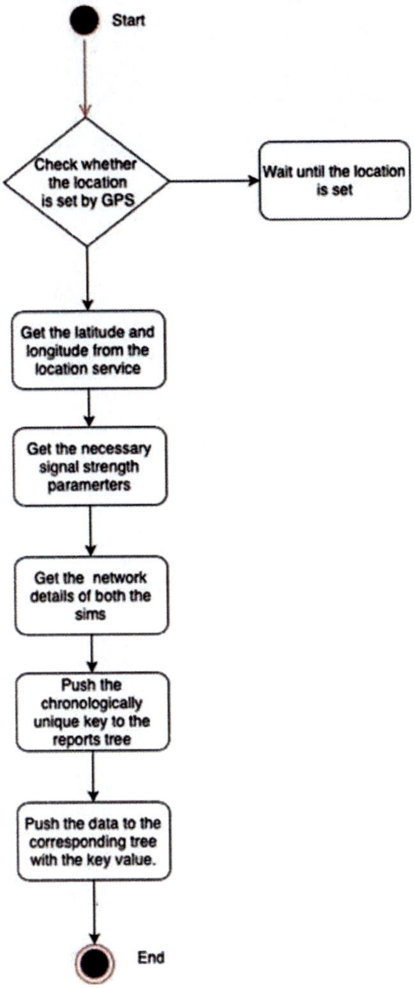

Fig. 7 Database upload module

The 500 m circle is highlighted in a color according to the coverage available in the area for the particular operator and the type of network.

The module fetches the current location as shown in Fig. 9. Using the current location it fetches the records which are within 500 m radius of the location. It selects only the records which correspond to the user's primary network. It plots the data in the map with the location data of each record with the signal strength value as a label for the marker.

Fig. 8 Process of map module

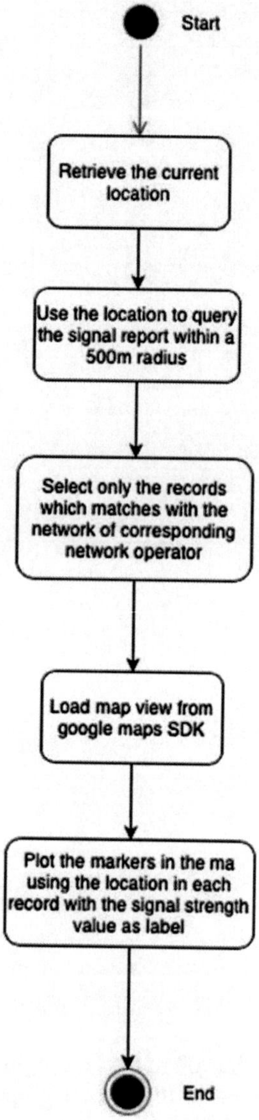

Test Cases

A test case is a set of conditions or variables under which a tester will determine if a requirement upon an application is partially or fully satisfied. The types of testing that are to be carried out on the system is as follows in Table 2.

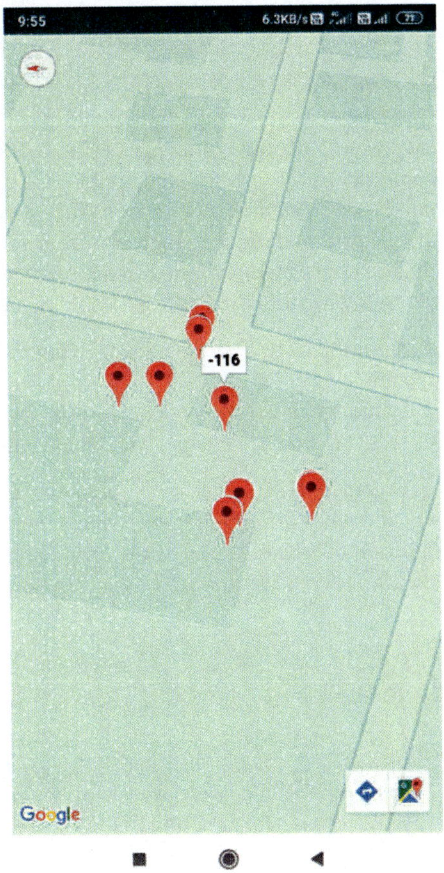

Fig. 9 Signal strength through map module

Table 2 Various test cases

Test case no	Description	Pre- conditions	Pass/Fail	Expected results
PTS_001	Permission request granting	Permissions not given already	Pass	Permission granted successfully
PTS_002	GPS service online	GPS service not turned on	Pass	GPS turned on successfully
PTS_003	Location set	GPS not set	Pass	GPS set successfully
PTS_004	Signal data upload	None	Pass	Upload successfully
PTS_005	Data query within specified radius	Location set and radius specified	Pass	Data query successful
PTS_006	Circle color change for signal strength	Average signal strength in a specific category	Pass	Color changed successfully

5 Conclusion and Future Work

Existing solutions each have different shortcomings. The proposed solution can overcome these shortcomings and provide a better solution for the problem. User privacy, easy access of huge amounts of data collected from the large amount of devices help in detecting poor connectivity. This also helps in promoting healthy competition between telecom operators as they can identify poor connectivity areas of their competitors and improve their own connectivity in the region and gain market share. This provides a win–win situation for both customers and telecom operators. In future the following things need to be implemented in this LIMBO system.

- An interactive heat map for viewing the connectivity strength in different areas and
- Rankings of network operators in a given location and radius.
- Stats on type of network which has highest penetration in a given location

Create an API for telecom operators so that they can use the API and fetch raw signal data of different regions to use it in their own way to improve their coverage in different areas. And to allow third party apps to use the API to create their own client for people to view the signal strength in their own areas.

References

1. Madariaga, D., Madariaga, J., Bustos-Jiménez, J., Bustos, B.: Improving Signal-Strength Aggregation for Mobile Crowdsourcing Scenarios. Sensors **21**, 1084 (2021). https://doi.org/10.3390/s21041084
2. Bhuvaneshwari, A., Sathyasavithri, T.: Comparative analysis of mobile radio path loss models for suburban environment in Southern India. In: 2013 International Conference on Emerging Trends in VLSI, Embedded System, Nano Electronics and Telecommunication System (ICEVENT), Tiruvannamalai, India, pp. 1–5. doi: https://doi.org/10.1109/ICEVENT.2013.6496544 (2013)
3. Ahamed, M.M., Islam, Z., Hossain, S., Faruque, S.: LTE network coverage prediction for multi-traffic users in an urban area. IEEE Int. Conf. Electro Inform. Technol. **2013**, 1–6 (2013). https://doi.org/10.1109/EIT.2013.6632703
4. Oughton, E.J., Frias, Z.: The cost, coverage and rollout implications of 5G infrastructure in Britain. Telecommun. Policy **42**(8), 636–652. ISSN 0308–5961 (2018)
5. Backman, W.: Signal level interpolation for coverage area prediction. In: Proceedings of the IEEE 60th Vehicular Technology Conference, vol. 60, no. 1, pp. 67–71. (2004)
6. Erceg, V., Greenstein, L.J., et al.: An empirically based path loss models for wireless channels in suburban environment. IEEE On Select. Areas Commun. **17**, 1205–1211 (1999)
7. Abou-zeid, H., Hassanein, H.S., Tanveer, Z., AbuAli, N.: Evaluating mobile signal and location predictability along public transportation routes. In: 2015 IEEE Wireless Communications and Networking Conference (WCNC), New Orleans, LA, USA, pp. 1195–1200. doi: https://doi.org/10.1109/WCNC.2015.7127639 (2015)
8. Choudhary, S., Sharma, A., Srivastava, K., Purohit, H., Vats, M.: ReadRange optimization of low frequency RFID system in hostileenvironmentalconditions by using RSM approach. Evergreen J. Novel Carbon Resour. Sci. Green Asia Strat. **7**(3), 396 403 (2020)

Path Extraction and Planning for Intelligent Battlefield Preparation Using Particle Swarm Optimization, Gravitational Search Algorithm, and Genetic Algorithm

Lavika Goel

1 Introduction

With the advancements in the modern era, it is critical to find the shortest path (obstacle free) to the enemy's base battle station to have a strategic advantage on the ground. Therefore, path planning and path extraction algorithms are hot topic of research. Bio-inspired techniques have found increasing application in path planning because unlike traditional techniques, these robust techniques are not limited by the pre-requisite of a well-behaved terrain. Hence, for the accommodation of diverse terrains, we have adopted the more robust–swarm optimization algorithms instead of the traditional techniques. The specific algorithms adopted are PSO, GSA, and GA [1–3].

Nature inspired swarm intelligence techniques have been recognized as some of the most robust search algorithms for complex and ill-behaved optimization problems. These techniques have found increasing application in path planning, both dynamic as well as static owing to their robustness and versatility in accommodating a wider variety of complex terrains.

The concept of anticipatory computing in battlefield preparedness has been a hot topic of research for many years. The importance of gaining an awareness of the enemy before the enemy gained a similar awareness was emphasized by Gilson et al. [4] in 1995. Automation in defense control systems comes as natural progression in defense technology which has made tracking the battlefield beyond human capabilities [5].

Input data to the algorithm includes a satellite image of an area with the coordinates of the starting (friendly) and the destination base stations (enemy). Through

L. Goel (✉)
Department of Computer Science and Engineering, Malaviya National Institute of Technology (NIT), Jaipur, Rajasthan 302017, India
e-mail: lavika.cse@mnit.ac.in

B. Agarwal et al. (eds.), *Proceedings of International Conference on Intelligent Cyber-Physical Systems*, Algorithms for Intelligent Systems,
https://doi.org/10.1007/978-981-16-7136-4_7

different layers, best path will be extracted from the image. First, PSO is used to generate a binary mask around the source station by classifying the area into obstacles and obstacle free areas. Following which, agents for the GSA are initialized in this obstacle free area. These agents fly through the hyper-plane from source to destination driven by the GSA algorithm. This movement of agents traces way-points at every iteration. This set of feasible way-points forms the solution space for GA. Finally, the path is optimized using GA by finding the best possible combination of way-points generated by GSA.

In the initial phase, PSO ensures that the agents initialized around the source station occupy valid via-points, i.e., they do not lie within obstacles. This ensures that there is control over the precise number of GSA agents. GSA has been preferred over local PSO for its inherent local exploration feature. Since masses are accelerated through the hyper-plane at a rate inversely proportional to the distance between the agents, clusters exploring different paths do not attract each other. This facilitates high exploration of the dataset. As the GSA agents iteratively migrate towards the destination, each position they occupy forms the initial population space for GA. GA iteratively converges toward the best possible combination of via-points from among this solution space. This method is preferred because, GSA, being a stochastic process involving random parameters, has an inherent stagger in the movement of the agents. Thus, the paths traced by GSA agents do not give the shortest possible path.

2 Related Work

In the last two decades, different conventional methods have been developed to solve the path planning problem, such as the cell decomposition, road map, and potential field. Most of these methods were based on the concept of space configuration. These techniques show lack of adaptation and a non-robust behavior. To overcome the weakness of these approaches, researchers explored variety of solutions.

Gravitational Search algorithm was introduced by Rashedi in 2009 as a new population-based meta-heuristic that uses universal law of gravitation to drift solutions towards the global optimum. Since its conception, GSA has found application in several fields one of them being path extraction and planning. Purcaru et al. [6] uses GSA for optimal static and dynamic path planning in a group of mobile robots. Since the environment remains static in the problem at hand, this paper aims to exploit the static path planning capability of GSA.

GA has been recognized as one of the most robust search algorithms for complex and ill-behaved optimization problems [7]. The basic characteristic which makes GA attractive in solving such types of problems is that it is inherently a parallel search technique and can search for optimal in dynamic environments [8–10]. GA was used to control a mobile robot moving in an environment which has static obstacles and to control robot motion with dynamic obstacles [11]. Some of the proposed techniques in [9, 11] suffer from many problems. They include (1) being computationally expensive (2) requires large memory spaces when dealing with dynamic and large sized environments, (3) being time consuming. In the last decade, genetic algorithms have been widely used to generate the optimum path by taking the advantage of its strong optimization ability. This research is motivated by earlier work presented in [8]. In this study, we provide an initial idea based on genetic algorithm to select the shortest path in predictable environment which will be able to handle static obstacles.

3 A Brief Review of PSO, GSA, and GA

This chapter briefly reviews various algorithms used in the development of our hybrid nature inspired path extractor/planner.

PSO is a population-based stochastic optimization algorithm. It comes under the broad category of Swarm Optimization that is based on social-psychological principles and provides insights into the social behavior, and has applications in engineering problems. The seminal paper on PSO by James Kennedy and Russell C. Eberhart was published in 1995. It optimizes the given problem using agents called particles. The dimension of the hyper-plane denotes the number of design variables to be optimized to get the highest fitness. These particles fly through hyper-plane with two prime reasoning capabilities: their memory of their own best position and knowledge of the global or the neighborhood's best position. So a particle has the information to accommodate a suitable response in its position and velocity [12–14].

GSA, like PSO, is a population-based optimization algorithm that was proposed by Esmat Rashedi in 2009. This technique is inspired by the Universal law of gravitation. GSA agents are particles that have masses. The performance or the fitness of the agent is measured by its mass. The higher the fitness, the heavier the agent. Each agent in the multi-dimensional solution space attracts every other agent by a force directly proportional to the product of their masses and inversely proportional to the square of distance between them. The net force on an agent is the vector sum of all forces due to all agents. Net force per unit mass of the agent gives the net acceleration of that agent. Hence, acceleration is a function of mass and therefore of fitness. Velocity and consequently position is updated using the acceleration thus computed. This interaction causes the poorer solutions (lighter masses) to drift toward the better solutions (heavier masses). Heavier masses which represent better solution move much slower than the lighter masses thereby guaranteeing good exploitation. Each mass is a candidate solution, and the algorithm is navigated by adjusting the gravitational and inertial masses. The algorithm iteratively converges to the heaviest mass in the search space which represents the optimum solution.

A Genetic Algorithm (GA) is a meta-heuristic inspired by the process of natural selection, which belongs to the larger class of Evolutionary Algorithms (EA). GA optimizes the solution space by simulating evolution through natural selection, crossover of traits and mutation. The GA solution space is comprised of bit strings where each substring represents the design variables. GA simulates evolution and optimizes the objective function through the principle of survival of the fittest. The fitter solutions of the parent generation are chosen randomly and subjected to genetic operators like crossover and mutation. Thereby they transfer their "favorable" traits to the next generation of solutions. This process continues iteratively till global optimum is attained [15, 16].

Candidate solutions together make up the solution space of the current generation. Fitness for each individual is computed using the objective function. The more fit individuals are stochastically selected from the current population for reproduction, and genome of each individual is modified (recombined and possibly randomly mutated) to form a new generation. The fitter solutions of the parent generation exchange sub-strings among themselves to produce the new generation of solutions. This genetic operation is called crossover. Mutation introduces an added degree of randomness to the optimization process. Individuals (solutions) of a generation are chosen randomly and their bits are mutated at random. Generally, the frequency of mutation is much lower than crossover which happens at every iteration.

4 Proposed Methodology

This section describes the path extraction and planning scheme adopted (Fig. 1).

Step 1: The first phase of the algorithm uses PSO to extract obstacles from histogram analysis of a 20 × 20 pixel area around the friendly base station. 50 GSA agents are initialized in the obstacle free area hence extracted.

Step 2: PSO agents are assigned random velocities and fitness for each agent is calculated as the inverse of distance to destination. The fitness thus calculated is used to assign masses to the agents with more fit agents having higher masses. Masses are assigned to GSA agents such that better solutions, i.e., agents with higher fitness values have higher masses.

$$mass(i) = \frac{fitness(i) - \min(fitness)}{\max(fitness) - \min(fitness)} \tag{1}$$

Step 3: The algorithm then checks if an agent is currently in an inaccessible area by comparing the pixel value of the grayscale value of the pixel it occupies. If the agent is found to be in an inaccessible region, a penalty is added to the fitness. This is done so as to prevent agents within obstacles from attracting more agents towards them.

Step 4: Forces on each agent due to the masses of the other are calculated subsequently. Each agent *i* is attracted toward every other agent *j* due to its mass by a force given by,

$$F_{ij} = \frac{G \cdot mass(i) \cdot mass(j)}{norm(pos(i) - pos(j))^2} \tag{2}$$

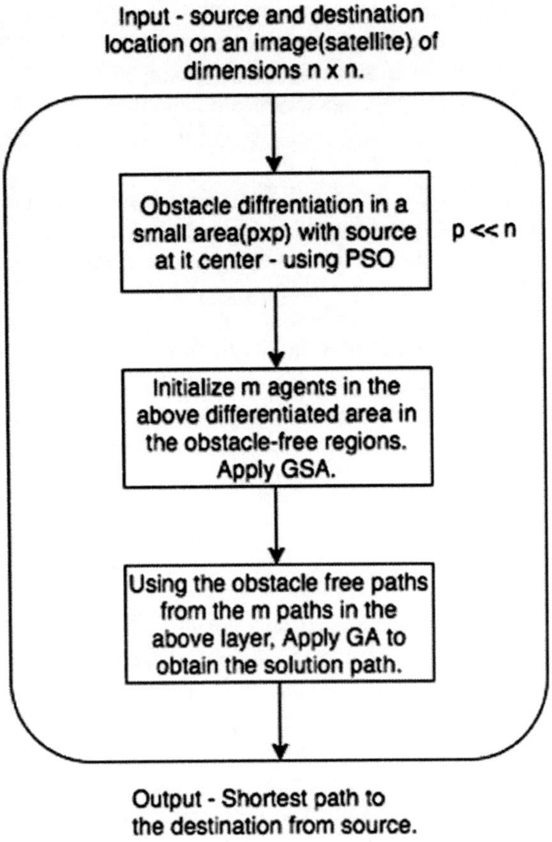

Fig. 1 Flowchart of the proposed approach

Following which, net acceleration of each agent is updated. Velocity and position are updated from acceleration calculated above. Thus, the particles the net acceleration of an agent "i" is given by

$$a(i) = \frac{\sum_{j=1}^{m} F_{ij}}{mass(i)} \tag{3}$$

The velocity and position of all the agents are given by

$$v_i(t+1) = \sigma \cdot v_i(t) + \gamma \cdot a_i(t) \tag{4}$$

$$pos_i(t+1) = pos_i(t) + v_i(t+1) \tag{5}$$

where σ and μ are weightage parameters to regulate the step-size of exploration.

Using the above equations, all the particles will explore the search space as they move toward the destination (enemy base station).This process is iterated till all agents are in an around the destination.

Each agent's trajectory is a potential path from source to destination. However, those paths wherein an agent was at least once within an obstacle are infeasible and hence must be eliminated.

Step 5: Once the collision free paths are extracted from each of the remaining agents, this vector space is passed to GA as its initial solution space. GA iteratively optimizes the path by choosing the best possible combination of substrings. This translates to the best possible combination of way-points. This is done by defining the GA objective function as the net path length of each path vector. GA finally converges to the shortest possible path.

GSA has been used for the purpose of path extraction in the algorithm. The path extraction phase requires high exploration for this problem statement. Therefore, a local best approach is to be adopted. Though this can be accomplished with local best PSO variant, GSA has this local interaction built in since

$$F(i, j) = \frac{G(t) \cdot M(i) \cdot M(j)}{r(i, j)^2} \tag{6}$$

Thus, cluster of particles interact with each other instead of drifting toward the global best and explore the individual paths to the destination from the friendly base station. Additionally, in the later stages of the path extraction phase, the exploitation can be improved for faster convergence by manipulating the universal gravitational constant G in Eq. 6. Universally gravitation constant has been set at 1 in the path extraction phase. A penalty of -2 is added to the fitness of agent *i* if the agent lies within an obstacle (this value can be increased or decreased depending upon the dataset characteristics like accessibility, relative position of the destination coordinates with respect to the source coordinates, possible shortest paths, obstacle identification criteria, etc. In our datasets of Alwar and Mussourie, the penalty was taken as -2 to identify the obstacle depending upon the dataset characteristics.).

$$f(i) = k + \frac{1000}{\sqrt{d^2 - r_i{}^2}} \tag{7}$$

where penalty *k* takes values,

$$k = \begin{cases} 0, iftheagentiisinaccessiblearea \\ -2, iftheagentiiswithinanobstacle \end{cases} \tag{8}$$

where f(i) is the fitness of agent *i*, *d* is the coordinate of the destination and represents the position vector of agent *i*. The introduction of penalty into the fitness equation adds the obstacle avoiding feature to the GSA phase. As the fitness of inaccessible agents decreases, their mass decreases making them lighter. The now lighter agents attract fewer agents and with lesser intensity than the heavier agents. Thereby ensuring that agents do not cluster around ones within obstacles.

GA converges to the optimal combination of way-points extracted by GSA to give the shortest possible path. This step is essential because GSA being a stochastic population-based scheme, causes some inherent stagger in the movement of the agents. Hence the path extracted by GSA isn't the one with smallest possible pixel length.

One of the key deciding factors of the effectiveness of GA in the proposed algorithm is the encoding of via-points of the path traced by a GSA agent. The size and complexity of the bit-string that contains the combination of way-points is critical to the success of the methodology. To reduce the size of the search space by shortening the bit strings, a coordinate transformation has been adopted. This has been done by projecting the way-points (two-dimensional data) to linear array of one dimension. The algorithm aims to determine the optimal combination of way-points generated by GSA that forms the shortest path from the source to the destination. The given workspace has been converted to a new coordinate space where each point i is uniquely identified by its index x_i in the one-dimensional array and its distance y_i to the line joining source to destination. Thus, y_i becomes the search space for each way point in the path and the candidate solution becomes one-dimensional data.

GA converges to the optimal combination of way-points extracted by GSA to give the shortest possible path. The pseudocode of the proposed approach is given below

```
initialize start, destination
    initialize agents in a sub matrix(subset of image)
    Threshold the sub-matrix into accessible or obstacle-prone areas using PSO
    Initialize m agents in the accessible area
    while (all agents not at destination)
    for all agents
    update position
    update fitness
    if particle in obstacle-prone area - add penalty to fitness
    update mass
    end for
    calculate distance, velocity and acceleration
    update velocity and position
    end while
    Let s be the set of all collision free paths extracted by GSA
    while (change in path length > ε ) where ε is minimum error
    for all agents
    if(s[i] is a collision-free path)
    fit1(i) = 1
    else
    fit1(i) = 0
    end if
```

```
for j=1: (no. of iterations for each agent)
fit2 = fit2 + norm(pos[i,j] - pos[i,j-1])
end for
path_fitness[i] = fit1(i) / fit2(i)
end for
crossover, mutate
path = s[k] where k is the index of the path with maximum fitness
end while
```

5 Results and Analysis

We demonstrate the implementation of the proposed PSO-GSA path extraction with GA path planning algorithm on two terrains, viz., Alwar area in Rajasthan (Plain/Desert) and that of Mussoorie region in Himachal Pradesh (Mountainous/hilly). The datasets chosen are satellite images taken through LISS-III sensor obtained courtesy of DRDO. The comparison of the pixel lengths obtained with different schemes is given in Table 1.

GSA agents successfully avoid obstacles even in ill-behaved terrains. This has been further reinforced by extracting paths on custom data sets. A set of custom datasets were generated by extending the obstacle in Figs. 4a and b for Alwar and Mussourie datasets. These datasets are shown in Figs. 4e and f. This helps demonstrate the obstacle avoiding nature of the GSA agents due to penalty incurred to fitness when inside an obstacle. As shown in the aforementioned figures, the GSA agents show promising obstacle avoiding behavior since they bend around the enlarged obstacle to reach the destination. Each cyan marker is the position of one of the aforementioned agents as they drift toward the destination. Figures 4c, d, e, and f show all positions of the GSA agents superimposed on the same image. The GSA agents extract paths by avoiding the obstacles in the custom datasets and thus the initial solution space generated by GSA is guaranteed to contain the shortest obstacle free path to the destination.

Nature inspired optimization techniques have been employed for path planning in the same datasets for the same set of start and destination coordinates [1, 7]. The results of it have been promising. The extended BBO-PSO-ACO model adopted in [1, 7] for the same input data, gave paths of lengths 351 pixels for Alwar and 310 pixels for Mussourie. As mentioned earlier, the proposed GSA-GA model gives paths 34% shorter for Alwar dataset and 17% shorter for Mussourie.

5.1 Mussourie Dataset

In the 258 $\times$ 203 pixel dataset, (250, 7) is the friendly base station and (16, 190) is chosen as the enemy base station which is the destination to which shortest path is

Table 1 Comparison of Results

	Hybrid PSO-GSA-GA	Hybrid BBO-ACO-PSO	ACO	PSO	ACO2/PSO	Stud GA	Evolutionary-strategy	Self organizing feature maps	Fuzzy inference	GA
Alwar	227	351	355	354	353	352	358	363	311	358
Mussourie	257	310	314	313	312	311	316	321	319	316

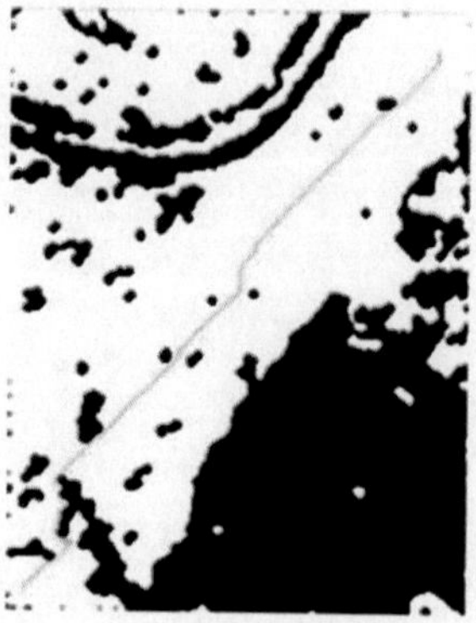

Fig. 2 Final path for Mussourie dataset

to be extracted. The collision free path extracted in the morphologically smoothened dataset is found to be 257 pixels in length. GA iteratively converges to a path 257 pixels in length from the set of all way-points created by GSA (Fig. 2).

5.2 *Alwar Dataset*

For the Alwar dataset, the friendly and enemy base stations are at (250, 7) and (26, 164). After morphological operations on the dataset, GSA extracts the set of all way-points which forms the initial solution space for GA. GA iteratively converges to a pixel length of 227 which is a 34% improvement on the path extracted by extended BBO-PSO-ACO in [1, 7] (Fig. 3).

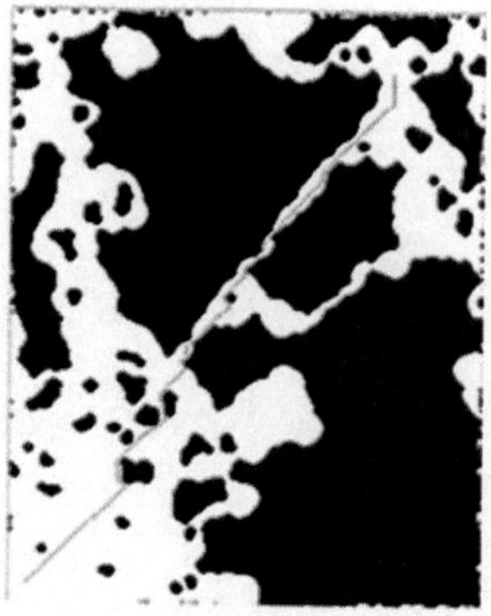

Fig. 3 Final path for Alwar dataset

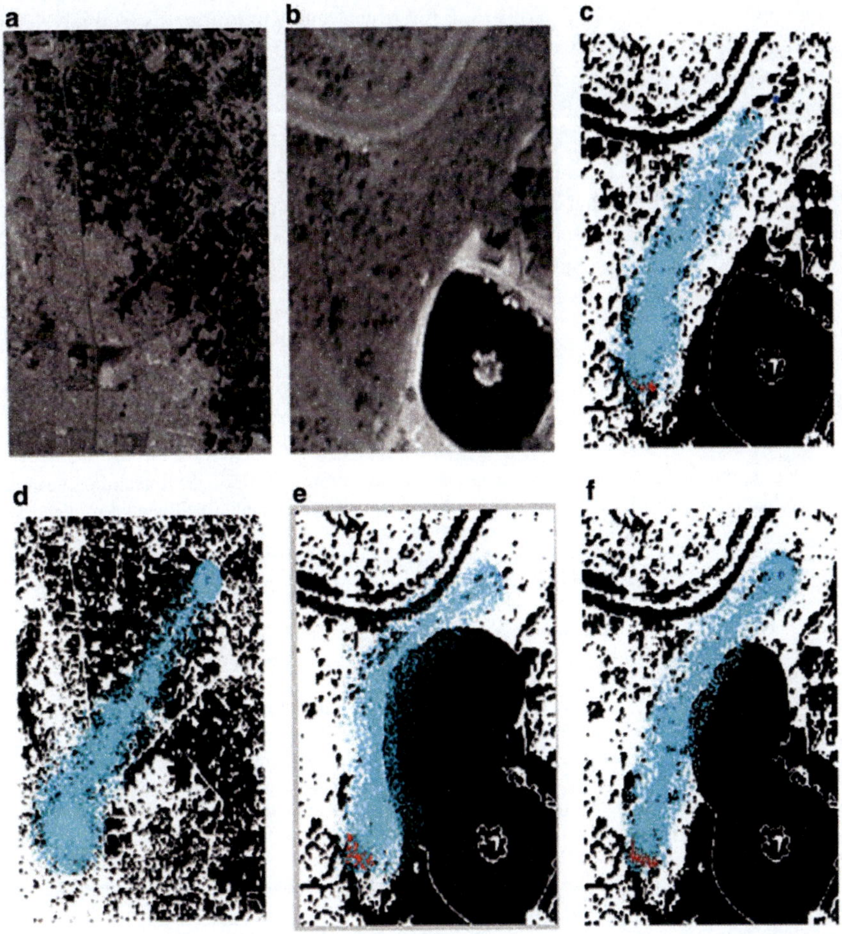

Fig. 4 **a** Alwar dataset **b** Mussourie dataset **c** GSA path extraction on Mussourie dataset **d** GSA path extraction on Alwar dataset **e** Path extraction in custom dataset I **f** Path extraction with obstacle avoidance in custom dataset II

6 Conclusion and Future Scope

We propose a hybrid of nature-inspired and evolutionary algorithms namely PSO, GSA, and GA for the problem of shortest path extraction in the given satellite datasets. PSO is used to initialize the agents in the obstacle free area around the friendly base station. This initialization ensures a check on the total number of valid agents, since all agents are in obstacle free areas. The agents then traverse from source to destination in incremental steps driven by GSA. This set of way-points created by GSA forms the initial population for GA which iteratively converges to the shortest path which are 227 and 257 pixels long for Alwar and Mussourie, respectively. This result is a significant improvement over other implementations of similar algorithms and previous studies in the area.

As this is a hot topic of research, new algorithms with better customization to the problem statement and improved response to a similar feature set can result in further improvement of results and ultimately a more powerful path planning tool. One of the possibilities that could be explored is increasing the penalty so as to give a negative fitness and consequently a negative mass to the GSA agent. This would lead to gravitational repulsion toward an inaccessible agent which could result in a smoother set of way-points for the GA phase.

References

1. Wu, J., Feng, S.: Improved biogeography-based optimization for the traveling salesman problem. In: 2nd IEEE International Conference on Computational Intelligence Applications (ICCIA 2017), pp. 166–171. China, https://doi.org/10.1109/CIAPP.2017.8167201 (2017)
2. Sörensen, K., Sevaux, M., Glover, F.: A history of metaheuristics. In: Martí, R., Pardalos, P., Resende, M. (eds.), Handbook of Heuristics, pp. 791–808. Springer, Cham https://doi.org/10.1007/978-3-319-07124-4_4 (2018)
3. Pijarski, P., Kacejko, P.: A new metaheuristic optimization method: the algorithm of the innovative gunner (AIG). Eng. Optimiz. **51**(12), 2049–2068. https://doi.org/10.1080/0305215X.2019.1565282 (2019)
4. Ghorbani, A., Shiry, S., Nodehi, A.: Using genetic algorithm for a mobile robot path planning. In: International Conference on Future Computer and Communication, pp. 164–166. IEEE Publications, Kuala Lumpur, Malaysia. doi: https://doi.org/10.1109/ICFCC.2009.28 (2009)
5. Goel, L., Gupta, D., Panchal, V.K.: Two-phase anticipatory system design based on extended species abundance model of biogeography for intelligent battlefield preparation. Knowl. Based Syst. **89**, 420–445 (2015)
6. Panigrahi, P.K., Ghosh, S., Parhi, D.R.: Comparison of GSA, SA and PSO based intelligent controllers for path planning of mobile robot in unknown environment. Int. J. Electr. Comp. Eng. **8**(10), 1633–1642 (2014)
7. Thomas, C.E., Pacheco, M.A.C., Vellasco, M.M.B.: Mobile robot path planning using genetic algorithms. In: Foundations and tools for neural modeling, vol. 1606/1999, pp. 671–679. Springer, Berlin/ Heidelberg, ISBN: 3–540–66069–0 (1999)
8. Lozano-Perez, T., Wesley, M.: An algorithm for planning collision-free paths among polyhedral obstacles. Commun. ACM **22**(10), 560–570 (1979)
9. Brank, J.: Evolutionary approaches to dynamic optimization problems-introduction and recent trends. [Online] In: Proceedings of GECCO Workshop on Evolutionary Algorithms for Dynamic Optimization Problems, pp. 2–4. Chicago, USA (2003)
10. Lu, J., Yang, D.: Path planning based on double-layer genetic algorithm. In: 3rd International Conference on Natural Computation (ICNC 2007) [Online], vol. 4, pp. 357–361. IEEE Publications, China. doi: https://doi.org/10.1109/ICNC.2007.546 (2007)
11. Purcaru, R.E., Precup, D., Iercan, L.O., Fedorovici, R.C., David, F.: Dragan: optimal robot path planning using gravitational search algorithm. Int J Artif Intell **10**(S13), 1–20 (2013)
12. Yarmohamadi, M., Javadi, H.H.S., Erfani, H.: Improvement of robot path planning using particle swarm optimization in dynamic environments with mobile obstacles and target. J. Adv. Stud. Biol. **3**(1), 43–53 (2011)
13. Mohajer, B., Kiani, K., Samiei, E., Sharifi, M.: A new online random particles optimization algorithm for mobile robot path planning in dynamic environments. Hindawi J. Math. Probl. Eng. **2**, 1–9 (2013)

14. Ahmadzadeh, S., Ghanavati, M.: Navigation of mobile robot using the particle swarm optimization. J. Acad. Appl. Stud. **2**, 32–38 (2012)
15. Goldberg, E.: Genetic Algorithms in Search, Optimization and Machine Learning, 1st edn. Addison-Wesley Longman publishing Co., Inc., Boston, USA (1989). ISBN: 978–0–201–15767–3
16. Lin, J.H., Huang, L.R.: Chaotic bee swarm optimization algorithm for path planning of mobile robots. In: Proceedings of 10th WSEAS International Conference on Evolutionary Computing, pp. 84–89. Wisconsin, USA (2009)
17. Elshamli, A., Abdullah, H.A., Areibi, S.: Genetic algorithm for dynamic path planning. In: Canadian Conference on Electrical and Computer Engineering 2004 (IEEE Cat. No.04CH37513), vol. 2, pp. 677–680, IEEE Publications, Canada. doi: https://doi.org/10.1109/CCECE.2004.1345203 (2004)

Smart Green Roof: A Prototype Toward Sustainable Smart Agriculture

Ramsha Siddiqui, Mohammad Muzammil Khan, Aqeel Khalique, and Imran Hussain

1 Introduction

To facilitate life and comfort, humans have been obliged to consume nonrenewable energies and fossil fuels as a result of economic advances and technological advancement. Excessive use of these resources, on the other hand, has resulted in issues which include deforestation and climate change. Heavy rains, storm surges, and drought periods are becoming more often as the Earth's natural condition and its forests deteriorate [1].

As a consequence, cities are growing increasingly polluted as constructions destroy trees and green spaces on a frequent basis, because of sources of air purification are trees and forests, which are mercilessly removed for commercial gain; sources of energy such as fossil fuels are now being utilized very aggressively while they are still limited and we come closer to the end of such, despite the fact that human beings are aware of the pollution generated by these resources. On the other hand, we cannot overlook the importance of oxygen and green space in meeting our physical and mental needs for ensuring our survival. As a result, we are confronted with two major issues: first, the green spaces that have been lost in cities must be preserved in order to meet each person's per capita requirement for green space while also reducing emissions by planting trees; and second, green spaces that are being lost in cities must also be reconstructed in order to satisfy each individual's per capita needs for green space while still decreasing pollution and reducing nonrenewable energy usage by reforestation.

Due to the fact that residential buildings use and waste the most energy, and because people are spending significant amounts of time in them, transforming green space on residential roofs into a rooftop terrace or green roof to minimize energy

R. Siddiqui · M. Muzammil Khan (✉) · A. Khalique · I. Hussain
Department of Computer Science and Engineering, School of Engineering Sciences and Technology, Jamia Hamdard, India
e-mail: mmkhan.sch@jamiahamdard.ac.in

I. Hussain
e-mail: ihussain@jamiahamdard.ac.in

B. Agarwal et al. (eds.), *Proceedings of International Conference on Intelligent Cyber-Physical Systems*, Algorithms for Intelligent Systems,
https://doi.org/10.1007/978-981-16-7136-4_8

consumption will be one of the finest and feasible solutions to the issues that have arisen; at this point, it's critical to make them aware of environmental elements that can help them consume energy more efficiently [2, 3].

This project is exploring a solution that incorporates the beneficial effects of a green roof with a water reservoir underneath it. These green roofs are designed to respond to current and future weather events while also communicating with one another. This aids in reducing the negative consequences of severe weather occurrences and improving the overall well-being of city dwellers.

In Sect. 2, we shall discuss the need for such systems followed by our proposed prototype in Sect. 3 with discussing outcomes in Sect. 4 and having a conclusion in Sect. 5.

2 Need for Smart Agriculture

To meet our physical and emotional demands, as well as to exist, we cannot disregard the importance of oxygen and green space. Even in this COVID-19 pandemic crisis, where the entire world is struggling for survival, especially countries like India, it is the wisest approach to take [3] (Fig. 1).

Fig. 1 Features of smart agriculture

However, there is a popular misconception that green roofs are self-sustaining and thus do not need to be maintained. This is not the case; green roofs need maintenance at least twice a year, which the HSE considers being a common occurrence [4]. In addition to maintenance of the vegetation such as removing weeds or replacing dead plants, drain outlets and fire breaks must be checked, components such as flashings, mastic and roofing membrane inspected, roof lights cleaned, sensors maintained, and general rubbish removed from the roof. However, this is worth mentioning that research is currently being done on Unmanned Vehicles for Smart Agriculture [5].

The age of smart agriculture has arrived, thanks to the deep integration of modern information technology and conventional agriculture [6]. The community's role in the long-term viability of Smart Green Roofs is critical since the community may effectively determine the feasibility of the productivity of green roofs. This is demonstrated by the prominence of abandoned green space infrastructure as a result of the population's lack of involvement in caring for and maintaining them. The community's most important responsibility in green-roof sustainability is community awareness, education, and active engagement in green-roof maintenance and care [7] (Fig. 2).

Fig. 2 Benefits of IoT-enabled smart agriculture

3 Prototype

With this project, Intelligent Green Roofs is our proposal for solving this problem. To start off, we shall investigate the various systems built for green roofs to assess the existing roofing systems. Price, schedule impacts, and constructability will all be considered when evaluating these systems. With the best system selected, the roof will be facilitated with a resilience network of IoT devices to make this roof an intelligent green roof. Lastly, the effects of the proposed green roof will be determined using data fetched by the IoT sensors located on the rooftop sent to the Ubidots' API. Figure 3 shows our prototype's circuit diagram consisting of Arduino, relays, sensors, etc.

For communication, HTTP is the major communication standard that will be used. Ubidots shall be treated as a data and user interface. All the sensors are placed at their specific place to monitor and send sensor data. Engineered systems typically consisting of a series of layers are included. The list of materials required consists as shown in Table 1 and Figs. 4, 5, and 6.

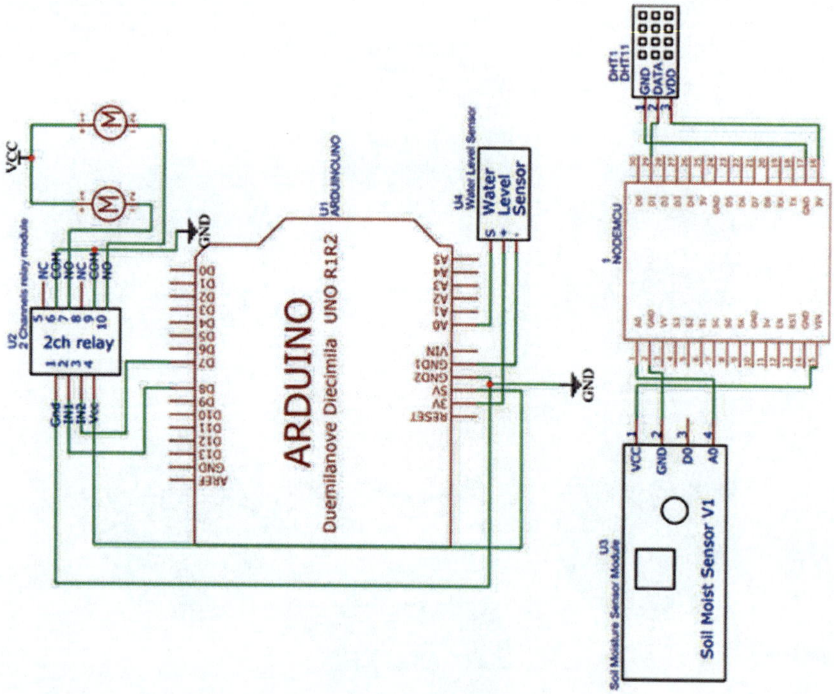

Fig. 3 Prototype's circuit diagram

Table 1 Requirements for the prototype

Material required	Hardware interfaces	Technological stack
A waterproof and root-proof membrane	Hard Disk—2 GB	Arduino
A filter fabric	Keyboard and Mouse	Tera Term
Growth medium—which is not soil but engineered lightweight granular medium typically consisting of expanded shales and clay minerals	Monitor Arduino UNO Board NodeMCU Board 5V Channel Relay module Water-level sensor Soil moisture sensor DTH22 Humidity and Temperature sensor Water motor Pipe	Ubidots (API) Windows XP C++ Embedded C# (small extent) Data analysis add-in

4 Results

A modular or smart green-roof system is a roofing design solution that is technically advanced as well as novel technological solution with multi-functional benefits and the ability to perform device integration. We were able to build our prototype as seen in the outputs in Figs. 7 and 8 with our proposed circuit diagram that can be seen in Fig. 3 to track the progress of plants in a wide range of real circumstances (e.g., soil quality and environmental conditions), as well as an ongoing condition of green roof. Variable irrigation and fertilizer/additives are used in these researches, which are performed in natural outdoor environmental conditions and locations where plants are growing [8]. By collecting appropriate time-series data from sensor networks, spatial data from imaging sensors, and human insights captured by smartphone apps, Internet of Things (IoT) technologies will reduce the cost and size of such studies. For example, IoT devices can provide scalability, and they help to monitor the temperature, humidity, water level in the tank, and much more data and share it among interested researchers and growers for further procedure. Wireless Sensor Networks have advanced to the point that they can now be used to track and regulate greenhouse parameters in precision agriculture [9–11].

The proposed model deals with predicting crop suitability by gathering information from sensors that present the following values:

- Temperature of the environment so that the exposure to climate can be adjusted as required;
- Humidity in the environment;
- Moisture content of the soil.

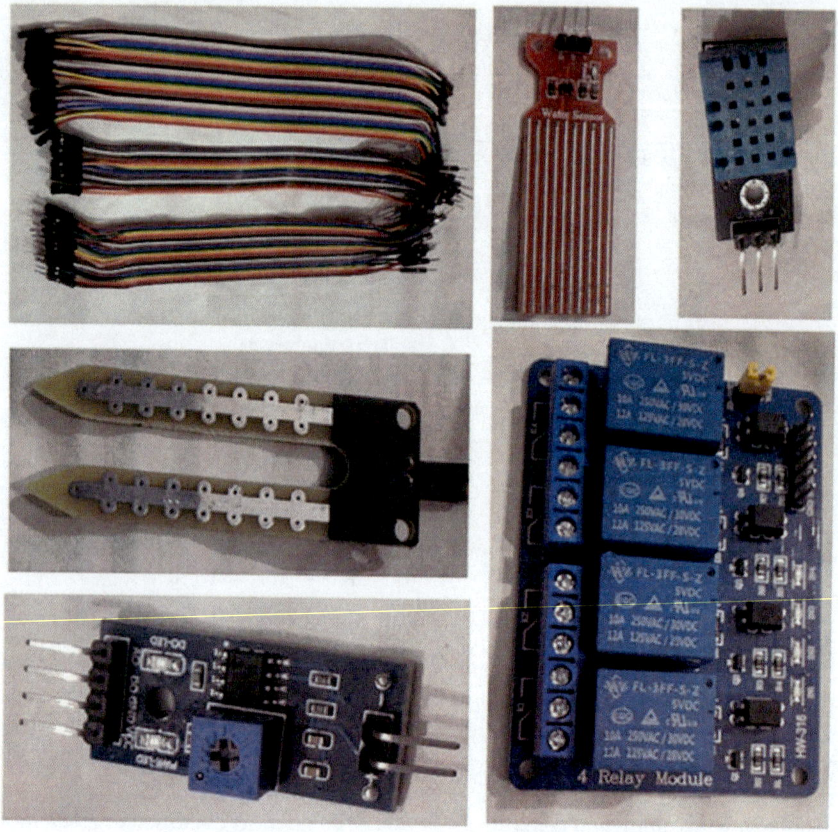

Fig. 4 IoT sensors and devices used in prototype

Fig. 5 Arduino Uno Board and NodeMCU Board

Fig. 6 The prototype

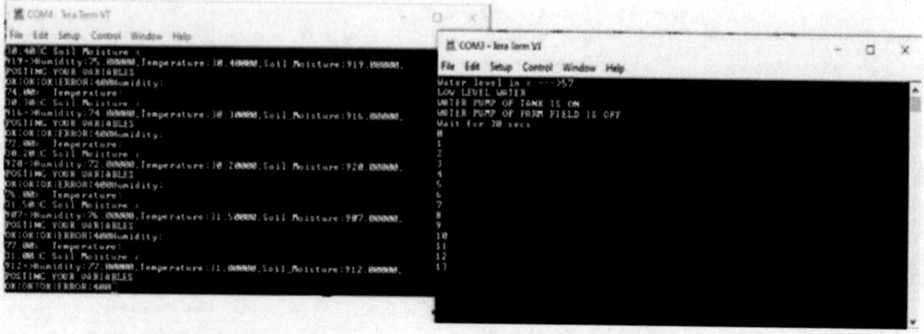

Fig. 7 Tera Term interface

5 Conclusion

The study concluded that the community's position in the long-term viability of Smart green roofs is critical because the community can effectively decide the viability of the green roof's operation. This is shown by the presence of abandoned green area facilities as a result of the community's lack of involvement in caring for and preserving them. The community's most important position in green-roof sustainability is community awareness, education, and active engagement in green-roof maintenance and care. Meanwhile, the community's function in providing green areas on building roofs will provide productive value to the occupied buildings, especially when the green roof is planted with productive plants. Because of the functionality and ease of access to carry out activities in productive green areas in occupied housing, universities, hospitals, and other places, the placement of green roofs on residential buildings is very strategic.

This also concludes that there is a very significant relationship of the role of the green area with intelligence (smart green roof) and another is a community with differences in economic status in support of green-roof sustainability. As a result, in order to introduce green roofing in both residential and non-residential areas, a pattern of mutually beneficial interactions is needed. Even though heavily developed urban areas have restricted green areas and residential land, the characteristics of low- to high-income people in dense residential areas appear to overlook the value of green roofs, whereas rural communities with various layers of the economy, low-, middle-, and high-income, have more support. The role of people who live in dense areas, such as metropolitan cities, provincial capitals, and cities, needs to be increased in stages through aspects of awareness, knowledge, and participation.

The role of the community can be managed by creating a gardening community network, which contributes to each other according to the economic capacity of the community. Worldwide, the gardening community network could be useful in other similar typical countries toward sustainable development. These findings of this study do not take into account the particular climatic area or types of roof materials, but rather concentrate on the characteristics that help smart green-roof applications.

Fig. 8 Ubidots interface for monitoring the green roof

This study suggests that further studies be done on management, monitoring, and community empowerment techniques that are relevant in green-roof sustainability. Further research may aid in the exploration of more issues in the field of smart green-roof implementation.

References

1. Bitaab, N., Golestani, N.: Smart green roof design in residential buildings approach to optimizing energy consumption. Int J Adv Mech Civil Eng (IJAMCE) **5**(1). IRAJ (2018)
2. Korol, S., Shushunova, N., Shushunova, T.: Innovation technologies in Green Roof systems. In: MATEC Web of Conferences, vol. 193, p. 04009. EDP Sciences (2018)
3. Babu Loganathan, G.: Smart agriculture system with E-carbage using Iot. Int. J. Modern Agricult. **10**(1), 928–931 (2021)
4. Safety on Green Roofs. https://specifierreview.com/2015/09/03/safety-on-green-roofs/. Accessed 12 July 2021
5. Maddikunta, P.K.R., Hakak, S., Alazab, M., Bhattacharya, S., Gadekallu, T.R., Khan, W.Z., Pham, Q.V.: Unmanned aerial vehicles in smart agriculture: applications, requirements, and challenges. IEEE Sens. J.
6. Yang, X., Shu, L., Chen, J., Ferrag, M.A., Wu, J., Nurellari, E., Huang, K.: A survey on smart agriculture: development modes, technologies, and security and privacy challenges. IEEE/CAA J. Autom. Sinica **8**(2), 273–302 (2020)
7. Yuliani, S., Hardiman, G., Setyowati, E.: Green-roof: the role of community in the substitution of green-space toward sustainable development. Sustainability **12**(4), 1429 (2020)
8. Jayaraman, P.P., Yavari, A., Georgakopoulos, D., Morshed, A., Zaslavsky, A.: Internet of things platform for smart farming: experiences and lessons learnt. Sensors **16**(11), 1884 (2016)
9. Devi, D.V.V., Kumari, G.M.: Real-time automation and monitoring system for modernized agriculture. Int J. Rev. Res. Appl. Sci. Eng. (IJRRASE) **3**(1), 7–12
10. Haseeb, K., Ud Din, I., Almogren, A., Islam, N.: An energy efficient and secure IoT-based WSN framework: an application to smart agriculture. Sensors **20**(7), 2081 (2020)
11. Manuel Ciruela-Lorenzo, A., Rosa Del Aguila-Obra, A., Padilla-Melendez, A., Jose Plaza-Angulo, J.: Digitalization of agri-cooperatives in the smart agriculture context. Proposal of a digital diagnosis tool. Sustainability **12**(4)

Identifying Predictors for Substance Consumption Pattern Using Machine Learning Techniques

Bijoy Chhetri, Lalit Mohan Goyal, and Mamta Mittal

1 Introduction

A drug is prescribed by the medical practitioner for remedies for some illness. Among the various classes of pharmaceutical drugs, the psychoactive drug contains a substance that influences mental functionalities. When these types of drugs are consumed above the prescribed dose, it is an abuse of a drug. Further, such drugs are an abuse-able psychoactive drug whose effects on the mental state are considerably high and gives a pleasant or interesting experience that some people choose to take them for a reason other than healing ailments. Such drugs of abuse have a chemical substance that influences part of mental functioning which is recreational, and enjoyable to cause preference to take multiple times leading to addiction. But, apart from substance consumption is enjoyable, the addiction intimate various kinds of diseases. These chemically or naturally available drugs are categorically called substances and as per Diagnostic and Statistical Manual of Mental Disorders (DSM) [1], the disorder caused by it is classified as a substance use disorder. Out of the various classes of drugs, depressants are those substances that include alcohol, amyl nitrite, benzodiazepines, tranquilizers, and opiates such as heroin and methadone. These classes of substances are used to reduce arousal and stimulations. The stimulants on the other hand consist of amphetamines, nicotine, cocaine powder, crack cocaine, and caffeine. These substances when taken speed up the movement of chemicals from the body to the brain making a person more energetic, alert, and too confident. The third broad category of drug classification is hallucinogens. These substances cause the person to hallucinate on sensations and images that appear to be real although they

B. Chhetri (✉) · L. M. Goyal
J C Bose University of Science and Technology, YMCA, Faridabad, India

M. Mittal
Delhi Skill and Entrepreneurship University, Dwarka,
New Delhi, India

B. Agarwal et al. (eds.), *Proceedings of International Conference on Intelligent Cyber-Physical Systems*, Algorithms for Intelligent Systems,
https://doi.org/10.1007/978-981-16-7136-4_9

are not in reality. Cannabis, ecstasy, ketamine, Lysergic acid Diethylamide (LSD), and mushrooms are the substances that fall under this category. Some other classes like Volatile Substance Abuse (VSA) which is used through glue-sniffing, inhalant abuse, and solvent abuse for the deliberate inhalation of volatile substances to achieve intoxication also exist. There are shreds of evidence on the use of these substances in high quantity for social or recreational purposes, including fun, stress suppressor, or to feel different. People of all age groups have been taking this for various reasons irrespective of gender or ethnicity. Several reasons are associated with early drug use including psychological, social, individual, environmental, and economic factors [2]. These aspects are likewise linked with several behaviors [3, 4] which are social and personal, but the consequences are highly problematic.

When data records of various international and national agencies are looked at, consumption of both illicit and licit substances has been so often that World Health Organization reports more deaths globally due to tobacco and opioids than because of hepatitis or AIDS [5]. National Survey in India [6] reported that of 14.6% alcoholics, 2.8% use cannabis and 1.14% use opioids either prescribed or illicitly accessed. The US reports the same figure as 43.4 million (17.9%) and the young are more influenced. The fatality rate is also very high due to opioid overdose, especially heroin consumption [7]. On a similar note, the UK too has a large number of cases, and various strategies have been put in place to see increasing patterns of drug use and its threats [8]. World Drug Report [9] highlights record-high use of prescription drugs and opioids along with enormous users of common drugs like alcohol and tobacco globally. To address this rate of consumption, several case-based researches are done and drug-dependent disorderly behaviors are studied and tackled. However, in the heterogeneity of substance consumption patterns [10], opiates for stimulants have some differences in the behavior of consumers. The cause of neurobehavior and even their consumption patterns have been very distinct [11, 12] in personality involvement. With each class of substance, the pleasure circuit is caused to move very differently in different individuals [13]. The substance-specific uses are fundamentally agreed [14] on the choice, and preferences of individuals like alcohol seekers will have a different profile than that of LSD. This apart, the initial onset of one of them could lead to the addiction of another and the seeking tendencies cause one to ON and the other one to OFF [15]. For example, substances opiate and stimulant addictions have opposite effects such that the former activates inhibitory and sedative circuit of the brain whereas the latter arouses and creates lots of excitement, thus, personality trait plays a major role in the selection. The initial onset of marijuana at a young age could lead to another substance intake [16] which is also dependent on how a person perceives the substance. Thus, the identification of layers of influencing factors that could justify variable consumption patterns is required. This phenomenon of seeking one substance after another has prevailing psychological thoughts which have been somehow disseminated by the Machine Learning (ML) model applied on the various clinical or non-clinical datasets revealing some of the remarkable facts on substance consumption as presented in Table 1.

Hence, knowing the fact that substance intake varies from person to person depending on their personality traits and consumption patterns, this article presents

Table 1 Various existing studies performed machine learning approach

Authors	Study and sample	Method used	Predictors	Outcome	Remark
[17]	Cohort/adolescents from Canada and Australia	Supervised learning	Personality, cognition, and alcohol attitude variables	Prevalence of alcohol drinking	AUC: 0.86
[18]	Cross-sectional/cocaine user and control	Supervised learning and regression	Impulsivity traits	Prevalence of cocaine and high impulsivity	AUC: 0.91
[13]	Cross-sectional/heroin user and Control	Supervised learning and regression with elastic net	Personality, impulsivity	Heroin (H) and Amphetamine (A) have distinct seeking and personality traits	AUC: 0.86(H) AUC: 0.71(A)
[19]	Cohort/smokers	CART supervised learning	Clinical, executive functions	Prevalence of cigarette and delay of execution	Accuracy: 0.80
[20]	Cohort/twitter users	Ensemble supervised	Tense level, arousal level, restlessness	Prevalence of alcohol more prominent	Accuracy: 0.89
[21]	Trail/cocaine dependent	Reinforcement learning	Deprivation	Sensitive to drug	R: 0.46
[16]	Cross-sectional/survey data	Supervised learning	Income, employment, early onset drug	Marijuana seeks to be driving the other substances	AUC: 0.89
[22]	Cross-sectional/drug user and non-user	Artificial neural network	Personality, demography	Alcohol user	Accuracy: 98.7

Label: AUC: Area under the Receivers Operating Characteristics

the use of the ML model to identify the baseline markers to understand the nature of substance intake and their behavioral association with an individual. Various reasons are likely to predict the initial onset of substance till it becomes an abuse. It is also the likelihood of one substance to induce another substance to cause poly-drug influence. The socio-economic parameters, demography, and location may also influence the cause. Although each of these aspects has a considerable amount of contribution to increase the threat of substance intake, neither one is the only origin of conclusion. Therefore, this research is aimed at analyzing the complex relationship among these dynamics and identifying patterns of consumption, their key predictor if any.

The relationships obtained are the measure of substance prevalence and consumption patterns which are the potential markers of risk assessment. Because of the fact that substance consumption may lead to psychological disorders in the later stage, therefore, these predictors along with potential clinical observation can help health workers and officials in planning out an appropriate intervention and precautionary approach.

The article is divided into various sections where Sect. 2 has a methodology explaining the approach used to carry out this research. Section 3 is a result section that highlights key findings obtained while carrying out this research. The primary findings are discussed in Sect. 4 which is followed by a conclusion and references.

2 Proposed Methodology

A systematic diagram that explains the methodology is presented in Fig. 1. The dataset acquired from the repository is subjected to rigorous pre-processing of data that are required. The first phase is data transformation where the categorical data is transformed into numerical values to facilitate the requirements of ML models. Data curation is also performed so as to obtain the required field for the classification problem. The subclass where similar substances are included is also prepared for association analysis. The insignificant columns are also removed.

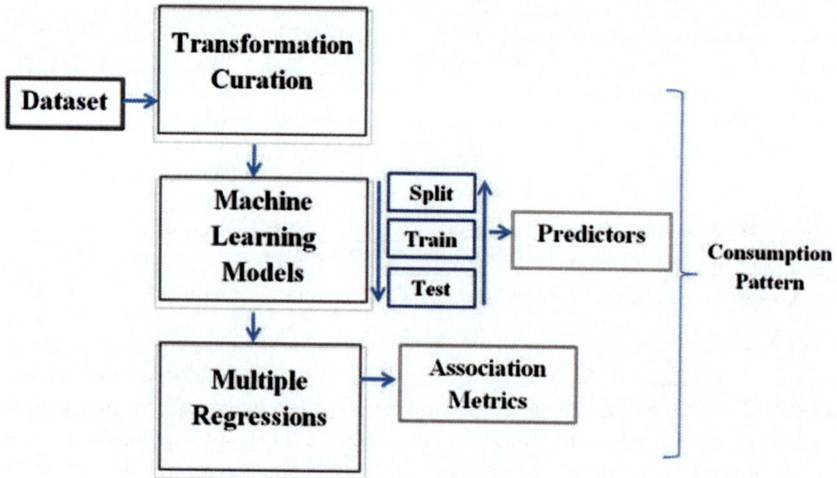

Fig. 1 Systematic diagram of proposed methodology

2.1 Dataset

The research is based on existing data [23] which is available in the UCI repository. The dataset contains 1185 × 31 values derived from the cross-sectional survey on personality trait parameters and substance consumption information of 18 licit and illicit substances. The categorical values are quantified. The personality traits are measured with 5 parameters to measure personality in terms of Neuroticism (N), Extraversion (E), Openness (O) to experience, Agreeableness (A), and Conscientiousness (C) which is popularly known as NEO-5 personality traits. It also contains a measure of Impulsivity (BIS-11) and Sensation Seeking measure (ImpSS) that is the standard battery of assessment used by psychologists to measure disinhibition, distraction, consciousness, and insufficiencies in decision-making. The dataset also contains little demographic information, i.e. level of education, age, gender, country of residence, and ethnicity. In addition, the dataset contains the response from the participants about their intake of 18 licit and illicit classes of substances. Alcohol, amphetamines, amyl nitrite, benzodiazepine, cannabis, chocolate, cocaine, caffeine, crack, ecstasy, heroin, ketamine, legal highs, LSD, methadone, mushrooms, nicotine, and VSA are included. A dummy drug (Semeron) is also used to rule out the over claim from the respondents. The information about consumption of each of these substances is collected through a questionnaire being asked to select one of the answers from the option of never used, used over a decade ago, or in the last decade, year, month, week, or a day suggesting the use patterns of the individual who has responded with the survey questionnaire. The dataset is curated and used in this study to suit the objective of the research. Out of 18 substances, the effect of VSA is not considered as it has insignificant statistics as compared to other substance use. The three classes of substances, namely depressants, stimulants, and hallucinogens, are made based on the standard composition of the substance. The outcome of multiple classifiers, multivariate analysis, and correlation analysis is presented with evaluation and comparisons with other literature. The paper takes a perspective to seek the likelihood of substance pattern due to one substance following the other in continuation. The outcome variables in terms of predictors signifying the likelihood are sought from the dataset.

2.2 Classifiers

The ML-based classifiers are employed to classify each of the respondents into substance user and non-user categories. Table 2 has a detailed list of the predictor variables used for classification on the ML models. The important features are extracted from the ML model to further understand their impact through multivariate analysis. The predictors' variable values are split into training and test cases. The training set is used for training the classifiers and test sets are kept for validation. The usual

Table 2 Summary of the predictor variables

Outcome: user/non-user			
Predictor variables	N	Mean	Levels
Gender Male Female	943 942	NA	2
Age 18–24 25–34 35–44 45–54 55–64 65+	643 481 356 294 93 18	42.44	6
Educational background Left school before 16 yrs Left school at 16 yrs Left school at 17 yrs Left school at 18 yrs Studied but No certificate Diploma Degree Masters and above	28 99 30 100 506 270 480 372	NA	8
Personality factors			
Neuroticism (N score)	1875	23.92	6
Extraversion (E score)		27.52	
Openness (O score)		33.64	
Agreeableness (A score)		30.87	
Conscientiousness		29.44	
Impulsivity and sensation			
BIS	1875		4
ImpSS	1875		
Potential drivers within the group of substances	NA		

percentage of training and test sets is 80–20 of all the responded values. The prediction showed low accuracy in the initial stage; it has a high number of non-users of any of the substances making it an imbalanced dataset. The authors, however, have used downsampling to balance the dataset, and classifications are performed. The datasets are used to train and test DT, RF, and K-NN classification algorithms. A Multivariate Regression (MR) analysis is also performed using Minitab to evaluate the association of various classes of substances. Python packages under the Scikit Learn library are used to build the classifiers model and perform classification. For all of the three classifiers, classification algorithms are run up to 100 times, and the average of the performance is measured using a 95% confidence level on the statistically significant parameters.

2.3 *Evaluation of Performance*

The results predicted by each of the classifiers are compared against the observed outcome to obtain the ground truth. The analysis is done through a confusion matrix which refers to relationships among the predicted and the observed true values. Here, true positive refers to the prediction of the substance user who is an actual user of the substance. False positive informs about substance abuse which is absent in the true case. Similarly, false negative is the predicted result which is incorrectly predicted as a non-user of substance which is a negative prediction as is false positive. However, certain cases of false predictions are considered and sensitivity percentage is computed. In addition, specificity which is a percentage of true negative cases is also taken as a measure to evaluate the classifiers. To visualize the performance, an Area under the Receivers Operating Characteristics curve is also assessed using various threshold values. In general, a range above 80% is considered as fair and good to predict the outcome, whereas 90% and above is targeted throughout this research. MR is evaluated using squared error measurement. The set of predictor variables, while they are identified, is evaluated statistically with a significance value of $p < 0.001$.

3 Result

This section presents the research findings in three modes; first, the classification algorithm is run to find the predictor variable among all the variables along with their importance. The ML classifiers are not very accurate in classifying the individual into their respective class. Therefore, an additional method of classifying the individual into each substance class using the ensemble method is also carried out and presented in a separate section. The third section has unique findings as to what extent the substances themselves drive the other substance consumption. To have a deeper understanding of every outcome, MR is performed with a statistically significant value of $p < 0.001$. The outcomes of classifiers as key predictor variables along with the importance of each variable are also presented along with the performance of ML models.

3.1 *Classification Based on Three Classes of Substances*

Based on the classification of substances [24], a total of 17 licit and illicit substances are categorized into three groups, namely Depressants, Stimulants, and Hallucinogens. The response value recorded under each category are taken all together to find correlation among the individual substance within the group like amphetamines, nicotine, cocaine powder, crack cocaine, caffeine, and chocolate falls under the same class of stimulants. The coefficient enabled the research to deepen down to find the most correlated preferences among the users. Further, a new target class is computed based on the preference of either HIGH or LOW. HIGH here signifies the respondents who have been taking substances for a month, but LOW is to signify that person is in abstinence for at least a month. For example, in a class of depressants, the three substances, namely benzodiazepine, alcohol, and meth are positively correlated with significance value ($p < 0.001$); thus a class of response variable (Depressant) is determined with the binary classification of HIGH or LOW as their indicators. The classifiers are trained and tested with the predictor variables considered, and the outcome evaluation metrics are presented in Table 3. Due to an imbalanced set of data in the stimulant class, where there is very minimal user classified as HIGH, the classifier was overfitted to predict with 99% of accuracy, which is not achieved in the other cases. Finally, the imbalanced set is managed using downsampling of the dataset to accord a result as shown.

Out of all the classifiers, RF seems to have scored high accuracy to predict the class of substances. The predictor variables as rated by RF have impulsivity factor on the top and relevance of education being very insignificant as far as substance intake likelihood is a concern. The predictor variables identified by the classifier(s) along with their percentage of importance are also presented in Table 4.

Table 3 Result of classifiers

Class of substance	Classifier	Precision	Recall	F1 score
Hallucinogens	K-NN	0.85	0.73	0.79
	RF	0.88	0.78	**0.83**
	DT	0.96	0.65	0.78
Stimulants	K-NN	0.65	0.81	0.72
	RF	0.74	0.81	**0.77**
	DT	0.85	0.65	0.74
Depressant	K-NN	0.87	0.67	0.76
	RF	0.75	0.85	**0.80**
	DT	0.89	0.65	0.75

Table 4 Relative importance of variables contributing to substance consumption

Predictor variables	Relative importance (%)
Impulsivity (ImpSS)	17
Impulsivity (BIS)	17
Openness	15
Extroversion	15
Neuroticism	12
Agreeable	5
Conscientiousness	4

MR analysis is performed on the target variables after removing less important predictor variables and considering only personality trait predictors (NEO) and impulsivity predictors (BIS and ImpSS) to test the regressed variables. The outcome tests the differences and associations to conclude an effect to the response variable or not and also validates the class of substances with a statistically significant p-value < 0.001 to support the assumption made at the beginning of this study.

3.2 *Classification Based on Individual Substance*

While the classification is done according to the class of substances, the viability of testing one's use of drugs cannot be mapped to the individual substances apart from the classification being made by the classifiers that has moderate accuracy; thus, it is difficult to find a final model of prediction. Therefore, to perform the classification synergistically, an ensemble method is adopted to combine the prediction of three ML models and lessen inconsistency in prediction as well as generality errors.

The training data is varied in every model using a random split of training and test data as well. The models are validated with a tenfold cross-validation technique to produce the performance as presented in Fig. 2 with performance metrics in Table 5.

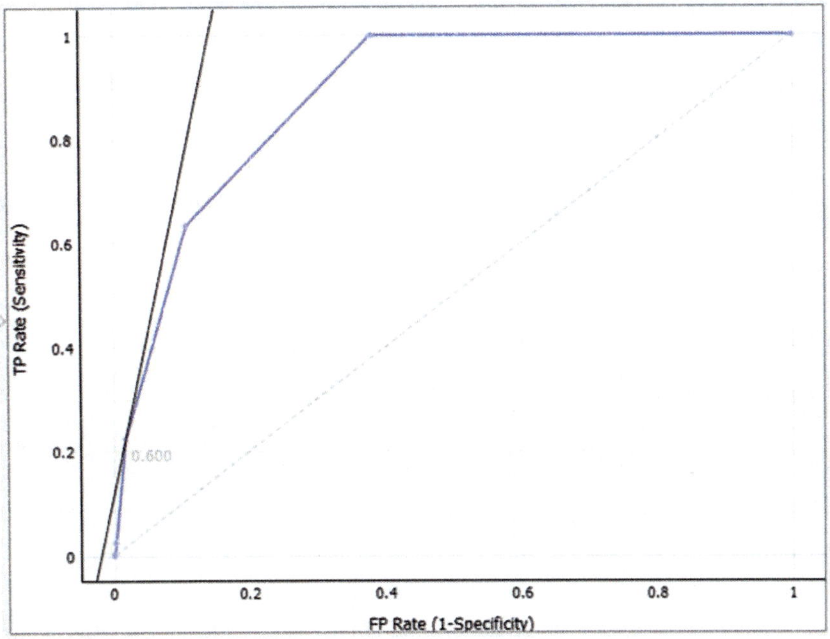

Fig. 2 Ensemble model AUC curve

Table 5 Performance of ensemble model

AUC	CA	F1	Precision	Recall
0.973	0.936	0.932	0.939	0.936

3.3 *Potential Drivers for Substance Consumption*

Various researches have concluded the fact that one's initial onset of driver substance could lead to their future increase in consumption thereby causing more harm. Therefore, to understand the consumption pattern and such driver substances, an analysis has been performed based on the objective to find a substance that could be a potential marker inherent within the group of substances driving an urge. It is found that both the statement of early onset and carrying further are significantly true with $p < 0.001$, and it was also surprisingly found that the drugs that belong to the same class have minimal correlation and association ($r << -0.01$) when they are analyzed within the group. However, the substance that belongs to a different class tends to have a higher association ($r > 0.21$). A MANOVA is performed with three classes of substances and the substance which is present in another class of substances. It is found that heroin is the leading driver in the group of depressants. It would lead to the consumption of other substances belonging to another group of substances like stimulants. Table 6 has a significant list of substances that falls under the potential driver to another class of substances with their significance value.

There are quite a few substances that are closely related to the hallucinogen class like cocaine, and amphetamines belonging to stimulants. Stimulants though have indicated a very limited association with depressants except with heroin. But, when analyzed along with hallucinogen class, cannabis and LSD has a high chance of leading to consumption of stimulants. Heroin also has been an active substance to have a strong association with hallucinogens. Methadone, an opiate medicine, is also of depressant class, which has a high influence on hallucinogens. Among the stimulants, crack cocaine has a high likelihood of driving it toward hallucinogens. Thus, there are some driver substances that are causing consumption patterns to

Table 6 Association of various driver substances in the pattern of consumption

Class	Name of substance	T statistics	F	p
Hallucinogens	Crack	0.97096	28.11	0.001
	Amphetamines	0.99295	6.64	0.001
	Heroin	0.96.73	38.54	0.001
	Methadone Opiate	0.9967	8.87	0.001
Stimulant	Cannabis	0.98818	11.23	0.001
	LSD	0.98303	16.97	0.001
	Heroin	0.9785	20.88	0.001
Depressant	Crack	0.98564	13.68	0.001

follow with the personality traits, as visually presented in Fig. 3a–c, and along with the key features of traits being the key predictors.

The regression model for various classes of substances highlights that the consumption pattern suggests sufficient association with the personality trait variables. The R2 measure of the regression model shows quite a high value with a significance measure of $p < 0.001$. As shown in Fig. 3a, it is a model building plot of the hallucinogen class and the predictor variables showing the impact of adding a new variable. The variation in the form of regression coefficient function is obtained, which is fitted on the curve to plot residual versus the fitted values. The final regression report suggests impulsivity being highly associated with the class of substances. Similarly, in Fig. 3b, stimulants are plotted against the predictors. There are some statistically insignificant values encountered in the analysis ($p > 0.001$), aside from the variable impulsivity which is evident in the variation bar plots, where R2 values are significant to fit the information to a model with low variation.

When the regression is plotted against the depressant class of the substance, each of the variables has a significant value ($p < 0.001$); thus the variation is significant as shown in Fig. 3c. The value of impulsivity is high while it is regressed with all the predictor variables, whereas whenever new information is to be predicted, the score of E contributes to the maximum in order to fit the model.

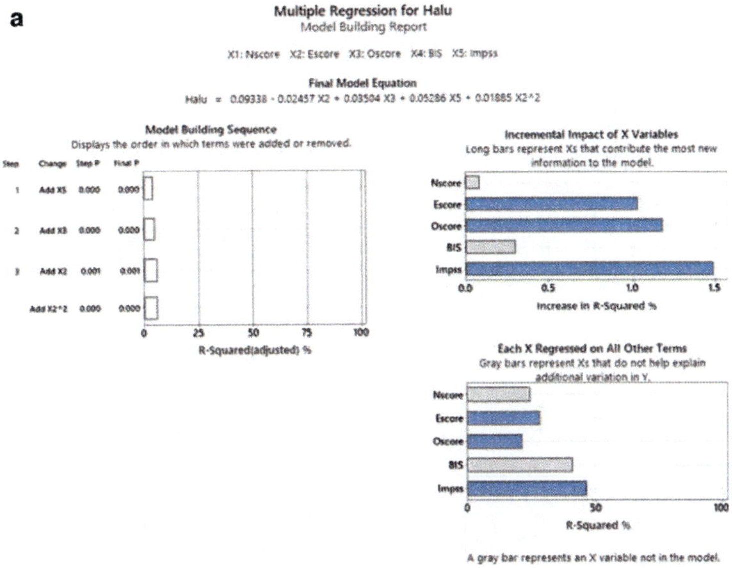

Fig. 3 **a** Multiple regression of halucinagens class. **b** Multiple regression of stimulant class. **c** Multiple regression of depressant class

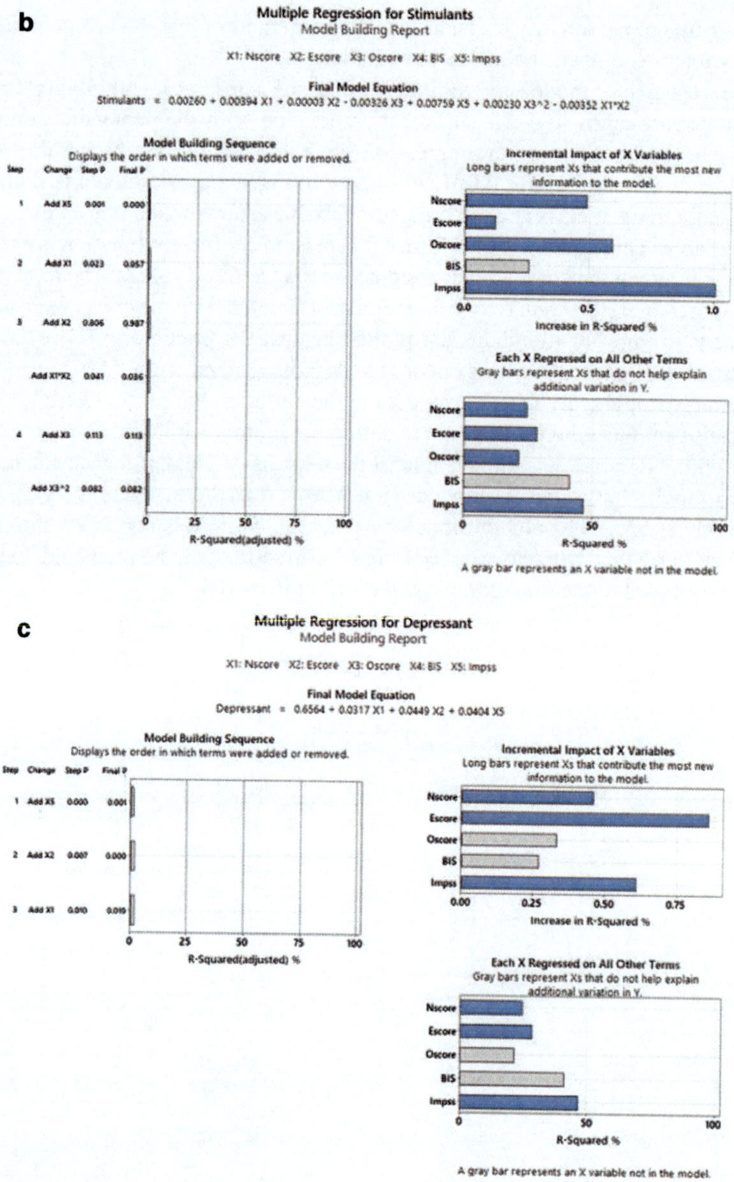

Fig. 3 (continued)

4 Discussion

This study presents a case on analysis of an existing dataset of substance use and its association with personality traits along with demographic variables. Findings show a positive association between consumption patterns and their predictor variables that influence an individual to consume substances. Initially, classification of substance usage as User/Non-User is performed using ML methods. Further, the substances are studied in association with personality traits and impulsivity along with some covariates as predictors to analyze multivariate kinds of patterns in their consumption, inter- and intra-class relationship as well as within individual substances. The existing literature [25, 26] has concluded the profile of personality traits to each Pleiades of substances. Similarly, a neural network applied to the same dataset has concluded results with the accurate classification of alcohol [22] consumption (AUC: 98%). This work has been able address major gaps that remained unaddressed, specifically (a) latent predictor variables which are addressed as drivers throughout the research are analyzed concurrently with multiple dimensions of personality traits within the substance-dependent individuals to gain insight into association among the various class of substance; (b) it identified a common predictor variable from multivariate patterns of consumption that classified them into user and non-user in new samples; and (c) it provided common patterns of evidence that an individual with certain endophenotypes correlates to substance intake.

The dataset contained a poly-drug assessment with personality traits; the only common predictor variable in all the categories of substance is impulsivity as in similar research [27]. It is strongly related to one psychological desire to perform any task without thinking about its result. When impulsivity goes too far, it strongly relates to aggressive behavior, restlessness, and most importantly gets easily distracted by influences, thus opening a wide scope of indulging in substance consumption [18]. Apart from this result, it also suggests that lifestyle personality factors like Neuroticism, Extraversion, and Openness has an equal stake in the development of substance consumption habits as well as their impact may lead to other comorbidities like Hypersensitivity or Conduct disorder, etc. [13]. The study also produces far-ranging evidence of one drug that has the chance of leading to another. ML models like K-NN, RF, and DT can understand substance consumption patterns, and relatively important variables are obtained. The consumption patterns show some of the inherent driving factors that are driving users toward the other class of substance. For example, someone is keen on prescribed opiates for some time; the desire of getting into illicit substances like LSD and Cannabis cannot be overruled and various profiles have already been identified in case of substance use disorder [28]. There is a prevalence of polysubstance consumption and such patterns even lead to mental illnesses like depression and anxiety [21, 29]. In contrast with other previous findings, the education and gender variables are not significant to drive people into substance consumption.

5 Conclusion

The ML approach has classified the substance user with acceptable accuracy and their relative predictors identified from the personality traits and demographical details to discover a consumption pattern. The study also produces some evidence on the inherent property of one or more substances to drive into the use of other substances. The dominant predictor variable is impulsivity along with the significant importance of personality trait variables. Depressant substances such as alcohol and its seeking sensation may not lead to heroin consumption but can have a strong association with hallucinogens such as cannabis. It can be established that variables like NEO that define personality trait evidence are indispensable for the estimated definition of substance consumption. The results are quite eye-opening. The truth of consumption patterns as revealed through data of substance intake seeks utmost attention. The multi way of influences in consumption pattern can be reduced while tapping the key predictors at the early stage using ML models and appropriate intervention is used to reduce the risks. Though the highest care is taken to carry out this research study, a few limitations exist: facilitation of individual consumption pattern with the development of the singular prevention method can be devised. The imbalance dataset can be corrected and can enrich the precision in the likelihood of time of consumption, which can also be taken up in future directions. In summary, the multiple variables are responsible for predicting markers that classify them to some substance consumption pattern or dependence and also that there are inherent capabilities of one substance to switch ON other substance intakes.

References

1. Pomeroy, E.C., Anderson, K.: The DSM-5 has arrived (2013)
2. Finn, K.: Why marijuana will not fix the opioid epidemic. Mo. Med. **115**, 191–193 (2018)
3. Zou, Z., Wang, H., Uquillas, F.D.O., Wang, X., Ding, J., Chen, H.: Definition of substance and non-substance addiction. Substance and Non-substance Addiction 21–41
4. Poria S., Gelbukh A., Agarwal B., Cambria E., Howard N.: Common sense knowledge based personality recognition from text. In: Castro F., Gelbukh A., González M. (eds) Advances in Soft Computing and Its Applications, MICAI 2013. Lecture Notes in Computer Science, vol. 8266. Springer, Berlin (2013). https://doi.org/10.1007/978-3-642-45111-9_42
5. Peacock, A., Leung, J., Larney, S., Colledge, S., Hickman, M., Rehm, J., Ali, R.: Global statistics on alcohol, tobacco and illicit drug use: 2017 status report. Addiction **113**(10), 1905–1926 (2017)
6. Ambekar, A., Agrawal, A., Rao, R., Mishra, A.K., Khandelwal, S.K., Chadda, R.K.: Magnitude of substance use in India. New Delhi: Ministry of Social Justice and Empowerment, Government of India (2019)
7. Abuse, S.: Mental health services administration. In: Key Substance Use and Mental Health Indicators in the United States: Results From the 2016 National Survey on Drug Use and Health (HHS Publication No. SMA 17–5044, NSDUH Series H-52). Rockville, MD: Center for Behavioral Health Statistics and Quality. Substance Abuse and Mental Health Services Administration (2017)
8. Drug misuse and dependence. www.assets.publishing.service.gov.uk (2017)

9. United Nations Office on Drugs and Crime (UNODC). World drug report 2017. Vienna: UNODC (2017)
10. Badiani, A., Belin, D., Epstein, D., Calu, D., Shaham, Y.: Opiate versus psychostimulant addiction: the differences do matter. Nat. Rev. Neurosci. **12**(11), 685–700 (2011)
11. George, O., Koob, G.F.: Individual differences in prefrontal cortex function and the transition from drug use to drug dependence. Neurosci. Biobehav. Rev. **35** (2010)
12. Vassileva, J., Paxton, J., Moeller, F.G., Wilson, M.J., Bozgunov, K., Martin, E.M., Gonzalez, R., Vasilev, G.: Heroin and amphetamine users display opposite relationships between trait and neurobehavioral dimensions of impulsivity. Addict. Behav. **39**, 652–659 (2014)
13. Ahn, W.Y., Vassileva, J.: Machine-learning identifies substance-specific behavioral markers for opiate and stimulant dependence. Drug Alcohol Depend. **161**, 247–257 (2016)
14. Clark, S.L., Gillespie, N.A., Adkins, D.E., Kendler, K.S., Neale, M.C.: Psychometric modeling of abuse and dependence symptoms across six illicit substances indicates novel dimensions of misuse. Addict. Behav. **53**, 132–140 (2016)
15. Peters, J., Pattij, T., De Vries, T.J.: Targeting cocaine versus heroin memories: divergent roles within ventromedial prefrontal cortex. Trends Pharmacol. Sci. **34**(12), 689–695 (2013)
16. Wadekar, A.S.: Understanding opioid use disorder (OUD) using tree-based classifiers. Drug Alcohol Depend. **208**, 107839 (2020)
17. Afzali, M.H., Sunderland, M., Stewart, S., Masse, B., Seguin, J., Newton, N., Conrod, P.: Machine-learning prediction of adolescent alcohol use: a cross-study, cross-cultural validation. Addiction **114**(4), 662–671 (2019)
18. Ahn, W.Y., Ramesh, D., Moeller, F.G., Vassileva, J.: Utility of machine-learning approaches to identify behavioral markers for substance use disorders: impulsivity dimensions as predictors of current cocaine dependence. Front. Psych. **7**, 34 (2016)
19. Coughlin, L.N., Tegge, A.N., Sheffer, C.E., Bickel, W.K.: A machine-learning approach to predicting smoking cessation treatment outcomes. Nicotine Tob. Res. **22**(3), 415–422 (2020)
20. Liu, J., Weitzman, E.R., Chunara, R.: Assessing behavior stage progression from social media data. In: Proceedings of the 2017 ACM Conference on Computer Supported Cooperative Work and Social Computing. IEEE, pp. 1320–1333 (2017)
21. Wang, J.M., Zhu, L., Brown, V.M., De La Garza II, R., Newton, T., King-Casas, B., Chiu, P.H.: In cocaine dependence, neural prediction errors during loss avoidance are increased with cocaine deprivation and predict drug use. Biol. Psychiatry Cogn. Neurosci. Neuroimaging **4**(3), 291–299 (2019).
22. Kumari, D., Kilam, S., Nath, P., Swetapadma, A.: Prediction of alcohol abused individuals using artificial neural network. Int. J. Inf. Technol. **10**(2), 233–237 (2018)
23. Machine learning repository. https://archive.ics.uci.edu/ml/datasets/Drug+consumption
24. Drug wheel: https://adf.org.au/
25. Fehrman, E., Egan, V., Gorban, A. N., Levesley, J., Mirkes, E.M., Muhammad, A.K.: Personality traits and drug consumption. Springer International Publishing (2019)
26. Fehrman, E., Muhammad, A.K., Mirkes, E.M., Egan, V., Gorban, A.N.: The five factor model of personality and evaluation of drug consumption risk. In: Data Science. Springer, Cham, pp. 231–242 (2017)
27. Argyriou, E., Um, M., Carron, C., Cyders, M.A.: Age and impulsive behavior in drug addiction: a review of past research and future directions. Pharmacol. Biochem. Behav. **164**, 106–117 (2018)
28. Zhang, X., Huang, R., Li, P., Ren, Y., Gao, J., Mueller, J.F., Thai, P.K.: Temporal profile of illicit drug consumption in Guangzhou, China monitored by wastewater-based epidemiology. Environ. Sci. Pollut. Res. **26**(23), 23593–23602 (2019)
29. Chhetri, B., Goyal, L.M., Mittal, M., Gurung, S.: Consumption of licit and illicit substances leading to mental illness: a prevalence study. EAI Endorsed Trans. Pervasive Health Technol. **6**, 21 (2020)

Web-Based Simulator for Operating Systems

K. Prajwal, P. Navaneeth, K. Tharun, Trupti Chandak, and M. Anand Kumar

1 Introduction

In the operating system field, we come across various concepts and solution to problems that have unique ways of being solved. Some of the various such concepts are CPU scheduling, memory management, disk scheduling, file system management, semaphores and some more. All the concepts mentioned above are fundamental when considering the operating system, as these algorithms that are suggested increase the operating system's speed significantly. Problems like deadlock and loss of information have been overcome by implementing these operating system's algorithms. Some of the concepts like Dining Philosophers and Consumer Producer are put forth as problems whose solutions have significantly affected how computers work today. Therefore, in light of this, we have decided to implement these concepts as a simulation.

To make the simulation more interactive and creative, we have combined the knowledge of web technologies and operating systems to come up with the following

K. Prajwal (✉) · P. Navaneeth · K. Tharun · T. Chandak · M. A. Kumar
Department of Information Technology, National Institute of Technology Karnataka, Surathkal 575025, India
e-mail: kprajwal.191it222@nitk.edu.in

P. Navaneeth
e-mail: navaneethp.191it132@nitk.edu.in

K. Tharun
e-mail: tharunk.191it255@nitk.edu.in

T. Chandak
e-mail: al.trupti@nitk.edu.in

M. A. Kumar
e-mail: manandkumar@nitk.edu.in

B. Agarwal et al. (eds.), *Proceedings of International Conference on Intelligent Cyber-Physical Systems*, Algorithms for Intelligent Systems,
https://doi.org/10.1007/978-981-16-7136-4_10

web application. For our project, we created a web application that simulates various CPU and OS concepts such as the ones mentioned above using graphical ways.

We have deployed our app at https://os-simulator.herokuapp.com/ which can be accessed by anyone, anywhere with internet access. Building the simulator was the main objective we were trying to achieve, a tool which can be used by anyone who wants to acquire knowledge on Operating System concepts, but reading text books or reference books is hard for them. This graphical way gives a whole new way of learning to either a student or a professor who wishes to use our web application.

Keeping the objective of building an educational tool in mind, we first decided to start with some basic yet essential algorithms to simulate with further research on advanced topics such as Socket Programming, Threading and System Calls.

2 Literature Survey

Operating System simulators simulate various Operating System concepts, and they vary from basic to complex real-time simulators. These simulators are used as educational tools at many top-tier institutions to make learning more interactive than just conventional reading from textbooks or reference books. Below, we have listed the popular and trending simulators available on the Internet with a short description. We have also tabulated this data, making it easier to see the differences and compare the available simulators.

SchedulerSim [1]: Here, a simulator of operating system was developed using various scheduling concepts. There was no simulation of the CPU processes. One of the main concepts here that was implemented in this project was the scheduling concept [1].

Sim. + assembler [2]: This project has a well-implemented CPU simulation. There is no OS simulation. Some of the important concepts that were implemented here were IO processing and interrupt handling [2].

MKit [3]: This project consisted of a very well-maintained OS simulator. No CPU simulation was implemented. Some of the concepts implemented were data path simulation as well as control unit management systems [3].

SOsim [4]: A simulator for an operating system was deployed in this project. There was no simulation on any CPU-based concepts. Some of the concepts implemented here are process management and memory management systems [4].

PsimJ sim [5]: It has an OS simulator set in place with various isolated OS component simulations that are present [5].

We have looked into each of these applications and felt like they could have included a few other concepts such as Memory Allocation Techniques, Page Replacement, Disk Scheduling which are also considered important in the field of Operating Systems. Hence, we decided to provide a graphical and interactive way of implementing the concepts including the other topics which were touched upon by the previous works such as CPU Scheduling algorithms. Using the concepts of operating system, to build an educational web app that simulates these concepts in an efficient and graphical way.

3 Proposed Work

We have implemented a simulation for six Operating System concepts.

1. Process Scheduling
2. Semaphores
3. Memory Allocation
4. File Systems
5. Disk Scheduling
6. Page Replacement.

We have built a web application using the following technologies which are available.

- Python3—used for scripting the algorithms
- JavaScript—used for scripting the algorithms
- Django—used as back-end for our web application
- HTML/CSS—used as frontend for our web application.

We have used python as to implement all the algorithms and concepts mentioned above and we used HTML, CSS and Javascript to implement, style and simulate them in a web environment.

We used Django framework to inter-link all these files into a web server which can be hosted on a server to make it available.

3.1 Process Scheduling

Pseudo Code for some of the scheduling concepts implemented have been written below. We have implemented the following algorithms.

- First Come First Serve (FCFS)
- Shortest Job First (SJA)
- Round Robin (RR)
- Priority Scheduling (PS)
- Shortest Run Time First (SRTF)
- Multi-level Queue Scheduling

$$\text{Turnaround Time} = \text{Completion Time} - \text{Arrival Time} \tag{1}$$

First Come First Serve: Here, we implement the processes in the increasing order of their arrival time. First come, first serve is a non-preemptive algorithm and therefore will only take the next job after it completes the implementation of the process that it is already implementing.

Shortest Job First: Here, the processes are implemented in ascending order of their burst time while keeping in mind their arrival time. Shortest Job First algorithm is a non-preemptive and therefore cannot implement any other process when a certain process is being implemented.

Round Robin: In the round robin algorithm, we take a quantum and continually keep doing different processes until all the processes are finished.

Priority Scheduling: In this algorithm, we implement the processes in the increasing order of their priorities keeping in mind the arrival time of each process. Priority scheduling can be made both preemptive and non-preemptive. Here, the pseudo code discussed below shows how to implement a non-preemptive priority scheduling algorithm.

Shortest Run Time First: Shortest run time first algorithm is inspired from the shortest job first algorithm that we have implemented above. But shortest run time first algorithm is a preemptive algorithm, implying that at any point of time if there is a process that has lesser burst time than that process that is being implemented, then, that process is implemented. Shortest run time first algorithm takes into consideration the burst time of all the processes along with their arrival time.

Multi-level Queue Scheduling: The modern CPU has processes in a queue known as Ready Queue. This ready queue is further partitioned or sub-divided into separate queues, namely, and not limited to:

- Foreground
- Background

Each queue has its own scheduling algorithm, you can assign any type of algorithm, hence the name, Multi-level Queue Scheduling.

Scheduling must be done between the queues

- The foreground queue will be run first and given higher priority over the background queue; the background queue will be run once the foreground has been exhausted.
- This scheduling method also requires a Time Slice or Time Quanta similar to that of Round Robin, which is the time after which the next process will be attended to by the CPU during process scheduling.

3.2 Semaphores

Semaphores are functions that can be used to solve various problems that are used to solve various critical section problems using wait and *signal* operations. Some of the concepts that use semaphores, that we implemented were producer consumer problem and dining philosophers.

Producer Consumer Problem: The following problem is a synchronization problem. There is a fixed size buffer that the producer can put his products into. The consumer takes the products from the buffer and consumes them. The problem occurs either when the buffer is empty and the consumer tries to take the product or when the buffer is full and the consumer tries to fill the buffer. This problem can be solved using semaphores.

Dining Philosopher: Dining Philosopher is a problem that is often used to display the synchronization problem that occurs in operating systems. It also shows us a technique for solving them. In our simulation, we have five philosophers that decide to dine together in round table. Each philosopher can either eat, think or wait. To eat, the philosopher needs two chopsticks, meaning that both the chopsticks beside him must be free. Since there are only five chopsticks on the table, two beside each philosopher, we need to time which philosopher eats when perfectly, so as to avoid deadlock or a situation where a philosopher starves forever. This too can be solved using semaphores.

```
Algorithm for Dining Philosopher (Output)
```

1. The array *Chopsticks* contain all the relation between the chopstick that each philosopher uses.
2. The number of philosophers is given by the number n.
3. fori in Philosophersdo
4. while true do
5. Think
6. *Callpickup* (chopsticks[i], chopsticks[(i + 1)%n])
7. Eat
8. *Callputdown* (chopsticks[i], chopsticks [(i + 1)%n])
9. end while
10. end for

3.3 Disk Scheduling

Disk Scheduling concept is a process that is implemented by the operating system to schedule and process the input output requests that it receives. It is similar to process scheduling. There are many requests that come at different times and have different execution times. Some of the algorithms that handle these kinds of requests are the following:

First Come First Serve: Here, the first request that arrives into the operating system is executed first.

Shortest Seek Time First: Here, we take into consideration all the request's seek time and their arrival time. The operating system then implements the requests in the increasing order of their seek time while taking into consideration their arrival time.

Scan Algorithm: In this algorithm, there is a head that traverses through the requests from the start to the end and executes them. Once the head reaches the end of the disk, the direction of the head is reversed and again the processes are implemented when the head reaches that request.

Look Algorithm: The look algorithm is similar to the scan algorithm, but here instead of going to the end of the disk the head goes only to the last request to be serviced in front of the head and then reverses its direction from there itself.

Circular Scan and Circular Look Algorithm: The pseudo code and the implementation for the circular scan and the circular look algorithm is quite similar to the scan and the look algorithm. The difference being that in the Circular Scan and Circular Look Algorithms, we jump to the beginning of the disk instead of reversing the direction.

3.4 Memory Allocation

These are the algorithms that helps the operating system in allocating spaces in the memory to programs that the operating system comes across. Memory allocation algorithms are spilt into two cases. One, when the partition size is fixed and when the partition size is varying.

The Fixed Sized Partition: Here, the partition is divided into equal or different sized partitions whose size is fixed and constant. Some of the algorithms that we implement to allocate memory to the blocks of programs are:

- *FirstFit*: The first fit algorithm finds the first block that fits the process/program and assigns the block to it.
- *Best Fit*: The best fit algorithm finds the smallest block that fits the program/process and assigns the block to it.
- *Worst Fit*: The worst fit algorithm finds the largest block that fits the program/process and assigns the block to it.

3.5 Page Replacement

When a new block of memory arrives and the old block of memory has to be replaced with the new one, the operating system uses some of the below mentioned algorithms:

- *First In First Out*: The first process/page that arrives first is the first one to be replaced. The operating system keeps a queue of all the process that arrives and the first process/page to arrive is at the front of the queue.
- *Optimum*: The page/program that is most likely to be not used in the future for a long time is replaced with the page that arrives. These page replacement algorithms are very hard to implement as the OS has no way of knowing whether a particular page/process may be used in the future.
- *Least Recently Used*: The page/program that is least recently used will be replaced with that of the incoming program or page.

3.6 File Structure

File structures are different ways of structuring the files in secondary memory that helps reduce the access time.

We have implemented a terminal style output of Linux commands which create, delete the files and folders. Hence, the user can also learn how files and folders can be created and deleted on a Linux terminal.

- *Single Level*: Here, there is only a big list of all the files that are present in the directory. The user can have only one directory and all the files must be present inside that. The implementation of this is very simple and the deletion and insertion is very fast and easy, but we cannot have two files with the same name and the protection of files are not guaranteed for multiple users.
- *Two Level*: In this File Structure algorithm, the user has two levels of files; the first level, being the directory and the second level being the files. Files cannot be created without being in a directory. Multiple folders having multiple files can be created.
- *Tree Structure*: Tree Structure is how modern operating systems implement file structure. A directory can have several files and sub-directories. These sub-directories can then further branch to contain additional files and sub-directories.

4 Result and Analysis

Given the process ID, arrival time and burst time, the process scheduling algorithms compute the Gantt chart, waiting time and the turnaround time. Additionally, the steps taken to arrive at the result and the time stamp in the simulation are presented at the bottom left (Fig. 1).

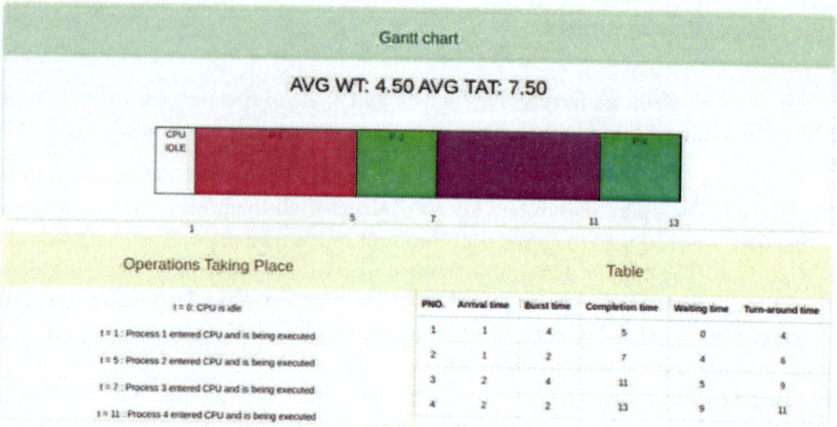

Fig. 1 Graphical representation of the data, along with the Gantt chart and the tabular format of experimental data

In the Reader Writer simulation. We need to select what the seven processes want to do, i.e., read, write or stay idle. Then on submitting using the "Submit Query" button, the result shows that if the process was permitted to continue, it is shown with blue indicating the process is allowed to read and with red indicating the process is allowed to write. We see that parallel reads are allowed but only one process can write at a time.

Given the cache size and a space separated list of requests, the cache hits represented by green tick and the cache miss represented by red cross can be visualized. The "Proceed" button processed the next request and displays if it is cache hit or miss. The algorithms FIFO, LRU and Optimum can be selected by the tabs provided.

Fig. 2 shows is the simulation of the file tree implemented step by step. The files and directories in the selected directory are shown at the top. The path to the current directory is shown above that. For every action, the equivalent terminal command to act is shown.

Fig. 3 shows the seek graph for the FCFS disk scheduling algorithm. The seek requests are given as a space separated list. Using the current position and the number of cylinders parameter, the order in which the seek happens is shown in the graph.

5 Conclusion and Future Scope

We successfully implemented all of the objectives, including concepts such as File Systems and Disk Scheduling, which have been entirely made into a graphical and interactive simulation, making it an immersive experience for the end-user, which helps understand how the concepts work.

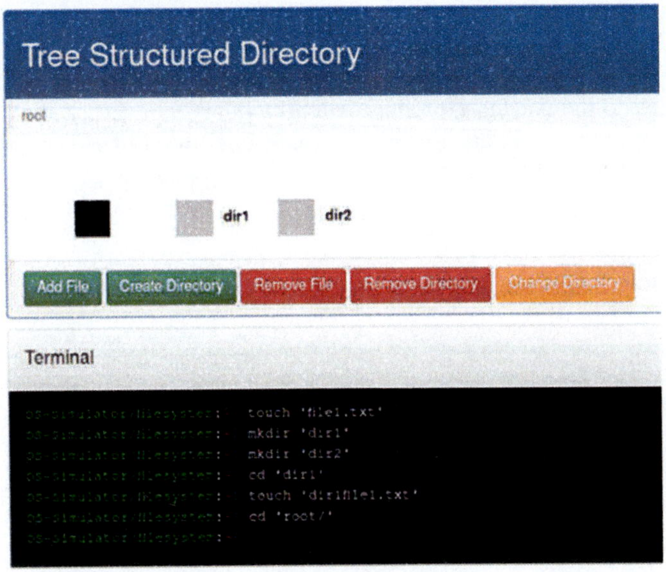

Fig. 2 Output of one of the algorithms used in file systems

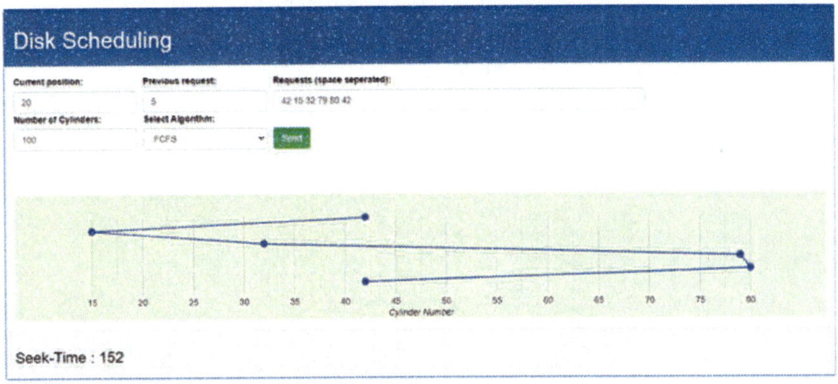

Fig. 3 Seek graph for the FCFS disk scheduling algorithm using experimental data

We have also deployed our app on Heroku web servers for easy access; it is deployed at https://os-simulator.herokuapp.com/.

One can open this using a device with internet access and a browser app. Just click on the link or copy-paste in the address bar to access the web page. There are no other requirements, making it very light and standalone, which is not the case with other simulators available, which requires various libraries and to run it locally on your machine.

The complete source code to our implementation can be found on the GitHub of one of the authors, using which we collaborate to put together this Operating System Simulator.

The link to the GitHub repository is https://github.com/navaneethp123/os_simulator/. Our future works are not limited to including advanced and latest Operating Systems such as Socket Programming and Parallel Programming\.

References

1. Chan, T.W.: A software tool in java for teaching CPU scheduling. JCSC **19** (2004)
2. Than, S.: Use of a simulator and an assembler in teaching input-output processing and interrupt handling. JCSC **22** (2007)
3. Nishita, S.: MKit simulator for introduction of computer architecture. In: 31st International Symposium on Computer Architecture, June 19, 2004. Munich, Germany
4. Maia, L.P., Pacheco, A.C.: A simulator supporting lectures on operating systems. In: 33rd ASEE/IEEE Frontiers in education Conference, November 5–8, 2003. Boulder, CO
5. Garrido, J.M., Schlesinger, R.: Principles of Modern Operating Systems (2008)

An Analysis Study of IoT and DoS Attack Perspective

Mahmoud Jazzar and Mousab Hamad

1 Introduction

In the coming years, everything will be connected to the Internet without the need for any human intervention. IoT technology let users control everything with smartphones and digital devices. As such, all devices and objects can communicate with each other and give users real-time data and control [1]. For example, users can receive notifications on smartphones about the status of washing machines, refrigerators, house oven, and more. Currently, lots of companies around the world make huge investments in IoT devices, and the growth of IoT industry is huge [2].

Denial-of-Service (DoS) is an attack used by cyber-criminals to not let users access their network or system for short time [3]. They shut down the network by pinging the main server with huge amount of spam requests so that it will not be able to respond. Correspondingly, DoS attack consumes the resources and bandwidth of legitimate users. As such, CPUs of main servers remain overloaded and may crash. In this paper, an overall view of IoT security is defined then an attack experiment with DoS is analyzed to determine conditions to shut down the Arduino sensor.

The rest of the paper is structured as follows. Section 2 introduces the background of IoT cybercrimes, overview of IoT security, and DoS attack when IoT as victim. Section 3 represents the experiment that is done on Arduino as victim of DoS attack. Section 4 demonstrates the process of DoS attack experiment. Section 5 discusses the results and findings. The concluding remarks are in Sect. 6.

M. Jazzar (✉) · M. Hamad
Faculty of Graduate Studies, Palestine Technical University – Kadoorie, Tulkarm, Palestine
e-mail: mjazzar@ptuk.edu.ps

M. Hamad
e-mail: m.a.hamad4@ptuk.edu.ps

B. Agarwal et al. (eds.), *Proceedings of International Conference on Intelligent Cyber-Physical Systems*, Algorithms for Intelligent Systems,
https://doi.org/10.1007/978-981-16-7136-4_11

2 Background

The new strategy to deal with IoT crimes aims to be composed in the early stages of the investigation process. According to [4], FBI initiated tips to be watchful for IoT systems. Collecting evidence of IoT systems is not easy or even on the same level as computers or phones, nonetheless, they are similar when it comes to classification. IoT cybercrimes are classified into three types.

IoT as a Target. These violations can be put within the modern period of wrongdoing. That is why organizations and people around the world are not prepared to battle them. This class of IoT crime includes exploiting vulnerabilities in smart devices, such as restorative implantation pump sand, etc. that empower malevolent instruments that posture a chance to human life [5].

IoT as an Apparatus. In such crime type, IoT devices play as tools that the cybercriminals use to harm. During this situation, recognizing and indicting the offenders may be part of the difficulty. This category of crime depends on the vulnerabilities that exist to the device itself as it needs minimal technical experience. Since security is frequently not the foremost focus of device producers, IoT devices are fabulous tools for offenders to make botnets to execute DoS and DDoS attacks. Attackers ordinarily abuse vulnerabilities such as mounted coding keys, default passwords, and failure to fix or overhaul device firmware [6].

IoT as an Eyewitness. These are not crimes of IoT. In this crime type, the IoT works for the police for investigation even though the crime is not related to cyberspace. Movement sensors, smart-light recorders, and smart CCTV systems can record the precise time of an offender. Smart locks can show whether the offender was brute-forced, hacked, or leveraged a legitimate code to enter the savvy domestic. Furthermore, Wireless Access Points (WAP) may contain historical logs of wireless connection endeavors and other nearby WAP exercises that will have incidentally associated associations from the intruder's device [7].

2.1 Overview of IoT Security

According to [8], the definition of things in IoT as appeared in Fig. 1 is exceptionally comprehensive and incorporates a wide assortment of physical components. This organizes a diverse sort of materials, which can bring parcel of challenges in creating applications and make existing challenges indeed greater, i.e., hard to bargain with.

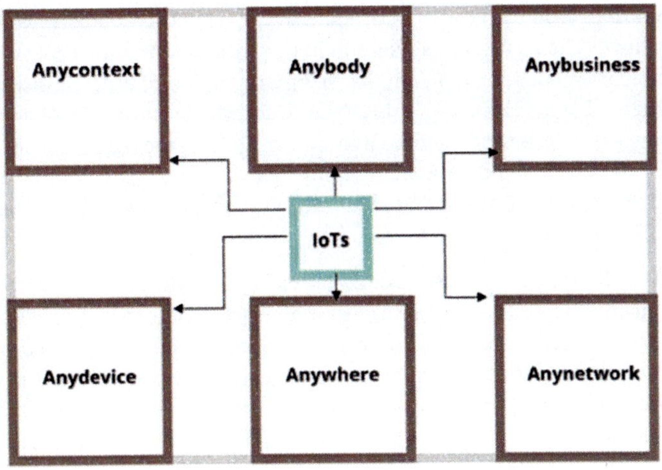

Fig. 1 IoT definition

2.2 *Security Characteristics Related to IoT*

Data confidentiality. Data security and reachability is a critical issue for a variety of digital systems. In IoT systems, there are two types of devices. Inside devices are important for the organization and they are part of it. The outside devices are not part of the organization as said in [9]. In IoT, there is a vitally important need for awareness when it comes to the customers, to ensure that the data are made and kept with the approved concerned party [9, 10]. Data created by the sensor nodes must be ensured against undesirable divulgence to the adversary [11]. To keep up the privacy of sensor data, the following areas of information at rest and in travel are identified: data generator, information capturer, and the communication links among these substances.

Even though, in principle, the encryption of all information may be an arrangement. As such, the unaltered application of standard encryption procedures, counting Rivest– Shamir–Adleman (RSA), information encryption standard (DES), Blowfish, and ElGamal strategies, is not completely viable on resource-constrained sensor hubs. Subsequently, data encryption must be lightweight as well as flexible against adversarial assaults [12].

Data Integrity. As stated in [9], the IoT system is built on the exchange of data between devices. As such, to ensure that data are not damaged during the transmitting process due to intentional or unplanned interruption. Besides, to ensure that data are accurate and came from the expected sender, such feature has to be employed when it comes to IoT. As known, data traffic is controlled by the use of firewalls, but that is not enough because the protocols do not guarantee security at the endpoint.

Data Availability. IoT's vision is to relate, however, many canny devices, which could be allowed. IoT customers ought to have all the information accessible whenever required. Nonetheless, data are not the main standpoint utilized inside the IoT; contraptions and organizations should hence be accessible and available as suitable in an ideal way [9]. Middleware devices may be modified by the adversary class to carry out dissent of benefit flooding attacks against particular targets within the arrangement, with the reason of disrupting scheduled organized operations and devastating critical sensor network services [12].

Data Authenticity. Authenticity is for checking the identity of users related to any item inside the IoT to distinguish and approve different devices. Regardless, this arranging might be troublesome when thinking about the presence of the IoT such as contraptions, individuals, organizations, producers, and handling units. In certain circumstances, things can be bound to multiple nodes, therefore, special arrangement needs to be made to authenticate IoT systems in every interaction between IoT devices [9].

2.3 *IoT with 5G Technology*

Tiburski in Ref. [13] clarified that security challenges of IoT middleware or information capturers will increase radically because IoT systems begin receiving 5G systems. Although 5G systems will give quick, solid, tall transmission capacity, and area mindfulness to IoTs, there stay numerous unaddressed security concerns. Attacks, such as Man-in-The-Middle, DoS attacks, and spying, will challenge the successful arrangement of the IoT on a 5G arrangement [13, 14]. With various associated devices and basic administrations depending on information created from billions of conclusions focuses, it is fundamental to have a proficient and strong security instrument to obstruct any malevolent endeavors to disturb administrations, particularly DoS attacks. Whereas customary procedures of the measurable investigation of network activity may be compelling in recognizing DoS assaults in sensor systems, the centralized examination of combined information at the data capture hubs encourages more precise recognizable proof of DoS network activity.

2.4 *IoT Common Attacks*

Over the last few years, IoT devices were exposed to the worst attacks such as the following.

Stuxnet. In 2010, Stuxnet was found by Kaspersky Lab [15]. It was introduced to a network via USB stick [16]. It targets both SCADA and PLC systems that are connected to machines and industrial control systems. Stuxnet harmed the control plants, fabricating businesses, atomic businesses, gas businesses, mining businesses employing exploit on their SCADA system. Stopped Iran's atomic program was the greatest Stuxnet assault up to so distant and that was the greatest offense at that time [15]. In 2014, other attacks on the German steel process too cautioned around the security significance within the execution of industry 4.0, trojan Lazoik made with the point to assault on the vitality segment [15].

Mirai Botnet. Mirai Botnet happened in 2016, one of the major attacks that are directed to IoT devices. In specific IP cameras and switches, exploit of those devices gives this Botnet the ability to surge DNS with a DDoS attack. The infected websites were GitHub, Netflix, Shopify, SoundCloud, Spotify, Twitter, and many other major websites. The vulnerabilities that let malicious code to take advantage of the system were the out of the data Linux version and most clients do not change the default usernames/passwords on their devices [17]. Mirai could be a frame of IoT Malware, particularly a worm, that has been appeared to cause a noteworthy risk to people that are arranged on the Web. It works based on the utilization of DDoS assaults against common targets through Botnet. The targets that Mirai has been utilized against are Deutsche Telekom (German ISP), which put 900 k German clients' offline, Krebson Security, and Dyn (DNS Benefit Supplier). Two of these targets are of critical standing inside the foundation of a nation and from the name, it is clear that they are Internet service provider companies so they are part of the Internet itself [18].

BrickerBot. As Mirai Botnet, it depends on doing DDoS to clients who do not change the default username/password of their gateway. The greatest difference between this is that BrickerBot exploits the gateway too [17]. BrickerBot was similar to Mirai according to BusyBox. It was discovered by analysts in 2017. BrickerBot attempted a permanent denial-of-service attack on the IoT devices. BrickerBot depends on exploiting the vulnerabilities that exist due to default configurations or misconfigurations [19]. A DDoS assault ended warming dissemination at least in two properties within the city of Lappeenranta, found in eastern Finland. In such DDoS attack, the organization is over-burden by activity from different areas with the point of causing the framework to fail [20].

2.5 IoT Architecture

Over the last few years, to employ the traditional security procedure based on ID network solutions is not enough. This is because IoT devices have a lot of types with different combinations, and there are numerous connection protocols in IoT system [21]. The attack vectors according to [21] which are announced by the open web application security project (OWASP) are about three layers of the IoT framework. The three-layer IoT architecture is demonstrated in Fig. 2.

Perception Layer. The perception layer is also known as the "Physical" layer in IoT. The main objective of the perception layer is to get the data from the environment with the help of sensors and actuators [9]. This layer also performs the IoT node collaboration in local and short-range networks [21].

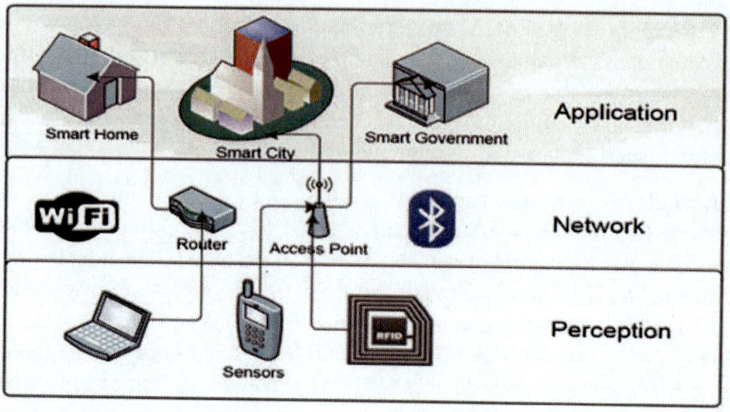

Fig. 2 Three-layer IoT architecture [21]

Network Layer. The network layer receives the collected and detected data from the physical layer. The role of the network layer in IoT according to [22] is for the data routing and transmission to different IoT hubs and devices. It runs by using very up-to-date technologies such as LTE, Bluetooth, 3G, ZigBee, and Wi-Fi. To use those technologies, this layer must have some components like cloud computing platforms, Internet gateways, switching, and routing devices [9].

Application Layer. This layer earns three of the four characteristics of security, which are confidentiality, integrity, and the authenticity of the data. The aim is to establish a smart environment by creating links and providing services [9].

IoT Protocol Stack. Protocols of IoT have been added with time due to prerequisites of new IoT devices as in Fig. 3 [23] (Table 1).

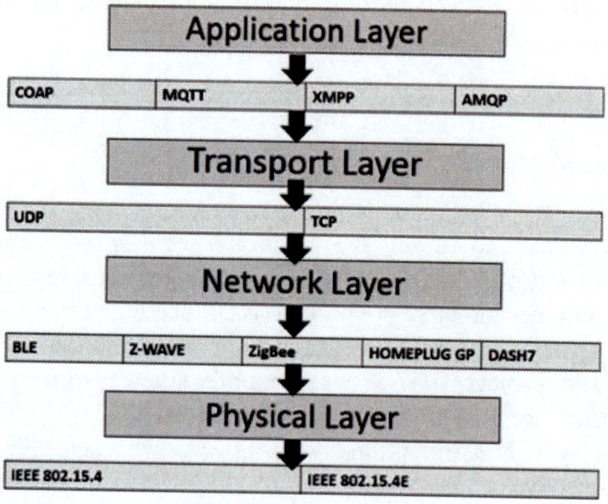

Fig. 3 IoT protocol stack

Table 1 Possible attacks on RFID and WSN in different layers [21]

Layer	RFID (Radio Frequency Identification)	WSN (Wireless Sensor Networks)
Perception/link	Jammers, replay attacks, Sybil, selective forwarding, synchronization attack	Passive interference, active jamming of temporarily disabling the device, Sybil, destruction of RFID readers, replay attacks
Network/transport	Sinkhole, unfairness, false outing, hello and session flooding, eavesdropping	Tag attacks: Cloning, spoofing Reader attacks: impersonation, eavesdropping, network protocol attacks
Application layer	Injection, buffer overflows	Injection, buffer overflows, unauthorized tag reading, tag modification
Multi-layer	The side-channel attack, replay attacks, traffic analysis, crypto attack	The side-channel attack, replay attacks, traffic analysis, crypto attack

2.6 DoS/DDoS Attacks

Denials of service attack is an open effort to make computing services inaccessible by inserting computer virus or by overflowing the network with unwanted traffic. Simply like thousands of people attempting to enter a room from the entry, finally wreaking havoc [24]. It can be done in different ways such as using botnets and buffer overflow vulnerability [5]. Other DoS attack involves the flooding of a huge amount of traffic to consume network resource, bandwidth (network attacks), target CPU time, and more. Some of the most common DoS attacks are SYN flood, DNS flood, Ping flood, UDP flood, ICMP broadcast [25]. Most of the victims of DDoS are government organizations and banking companies. Such high-profile entities have confidential information. New DoS attacks are being constantly introduced due to ever-growing advancement even though there are many software fixes to stop those attacks [25] (Fig. 4).

The DDoS assault is an endeavor to hinder a web foundation and make it blocked off by burdening it with multiple source activities. Frequently, few tainted systems influenced by a Trojan infection are utilized to assault a single web location or gadget that produces it blocked off. A large-scale DDoS assault (run-up to 400 Gbps) will affect the web network of the whole geological range.

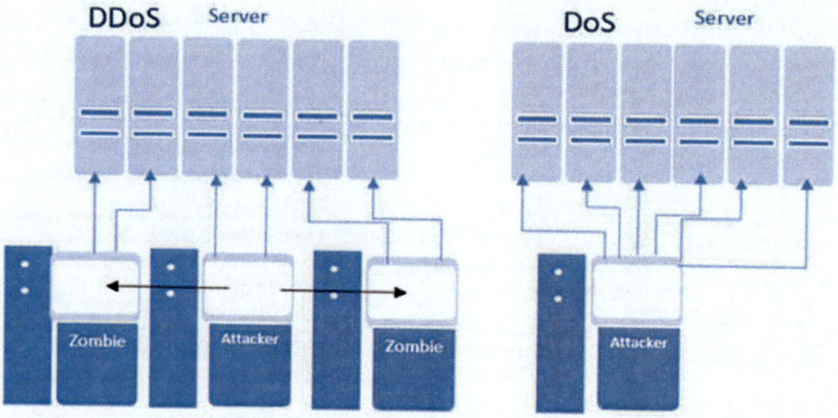

Fig. 4 DoS versus DDoS

2.7 DoS/DDoS Attacks on IoT

A. **DDoS on Perception Layer**

- **Jamming.** In this attack type, the tags cannot communicate with the reader [24]. This leads to wireless channels. It can be done using two ways, the first way is done by transmitting the power higher than the received power maximum value. The second way aims to make the quantized Received Signal Strength Indicator (RSSI) value constant in time. The inverse of this can be done with transmit power at high, and the received power is constant [26].
- **Kill Command Attack.** Because of low memory, this attack leads to disabling tags. The tags are protected with the password in writing mode. Ruggedly doing some huge load can lead to corrupt the memory [24].
- **De-synchronizing Attack.** This attack demolishes the sync between the tag and the RFID reader so that will enduringly turn off the capability of the authenticity of the RFID [24].
- **Bootstrapping Attacks**. Configuring both nodes requires some time to securely join (bootstrapping) during the initial network setup. In the most resource-restricted nodes, it can have two buttons on each node, ever press on any button causes changing the mode of it. If the attacker is not existing in the initial configuration or booting, it will be safe for the system such as remote controls [24]. The user needs some assurance that the device can be trusted because sharing confidential information with any IoT device can lead to this attack and let some intrusions happen [26].

B. DDoS on Network Layer

- **Flooding Attacks.** It works by unsettling the user's connection by drowning the bandwidth of the victims' network such as UDP flood, ICMP flood, and DNS flood [26].
- **Reflection-based flooding Attacks.** This type of attack sends fake response requests to responders instead of the original direct request; therefore, those responders send responses to the victims and drain the victim's resources [26].
- **Protocol Exploitation flooding attacks.** This type of attack uses certain features or implementation errors of the victim's protocol to maximize the victim's resources. For example, SYN flood, TCP SYN-ACK flood, ICMP flood [26].
- **Amplification-based flooding attacks.** This type of attack tries to use the app to generate news or multiple messages received to increase traffic toward the victim. Botnet is widely used for both amplification and reflection purposes [26].

3 The Experiment Model

According to [5], modeling an attack incorporates the attacker, the victim, and the monitoring device to measure the attack severity. Therefore, for establishing an experimental model, the experiment target needs to be defined. The proposed experiment targets the impact of DoS on IoT devices, and hence to analyze the cause and severity of such attacks. Figure 5 shows the system design for experimentation.

The experiment platform is composed of the following components.

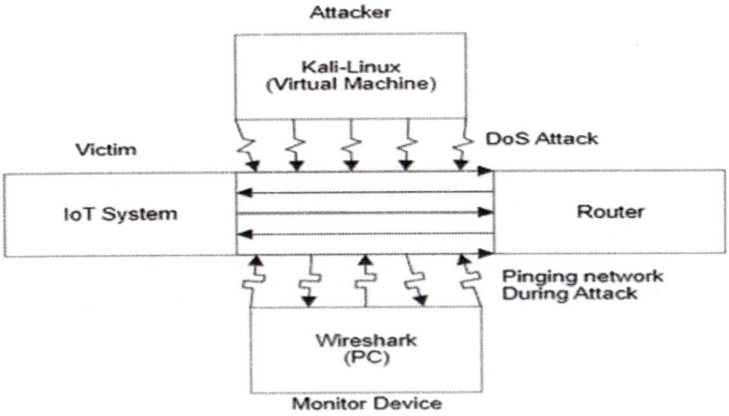

Fig. 5 System Design

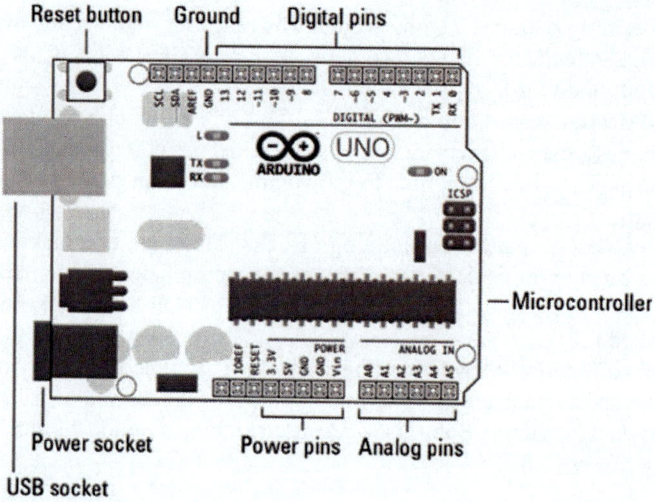

Fig. 6 Arduino Pinout

- PC connected to Router to obtain information from the ESP8266 Wi-Fi module.
- Establishing a WLAN (Wireless Local Area Network) by connecting the PC, Arduino nodes, and the virtual machine.
- ESP-8266 Wi-Fi module (connected to Arduino) to recode the data from the Arduino and the sensors.
- Kali-Linux/attacker system installed in the virtual machine of the PC.
- Monitoring device (PC) using Wireshark software for packet analysis between the router and ESP modules during the DoS.

Arduino itself is just an open-source hardware/software company that puts out single-board microcontroller kits for electronics development. The company has many products, however, Arduino Uno R3 was used for this experiment. ESP-8266 is a low-cost Wi-Fi microchip developed by a third-party manufacturer called AI-Thinker. It has an onboard MCU (Microcontroller Unit), which allows users to control I/O digital pins directly via the Arduino IDE (Figs. 6 and 7).

DoS attacks can perform easily using various tools like (Nemesy, RUDY, GolodenEye, UDP flood, PyLoris, HULK, pentmenu, Hammer, xerxess, LOIC, HOIC, and Metaspoilt). Pentmenu was used in this experiment, which can do many types of DoS attacks on any system [27] (Fig. 8).

4 Process of DoS Attack

The following diagram demonstrates the attack platform. Initially, the pentmenu on the Kali-Linux and the IoT system (victim) is set. In this part of the experiment, all the procedures were followed as mentioned in Ref. [28]. In addition, there is an

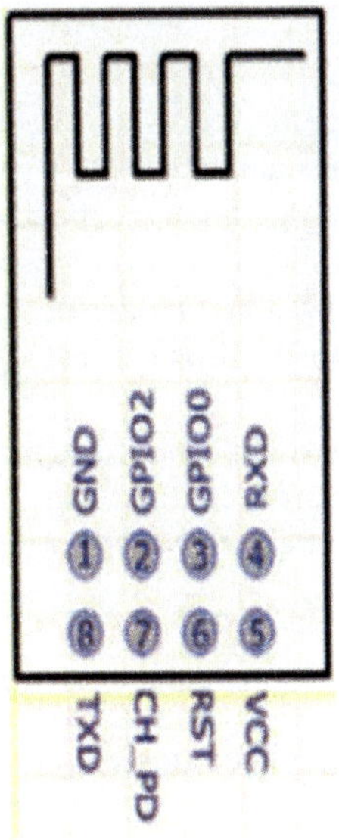

Fig. 7 ESP Pinout

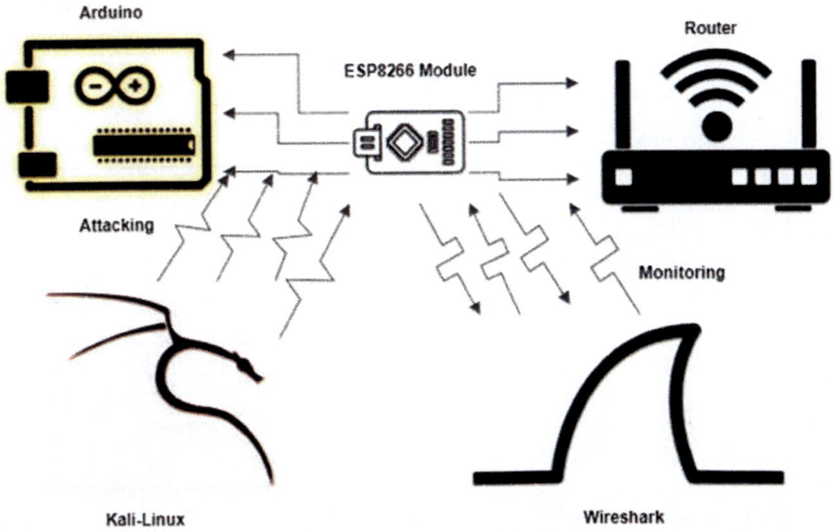

Fig. 8 Attack platform

essential need to understand how a local server works. To set up a local server, finding a way to send AT commands to the ESP-01 is required. The commands come from the pre-installed AT firmware of the ESP-8266.

The ESP-8266 AT commands allow users to perform operations like testing the connection, setting the mode of operation, connecting to Wi-Fi, determining the IP address, and more. Once confirming that the chip is working, the mode of operation is set. Generally, ESP8266-01 has three modes of operation: Station (STA), Access Point (AP), and Both. Once the Wi-Fi is successfully connected, an OK response shall be received.

ESP-01 can be used as a server and hence to support multiple connections. Single and multiple connections range between 0 and 1 for AT plus CIPMUX command. AT plus CIPSERVER, CIPCLOSE to start the server, and to close the communication channels.

5 Experiment Result and Discussion

Once the circuit is built and the code is uploaded. Figure 9 shows the normal operation in the Arduino serial monitoring window. At first, the ESP tries to connect to the WAP and then the HTTP server starts with IP address that has been checked in setting the local server section. Then, the temperature and humidity are received from the sensor and sent to the server.

The following Fig. 10 demonstrates the normal webpage of the local server.

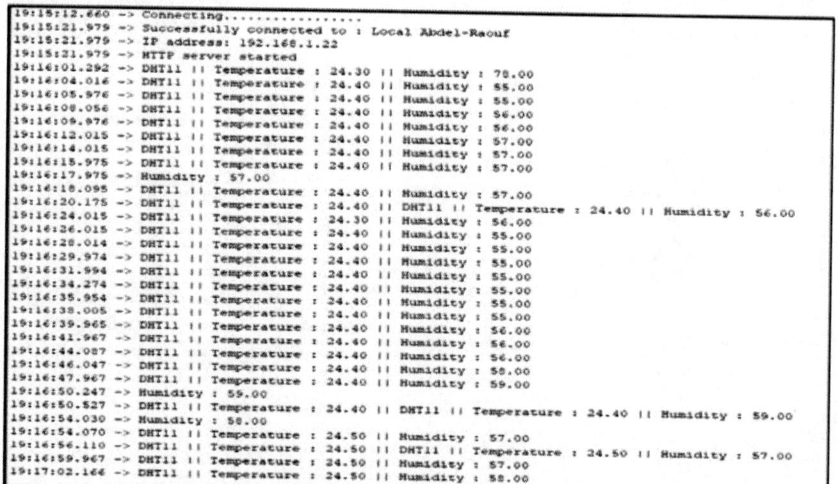

Fig. 9 Normal Operation Serial Monitor Arduino

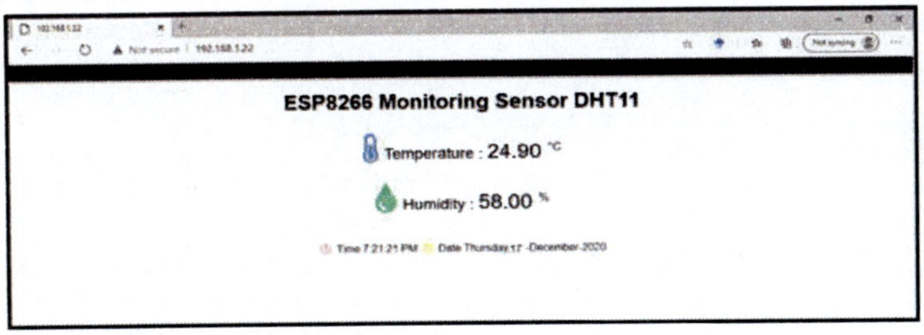

Fig. 10 Normal Webpage—Local Server

ICMP Echo Flood attack. This attack drowned the system with 852,842 packets. The local server was unreachable through the attack. There was no packet loss in the attack. The details are evident in the figures and Table 1. The attack did not shut down the IoT system.

ICMP Blacknurse: This attack drowned the system with 323,881 packets and it did shut down the IoT system after 2:30 min of backing to get data from the sensor. The local server was not available to reach during the attack. The packet loss to the IoT system was 100%, which means the attack had full success.

TCP-SYN flood: It did less damage as compared with the previous attacks; it did not harm the IoT system physically. The local server was still available to connect through the attack. During the entire minute, it just sent 1724 packets because it related to the original SYN TCP packet.

TCP-ACK flood. Similar to TCP-SYN but it did make the local server unavailable during the attack. Even though the TCP-ACK did send fewer packets (1550) during the attack to the IoT system but that is not related to the damage because as clarified before the TCP flood attacks are connected to the packets with the same type of flag.

Figure 11 illustrates data from different phases during 1 minute of operation, and the number of received data from the sensor node. Table 2 reveals the attack type and

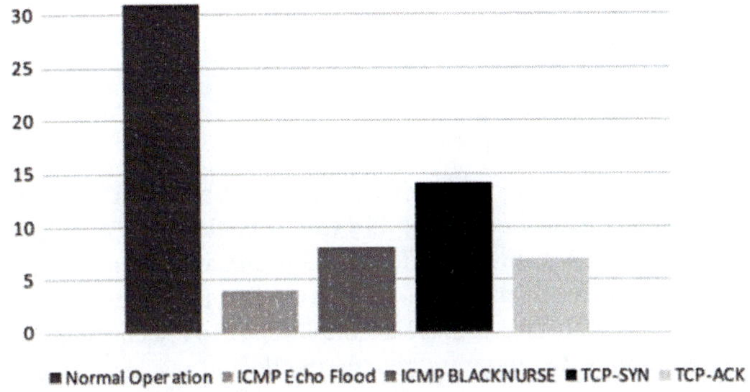

Fig. 11 Number of data received by IoT system in different phases

Table 2 Flood attacks over IoT system

Type of the attack	Number of packets (one-minute interval)	Damage
ICMP echo flood	852,842	Complete Paralyze IoT during attack but the system was able to response after the attack
ICMP blacknurse	323,881	Complete Paralyze IoT during attack but the system was not able to response after 2:30 from the attack
TCP-SYN flood	1724	The system could receive some data during the attack and local webserver remained up
TCP-ACK flood	1550	The system could receive some data but the local webserver went down during the attack

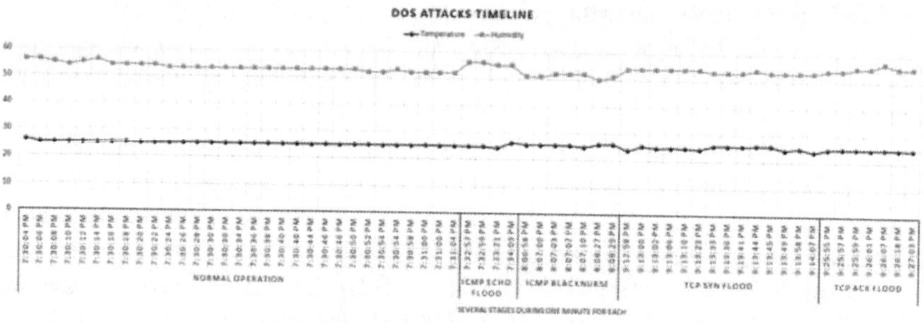

Fig. 12 One Minute DoS Attack Timeline

relevant damage, the analysis of the concepts of these attacks was explained clearly and simply, and in a way that can be understood. The drawn timeline shows all the data obtained during the four attacks (Fig. 12).

6 Conclusion Remarks

The study outlines IoT security, IoT architecture, and DoS attacks. DoS/DDoS attacks on IoT devices are explained according to every specific layer. An experiment of simple IoT system is exposed to four different flood DoS attack types, and the results were analyzed to reveal the damage level of each attack type. In general, IoT systems and device users can contribute to increase the level of IoT security and safety. As such, the consumer has the option to change the default password and the Wi-Fi key on first use to perform robust layer security. In addition, the access control lists donate to counteract brute-force attacks from happening outside of the network.

In conclusion, the awareness of the IoT components and relevant security training remains indispensable and vital to measure the risk of any future intrusions or attacks on such devices.

References

1. Peng, S., Pal, S., Huang, L.: Principles of Internet of Things (IoT) ecosystem: insight paradigm. ISRL **174**, 3—100, Springer, Heidelberg (2020)
2. Herrera Silva, J.A., Barona, L.I., Valdivieso, Á.L., Hernández-, M.: A survey on situational awareness of ransomware attacks—detection and prevention parameters. Remote Sens. **11**, 1168 (2019)
3. Liang L., Zheng K., Sheng Q., Huang X.: A denial of service attack method for an IoT system. In: 2016 8th International Conference on Information Technology in Medicine and Education (ITME), pp. 360–364. https://doi.org/10.1109/ITME.2016.0087 (2016)
4. Cyber Tip: Be Vigilant with Your Internet of Things (IoT) Devices | Federal Bureau of Investigation. Retrieved June 2021, https://www.fbi.gov/news/stories/cyber-tip-be-vigilant-with-your-internet-of-things-iot-devices
5. Cui, Y., Liu, Q., Zheng, K., & Huang, X.: Evaluation of several denial of service attack methods for IoT system. In: 9Th International Conference on Information Technology in Medicine and Education (ITME). https://doi.org/10.1109/itme.2018.00179
6. Kebande, V. R., Ray, I.: A Generic Digital Forensic Investigation Framework for Internet of Things (IoT). In: IEEE 4th International Conference on Future Internet of Things and Cloud (FiCloud), pp. 356–362 (2016)
7. Bouchaud, F., Grimaud, G., Vantroys, T., Buret, P.: Digital investigation of iot devices in the criminal scene. J. Univ. Comput. Sci., Graz University of Technology, Institut für Informationssysteme und Computer Medien **25**(9), 1199–1218. hal-02432740 (2019)
8. Vashi, S., Ram, J., Modi, J., Verma, S., Prakash, C.: Internet of Things (IoT): A vision, architectural elements, and security issues. In: 2017 international conference on I-SMAC (IoT in Social, Mobile, Analytics and Cloud)(I-SMAC) (pp. 492–496). IEEE Press (2017)
9. Mahmoud, R., Yousuf, T., Aloul, F., Zualkernan, I.: Internet of things (IoT) security: Current status, challenges and prospective measures. In: 10Th International Conference For Internet Technology And Secured Transactions (ICITST) (2015)
10. Roman, R., Najera, P., Lopez, J.: Securing the Internet of Things. Computer **44**(9), 51–58 (2011)
11. Baig, Z.A., Sanguanpong, S., Firdous, S.N., Nguyen, T.G., So-In, C.: Averaged dependence estimators for DoS attack detection in IoT networks. Futur. Gener. Comput. Syst. **102**, 198–209 (2020)
12. Johnson Singh, K., De, T.: Mathematical modelling of DDoS attack and detection using correlation. J. Cyber Secur. Technol. **1**(3–4), 175–186 (2017)
13. Tiburski, R., Amaral, L., Hessel, F.: Security challenges in 5G-based IoT middleware systems. Internet Of Things (Iot) In 5G Mobile Technologies, 399–418 (2016)
14. Li, S., Xu, L., Zhao, S.: 5G Internet of Things: a survey. J. Ind. Inf. Integr. **10**, 1–9 (2018)
15. Patel, C., Doshi, N.: Security Challenges in IoT Cyber World. Security In Smart Cities: Models, Applications, and Challenges, 171–191 (2018)
16. Farwell, J., Rohozinski, R.: Stuxnet and the future of cyber war. Survival **53**(1), 23–40 (2011)
17. Whittaker, Z.: Homeland security warns 'BrickerBot' can destroy unsecured internet-connected devices. ZDNet. Retrieved November 2020, from https://www.zdnet.com/article/homeland-security-warns-of-brickerbot-malware-that-destroys-unsecured-internet-connected-devices/.
18. Whitter-Jones, J.: Security review on the Internet of Things. In: Third International Conference on Fog and Mobile Edge Computing (FMEC) (2018). https://doi.org/10.1109/fmec.2018.8364059

19. Kolias, C., Kambourakis, G., Stavrou, A., Voas, J.: DDoS in the IoT: Mirai and other botnets. Computer **50**(7), 80–84 (2017). https://doi.org/10.1109/mc.2017.201
20. Kumar, M.: DDoS attack takes down central heating system amidst winter In Finland. The Hacker News. Retrieved November 2020, from https://thehackernews.com/2016/11/heating-system-hacked.html
21. Mohamad Noor, M., Hassan, W.: Current research on Internet of Things (IoT) Security: a survey. Comput. Netw. **148**, 283–294 (2019)
22. Mahalle, P.N., Anggorojati, B., Prasad, N.R., Prasad, R.: Identity authentication and capability based access control (iacac) for the Internet of Things. J. Cyber Secur. Mobility **1**(4), 309–348 (2013)
23. Sharma, C., Gondhi, N.: Communication protocol stack for constrained IoT systems. In: 3Rd International Conference On Internet Of Things: Smart Innovation And Usages (Iot-SIU) (2018)
24. Sonar, K., Upadhyay, H.: A survey: DDOS attack on Internet of Things. Int. J. Eng. Res. Dev. **10**(11), 58–63 (2014)
25. Herrera Silva, J., Barona, L., Valdivieso, Á., Hernández-, M.: A survey on situational awareness of ransomware attacks—detection and prevention parameters. Remote Sens **11**(10), 1168 (2019)
26. Danish, S., Nasir, A., Qureshi, H., Ashfaq, A., Mumtaz, S., Rodriguez, J.: Network intrusion detection system for jamming attack in LoRaWAN join procedure. In: IEEE International Conference on Communications (ICC) (2018). https://doi.org/10.1109/icc.2018.8422721
27. GinjaChris/pentmenu, GitHub, Retrieved December 2020, https://github.com/GinjaChris/pentmenu
28. Ramirez, R.: How to show arduino sensor data on a web page. Circuit Basics. Retrieved November 2020, https://www.circuitbasics.com/how-to-set-up-a-web-server-using-arduino-and-esp8266-01/

The Effect of Sampling in the Machine Learning-Based Malware Analysis

K. Sakshi Thimmaiah, Lakshmi S. Raj, Prasanthi Bolimera, and M. Anand Kumar

1 Introduction

Android Operating System has become very popular over the years, and it is a Linux-based operating system. It has been designed primarily for touchscreen mobile devices and tablets. They are increasingly used to access services, such as messaging, video/music sharing and e-commerce transactions that have been previously available on PCs only. Subsequently, it has attracted several Malware developers who target these mobile users [3, 10].

We must detect this malicious software that tampers with the device performance and steals personal data, such as accessing contacts, media and personal messages, without the user's knowledge. Machine Learning techniques can be used to classify software into two categories: malware and safeware. This classification can be done by using the XML file called "Android Manifest" to present in each Android application. It provides essential information to the operating system, like the first class to use when starting the app or the type of permissions used in the application [3].

Only permissions provided in the file will be used in the application, and this is done only after asking the user to grant these permissions. If the application tries to use some other permissions which were not allowed in the Android Manifest file, the execution fails. Unfortunately, many users tend to grant permissions to unknown applications, which is why malicious software infects the device [3].

K. S. Thimmaiah (✉) · L. S. Raj · P. Bolimera · M. A. Kumar
Department of Information Technology, National Institute of Technology Karnataka, Surathkal, India
e-mail: ksakshithimmaiah.191it124@nitk.edu.in

L. S. Raj
e-mail: lakshmisraj.191it225@nitk.edu.in

P. Bolimera
e-mail: prasanthibolimera.191it240@nitk.edu.in

M. A. Kumar
e-mail: m_anandkumar@nitk.edu.in

B. Agarwal et al. (eds.), *Proceedings of International Conference on Intelligent Cyber-Physical Systems*, Algorithms for Intelligent Systems,
https://doi.org/10.1007/978-981-16-7136-4_12

Thus, users must be made aware of the type of software they are installing so that they do not fall prey to malicious software and lose essential data from their mobile devices. For this particular project, we are making use of the DREBIN dataset. It contains 5,560 applications from 179 different malware families. The samples have been collected from August 2010 to October 2012 [5].

2 Literature Survey

There are two approaches to detecting malware in Android operating systems. The first one is a signature-based approach which generates a signature for every kind of malware and compares it with the application [1]. Typical antivirus software (e.g., Norton and McAfee) use signature-based methods to identify malware. However, this can be easily evaded by attackers. Example methods involve changing signatures using code obfuscation or repackaging [11]. The second is behavioural detection. The behaviour of an application is compared at runtime to identify malicious intent [1, 3].

In recent years, there has been an increasing trend using machine learning to overcome the challenges mentioned above to develop automatic and intelligent malware detection methods. These techniques are capable of discovering certain patterns to detect previously unseen malware samples and identifying the malware families of malicious samples. These systems can be classified into two categories: Dynamic analysis and Static analysis [3, 11].

Dynamic analysis [12–14] involves accumulating information regarding API calls, environmental variables and data transmission during the execution of an application. Dynamic analysis gives precise predictions and has lower false positive rates [3, 11].

Static analysis involves two parts—feature extraction and classification. The features are first extracted from the source file and a model is created to identify the malware families. A number of known datasets are used for feature extraction. DroidMat is used for static analysis using the manifest file and source code to extract features and k-means clustering and k-NN classification [7]. DREBIN uses the manifest file to extract features from 5,560 applications and SVM as a classifier [3, 5, 11].

Some effort has been to integrate static and dynamic analyses for better performance. Dynamic analysis could be used to reduce false positives obtained after static analysis but doing so could in turn increase the false positive if a particular path is not executed during the dynamic analysis [3, 8, 9, 11].

Permissions accessed by Android Applications have been significantly studied to understand malicious intent. The Android operating system provides a coarse-grained mandatory access control (MAC). A permission-based classifier can identify more than 81% of malicious software. It can be used for preliminary malware check before a complete second analysis [3, 3].

These applications are classified as malicious or benign by the combination of permissions required by them. The DREBIN dataset contains 5,560 applications and the respective permissions from 179 different malware families were making it a sufficient dataset [3, 5].

DREBIN database, in comparison with older datasets, gives a better False Positive Ratio parameter overall. Most Machine Learning algorithms provide high accuracy rates, which are more significant than 85% using the dataset. DREBIN performs better than older datasets and 9 out of 10 popular anti-virus scanners [2]. The analysis of the DREBIN dataset is speedy, usually taking lesser than a second on computer systems and lesser than a few seconds on a smartphone [3, 5].

On this dataset, Random Forest Classifier shows better results than Naive Bayes and Logistic Regression [2]. The Random Forest Classifier is most suited for high-dimensional data modelling. It is easy to use as it can handle all data types and manage dataset inconsistencies easily [4]. Support Vector Machine is a good choice for a classifier as it gives high precision and recall values. DREBIN data has embedded feature sets that make it suitable to run the SVM algorithm. Compared to the other two methods, the SVM algorithm takes longer to run but gives accurate results [6].

3 Dataset Description

The DREBIN approach made in this paper for malware detection and classification is to gather as many features as possible from the application's manifest and code and embed these features into a joint vector space where each feature is grouped into sets. Some machine learning techniques are used to identify patterns in these features, which were gathered earlier. These features, each collected from each application, have the following properties, which are further grouped into sets: feature as Set S1 (Hardware Components) permission as Set S2 (Requested Permission) activity, service receiver, provider, service as set S3 (App Components) intent as set S4 (Filtered Intents) api call as set S5 (Restricted API calls) real permission as set S6 (Used Permission) call as set S7 (Suspicious API Calls) url as set S8 (Network Addresses) [3].

Due to the large size of features, the actual contents have not been used. Some have different values running into thousands, and not many are the same across other application files. Therefore, building one hot encoder and exponential growth in the feature vectors has been used in the algorithms. The number of feature properties of each feature set has been counted and stored in the dataset. A feature vector of size eight was used where each feature has count values and the output being True (malware) or False (not malware). The input vector looks in the following way (Fig. 1):

	sha256	S1	S2	S3	S4	S5	S6	S7	S8	output
0	00002d74.	2	11	5	3	7	6	11	26	TRUE
1	000068210	1	2	9	5	2	1	2	0	FALSE
2	00007647:	5	11	4	4	6	5	6	3	TRUE

Fig. 1 A snapshot of the feature_vectors_dataċsv file

4 Proposed Work

Using the dataset, the machine learning techniques are used to classify the applications as malware or non-malware. As there is much advancement in the usage of Android apps, it becomes a need for us to detect the malicious behaviour of Android apps for users' security, privacy and safe usage. Our approach is detecting malware systems using machine learning techniques that classify the apps as malicious and benign and suggest a better detection method.

4.1 Sampling Data

A graph between the count of malware and safeware+malware is plotted (Fig. 2), and it can be seen that the number of malware are 5560 and safeware+malware are 129013. It can be observed that there is a huge difference between the values, and this implies that the data is imbalanced. To balance this data, we use the methods of upsampling and downsampling it. Since the malware is very less in number, it is made as a minority class, and safeware+malware are made as a majority class.

- Upsampling: It is the process of inserting zero-valued samples between original examples in order to increase the sampling rate. In this dataset, the minority class is upsampled using resample method of the scikit-learn library with the number of samples set to 123453 and a random state of 123. The number of malware now is 123453, and the number of malware+safeware is also the same (Fig. 3).
- Downsampling: It is the method of removing samples of a disproportionately low subset of the majority class examples, decreasing the sampling rate. In this dataset, the majority class is downsampled using the resample method with the number of samples set to the length of the minority class (i.e., 5560) and random state to 123. The number of malware now is 5560, and the number malware+safeware is also the same (Fig. 4).

For both upsampled and downsampled data, the data is preprocessed, standardised and split into 70% as training set and 30% as testing set.

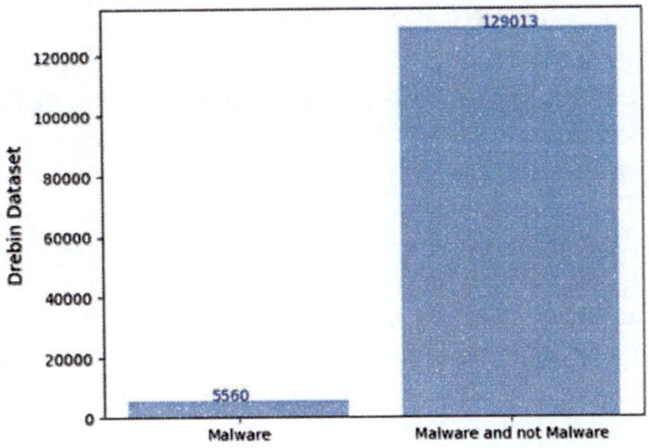

Fig. 2 A graph between the count of malware and safeware+malware in the Drebin dataset

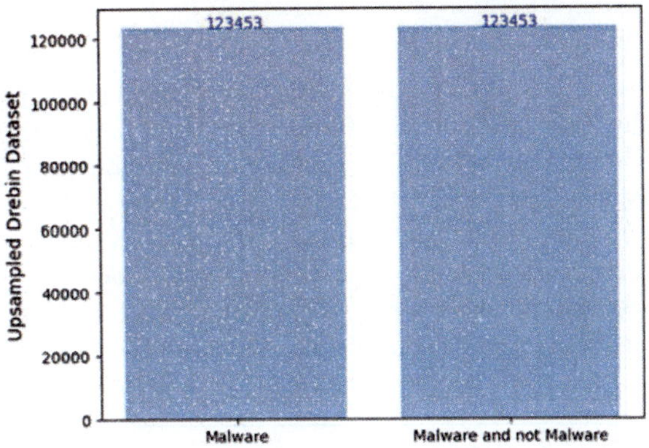

Fig. 3 A graph between the count of malware and safeware+malware after upsampling the DREBIN dataset

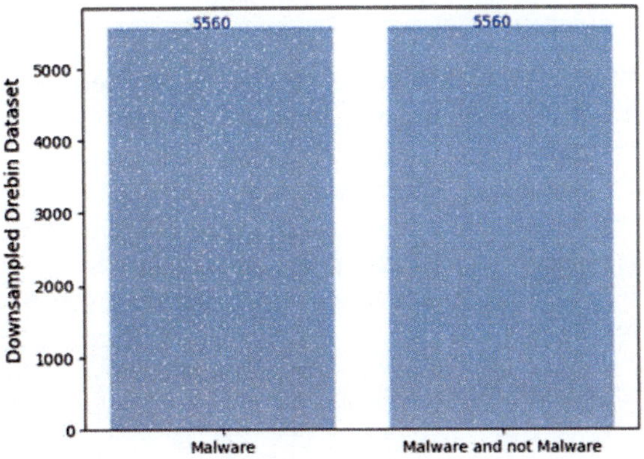

Fig. 4 A graph between the count of malware and safeware+malware after downsampling the DREBIN dataset

4.2 Logistic Regression

Logistic regression is a classification algorithm that is used to assign observations to a discrete set of classes. In this project, binary classification has been employed as we are classifying the software into two categories, viz. malware and safeware. The model, in which the data was split into training and testing sets, is trained with the logistic regression method from the linear model of scikit-learn. Then the labels of test data are predicted, and metric methods are used on these labels to calculate accuracy, precision, model recall and F1 score of the model. The same procedure was followed for unsampled, upsampled and downsampled data to calculate the results.

4.3 Random Forest Classifier

A random forest classifier is a classifying method that combines many decision trees by recursively selecting subsets of datasets to build different decision trees. It does so by building multiple decision trees and then merging them to get a more accurate and stable prediction. The model is trained with the Random Forest Classifier method from the linear model of the scikit-learn. The test data labels are then predicted, and metric methods are used to calculate accuracy, precision, model recall and F1 score of the model. The same procedure is followed for unsampled, upsampled and downsampled data to calculate the results.

4.4 Support Vector Machines

Support Vector Machine is a popular Supervised Learning algorithm used for classification. This algorithm aims to create the best decision boundary or line (hyperplane), which can segregate n-dimensional space into classes to make it easier for us to put the new data point in the correct category in the future. The algorithm involves choosing extreme points, called support vectors, in creating hyperplanes. The model is trained with the Support Vector Classifier (SVC) method from the SVM of scikit-learn. Prediction on the model and metric calculations on unsampled, upsampled and downsampled are similar to other classification methods.

5 Results and Analysis

The output plots, along with the results, are given below. We notice that all the three classifying methods have successfully classified the software in the given DREBIN dataset to Malware and Safeware, respectively.

(Formulae Used For calculations:)

$$TruePositives\% = \frac{TP}{P}$$

$$TrueNegatives\% = \frac{TN}{N}$$

$$Accuracy = \frac{TP + TN}{P + N}$$

$$Precision = \frac{TP}{TP + FP} = TruePositives\%$$

$$Recall = \frac{TP}{P}$$

$$F1 = \frac{2 * Precision * Recall}{Precision + Recall}$$

(TP = True Positives, TN = True Negatives, FN = False Negatives, FP = False Positives, P = Positives = TP+FN, N = Negatives = FP+TN)
Here, Positives are Malware and Negatives are Safeware.
Confusion matrix representation (Fig. 5):

Below are the confusion matrices of all three classications performed on unsampled, upsampled and downsampled data:

From the results given in Table 1, we can observe that the classification performed on the initial form of the dataset, which has not undergone resampling, produces inconsistent results. It is also clear that the three classifying methods perform better when the dataset has been upsampled or downsampled [3] (Fig. 6, 7, 8, 9, 10).

We can see that the Random Forest Classifier gives us the best results with an accuracy of 0.993 with the upsampled dataset, whereas with the downsampled dataset, it gives slightly lesser accuracy. It produced 99.7% true positives and 78.8% true negatives. The Logistic Regression method, on the other hand, gives a good accuracy of around 0.82. And on the other hand, the Support Vector Machine gives an accuracy of 0.866 when upsampled and 0.863 when downsampled, but these values are much lesser than the Random Forest Classifier method. Hence, we can conclude that the Random Forest Classifier method is a better method for malware classification among the three (Fig. 11, 12, 13, 14).

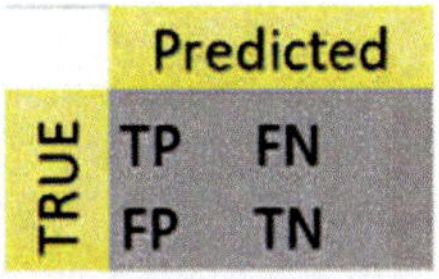

Fig. 5 Representation of a confusion matrix

Table 1 Results of classifiers on unsampled, upsampled and downsampled data

Classifier	Accuracy	Precision	Recall	F1 Score	TruePositives %	TrueNegatives %
Logistic regression (without sampling)	0.961011	0.671532	0.216853	0.327839	99.51	21.68
Logistic regression (upsampled)	0.819756	0.852773	0.773277	0.811082	85.3	77.3
Logistic regression (downsampled)	0.821342	0.854838	0.78125	0.816389	86.28	78.12
Random forest (without sampling)	0.989561	0.935353	0.818503	0.873035	99.73	81.73
Random forest (upsampled)	0.993627	0.989014	0.998354	0.993662	99.01	99.73
Random forest (downsampled)	0.940647	0.938524	0.9451650	0.941833	93.6	94.45
Support vector machine (without sampling)	0.956076	0.454545	0.008839	0.017341	99.95	0.0088
Support vector machine (upsampled)	0.866656	0.989275	0.850039	0.864488	88.32	85.00
Support vector machine (downsampled)	0.863609	0.879047	0.848466	0.863486	87.92	84.84

Random Forest Classifier

1. Without sampling

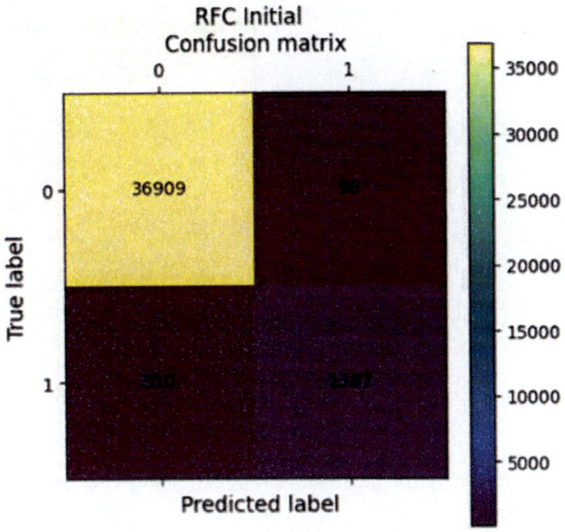

Fig. 6 Confusion matrix obtained for Random Forest Classifier method performed on the non-sampled data

2. For Upsampled data

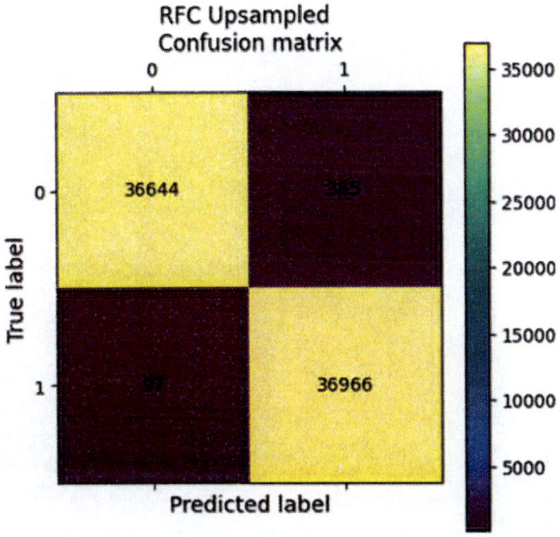

Fig. 7 Confusion matrix obtained for Random Forest Classifier method performed on the upsampled data

3. For Downsampled data

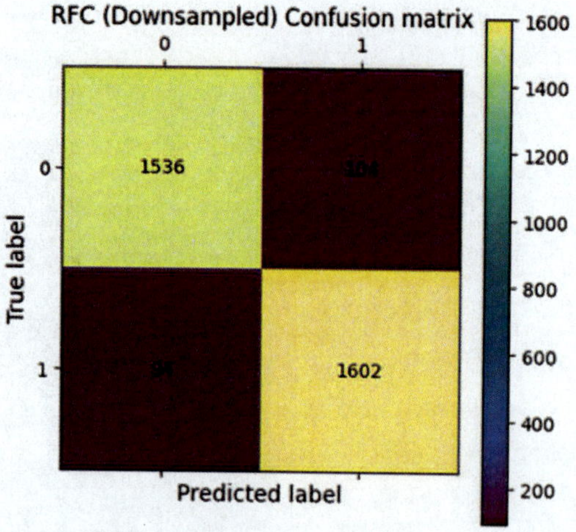

Fig. 8 Confusion matrix obtained for Random Forest Classifier method performed on the down-sampled data

Logistic Regression

1. Without Sampling

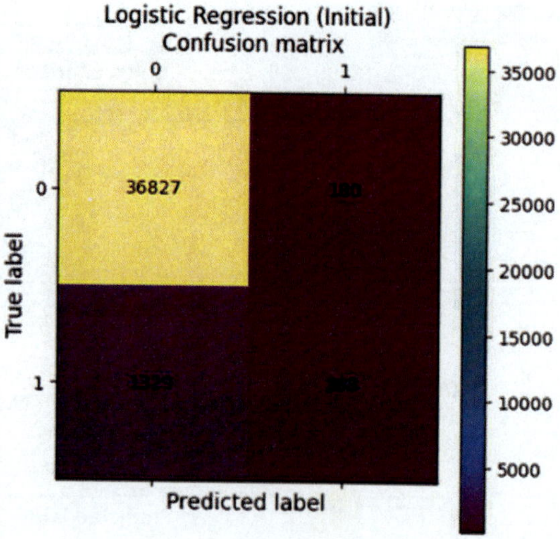

Fig. 9 Confusion matrix obtained for Logistic Regression method performed on the non-sampled data

2.For Upsampled data

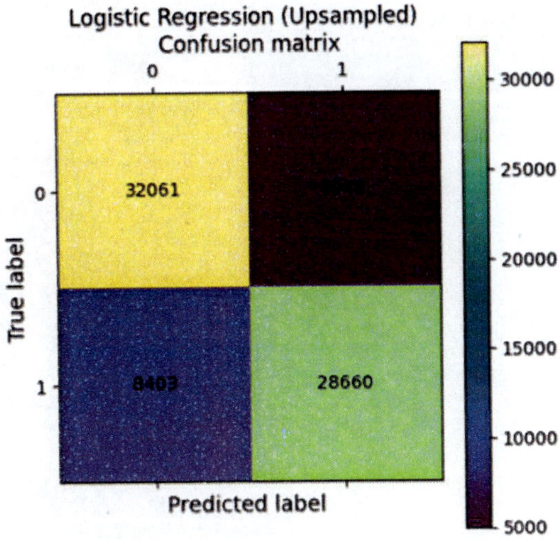

Fig. 10 Confusion matrix obtained for Logistic Regression method performed on the upsampled data

3.For Downsampled data

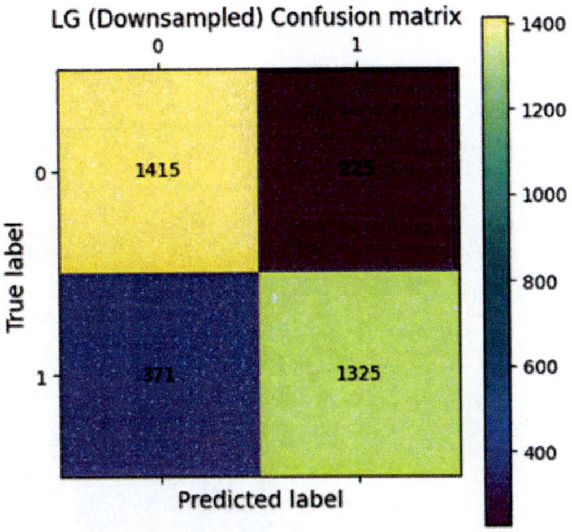

Fig. 11 Confusion matrix obtained for Logistic Regression method performed on the downsampled data

Support Vector Machine

1. Without sampling

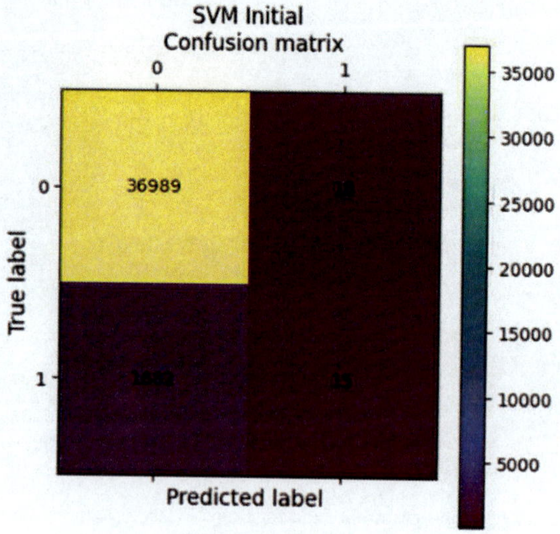

Fig. 12 Confusion matrix obtained for SVM method performed on the non-sampled data

2. For Upsampled data

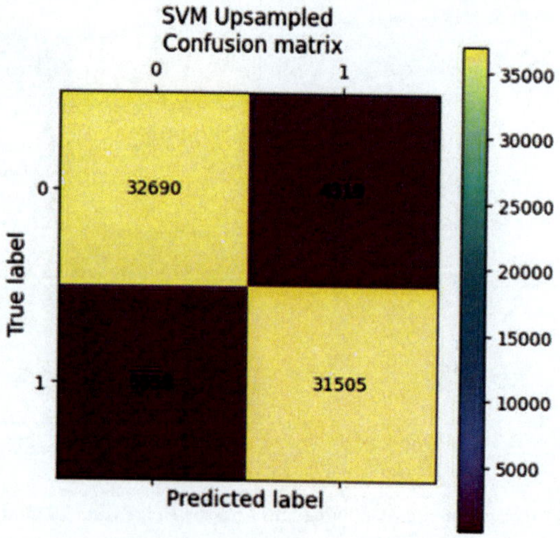

Fig. 13 Confusion matrix obtained for SVM method performed on the upsampled data

3. For Downsampled data

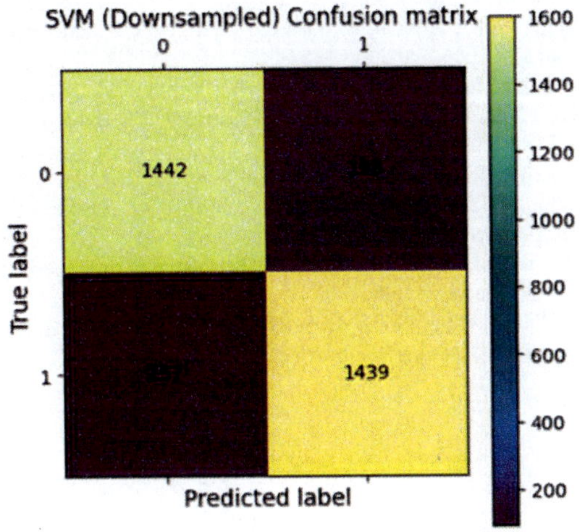

Fig. 14 Confusion matrix obtained for SVM method performed on the downsampled data

6 Conclusion and Future Work

In this paper, safeware static analysis has been carried out using the AndroidManifest.xml file to extract the features such as "permissions" and "API calls" after which the results of the classification carried out by the three methods have been analysed and compared. We have concluded from the results that the Random Forest Classifier method is more effective in malware classification among the three. For future work, the software can be classified using dynamic analysis by extracting system calls.

References

1. Patel, Z.D.: Malware Detection in Android Operating System, Department of Computer Engineering, Sarvajanik College of Engineering and Technology, Surat, India
2. de la Puerta, J.G., Sanz, B., Grueiro, I.S., Bringas, P.G.: The Evolution of Permission as Feature for Android Malware Detection
3. Huang, C-Y., Tsai, Y-T., Hsu C-H.: Performance Evaluation on Permission-Based Detection for Android Malware
4. Jehad Ali, J., Khan, R-U., Ahmad, N., Maqsood, I.: Random Forests and Decision Trees
5. Arp, D., Spreitzen-Barth, M., Hubner, M., Gascon, H., Rieck, K.: Drebin: Effective and Explainable Detection of Android Malware in Your Pocket
6. Rana, M.S., Sung, A.H.: Malware Analysis on Android Using Supervised Machine Learning Techniques, University of Mississippi
7. Wu, D.J., Mao, C-H., Wei, T-E., Lee, H-M., Wu, K-P.: Droidmat: Android Malware Detection Through Manifest and Api Callstracing

8. Ge, X, Taneja, K., Xie, T., Tillmann, N.: Dyta: Dynamic Symbolic Execution Guided with Static Verification Results
9. Jiang, Y.Z.X., Xuxian, Z.: Detecting Passive Content Leaks and Pollution in Android Applications
10. Bose, A., Hu, X., Shin, K.G., Park, T.: Behavioral Detection of Malware on Mobile Handsets
11. Li, C., Zhu, R., Niu, D., Mills, K., Zhang, H., Kinawi, H.: Android Malware Detection Based on Factorization Machine
12. Enck, W., Gilbert, P., Han, S., Tendulkar, V., GonChun, B., Cox, L.P., Jung, J., McDaniel, P., Sheth. A.N.: TaintDroid: An Information-flow Tracking System for Realtime Privacy Monitoring on Smartphones
13. Tam, K., Khan, S.J., Fattori, A., Cavallaro, L.: CopperDroid: Automatic Reconstruction of Android Malware Behaviors
14. Wu, W.-C., Hung, S.-H.: DroidDolphin: A Dynamic Android Malware Detection Framework Using Big Data and Machine Learning

Operating System Fingerprinting Using Machine Learning

Achintya Kumar, Ishan Soni, and M. Anand Kumar

1 Introduction

Since everyone is linked to the Internet these days, being safe from breaches and incursions is crucial. For businesses, this external danger causes them to investigate various security alternatives, such as firewalls and intrusion detection mechanisms, to protect themselves against hackers. Operating system fingerprinting is a much-needed approach for spotting and identifying a target machine's identity by looking at the TCP/IP packets it generates consistently. The most generally used technique in the market is to employ rule-based matching methods to identify the OS. Unlike machine learning, this approach does not require a significant quantity of data and the speed for identification to take place is also very quick. In cases of insufficient information from the packets received for identification due to network settings, newer versions, or other factors, the method will not recognize the operating system, and the resulting accuracy will be low.

Operating System fingerprinting techniques are categorized into two categories, active and passive. In Active fingerprinting, packets are sent to a target and received packets are analyzed. Nmap is a vital tool in this regard and is generally used by network admins for security and testing purposes. Using Nmap [13], one can ensure that all of the firewalls in their network are appropriately configured, and the TCP/IP stacks are not malfunctioning. Passive fingerprinting works by sniffing the TCP/IP

A. Kumar (✉) · I. Soni · M. Anand Kumar
Department of Information Technology, National Institute of Technology Karnataka Surathkal, 575025 Mangalore, India
e-mail: achintya.191it203@nitk.edu.in

I. Soni
e-mail: ishu.191it121@nitk.edu.in

M. Anand Kumar
e-mail: m_anandkumar@nitk.edu.in

B. Agarwal et al. (eds.), *Proceedings of International Conference on Intelligent Cyber-Physical Systems*, Algorithms for Intelligent Systems,
https://doi.org/10.1007/978-981-16-7136-4_13

ports rather than using extra bandwidth for requests. Passive OS detection has gained recent interest in identifying the host with no trace left behind. For this detection technique, the primary focus is upon the different parameters of packet headers, some of which are window size, do not fragment bit, time to live (lifetime), and TCP flags.

In this paper, we use the Passive Fingerprinting method by analyzing TCP network packet header info as well as info from HTTP header using several machine learning methods, such as K-nearest neighbours (KNN), Artificial Neural Network, Decision Trees, Naive Bayes, and Random Forest.

The motivation for this study was to decide the most suitable OS fingerprinting approaches as the present tools in use for OS fingerprinting were not accurate enough and were unable to detect dissimilarity in many cases. Moreover, many modern operating systems have default policies and firewalls that messes with the network services which in many cases might result in a lack of data for proper identification.

The rest of the paper is broken into six sections. Section 2 deals with the literature survey, Sect. 3 explains the framework proposed, Sect. 4 discusses the dataset used for the study, and Sect. 5 describes the methodology used. In Sect. 6, the outcomes of the experimental work are analyzed. At last, the conclusion is discussed in Sect. 7.

2 Literature Survey

There has been some research for active and passive fingerprinting techniques [3, 17] in the last 12–15 years. Spitzner [6] was the first to identify what passive OS fingerprinting was, how it worked, and the use cases. They also extensively compared both fingerprinting techniques using a wide array of tools. Al-Shehari et al. [1] had proposed machine learning techniques combined with traditional tools to build a system that can set up TCP/IP communication between different machines and then capture and inspect the TCP/IP packets for significantly better OS detection. Similarly, Matsunaka et al. [16] used DNS Traffic Analysis by analyzing the data sent by each OS and extracting the characteristics for OS fingerprinting such as interval time pattern of DNS queries and OS-specific query. Then, examine the estimation method by using DNS traffic in their own intra-network.

Lippmann et al. [5] had an interesting approach for near-match fingerprints, where they used machine learning classifiers to determine the most detectable OS categories that used fingerprinting. Tyagi et al. [4] also had a similar approach of using TCP/IP communication for identifying prohibited operating systems on private internal networks. For optimization and quick results, Gu et al.[6] focussed on using only memory for fingerprinting and caching the code hash of the kernel from the guest Operating System for faster results.

Song et al. [2] analyzed the identification capabilities of several ML methods with each having a unique approach for classifying. The models were based upon Decision Trees, K-nearest neighbours, and Artificial Neural Networks and showed a 94% probability of getting the prediction correct. They found ANN to perform best

Table 1 Summary of literature survey

Authors	Methodology	Merits
Taher Al-Shehari et al. [1]	TCP/IP header packet info for ML and new extended tool	Simple algorithm and use in real life
Martin et al. [10]	Used TLS Fingerprints for OS Identification	High accuracy as TLS, TCP/IP and HTTP headers are used
Song et al. [2]	Analysis of OS identification using ML techniques	Employed many ML models for higher accuracy
Yufei et al. [7]	Memory-Only Operating System Fingerprinting in the Cloud	Very quick results and wide range
Tyagi et al. [4]	TCP SYN packets for OS fingerprinting	Easy to compute and gather data
Takashi et al. [16]	Analysis of DNS traffic	Novel approach, works better in some cases

of three when the dataset was large and adequately trained. KNN was second in line with no bounds to data size and performed consistently. In the meanwhile, Beverly [11] also used the Naive Bayes Classifier for the same approach.

This paper draws inspiration from such authors' Machine Learning approach to OS fingerprinting and has employed many such methods from different authors' research (Table 1).

3 The Proposed Method

This paper deals with the following problem statement: To Determine the best ML classifier and the most influencing parameter.

The problem statement can be broken further into two parts:

- How to analyze all the different classifiers to find the most suitable classifier taking size, time taken, and other costs into consideration.
- How to determine which parameter(s) play a major role in the identification of operating systems and are necessary for fingerprinting.

Many authors and researchers use TCP/IP and HTTP features [14] for passive OS fingerprinting and TLS features [8, 11, 12] for fingerprinting specific browsers [9]. We propose a system that combines both of these features to identify all kinds of desktop and mobile (handheld) operating systems.

We have used conventional machine learning algorithms such as Decision Trees, K-nearest neighbours, and Artificial Neural Networks from [2] and tried some of the newer algorithms such as Random Forest and Bayes algorithm too in the proposed

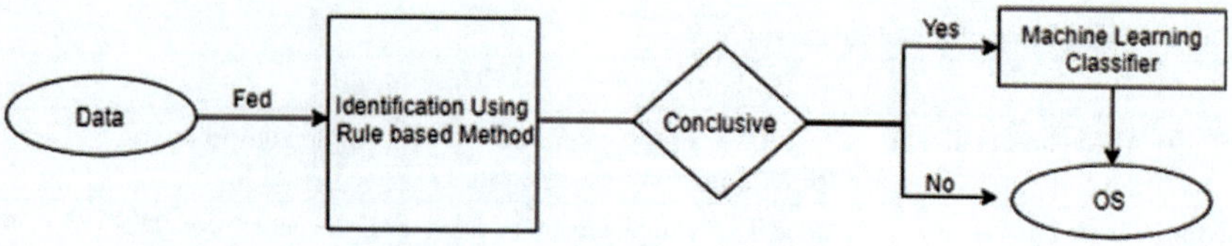

Fig. 1 Proposed model

implementation. We have also analyzed the probability of correct identification across different operating systems and the role of various parameters involved in the process.

Figure 1 shows the architecture of OS fingerprinting by combining Machine Learning Classifier and Conventional rule-based matching methods. Considering the cost of computation and time taken, the proposed model first uses a cheap and quick method of identification using Rule-based matching method, and in case of an inconclusive or partial match, the data from TCP/IP packet headers and HTTP headers go through ML classifier and then the results of both methods are compared. In case of no discrepancy, the output is given as a result.

4 Dataset and Model Setup

The dataset was acquired from [4] paper where authors had posted the dataset on Zenodo.org. It contains data from TCP/IP network, HTTP connection, and other metadata from the connection. The following are the different fields available in the dataset.

- Metadata about network
  - Beginning of connection
  - End of connection
  - Port used
  - Src IPv4 (address)
  - Dst IPv4 (address)
  - Receiver port
- Features of HTTP
  - HTTP UA OS, MAJ, MIN, BLD (information about major/minor versions of OS)
  - HTTP Hostname
- TCP/IP features
  - SYN size (packetsize)
  - TCP SYN TTL (Time to live)
  - TCP win (size of window)

After careful consideration and testing, some of the parameters were used for training and testing the model. Unique identifying data fields were removed and the rest were considered for the study.

- SYN size
- TCP win
- TCP SYN TTL
- HTTP UA OS
- HTTP UA OS MAJ
- HTTP UA OS MIN
- HTTP UA OS BLD
- Ground Truth OS

4.1 Pre-Processing

As shown in the Dataset Description, the data consisted of HTTP Features, network packet information, and other metadata. Each attribute had some blanks for specific instances, so the dataset was trimmed and prepared for the experiment. We took 232391 instances of data where approximately 80% of the data is for training the model, and the rest 20% is used to test it.

4.2 Model Setup

In the mentioned study, the model incorporates one layer each for input, hidden, and output. Seven attributes values were used for the input layer: SYN size, TCP win, TCP SYN, TTL, HTTP UA OS, HTTP UA OS MAJ, HTTP UA

OS MIN, HTTP UA OS BLD. The output layer was configured to conceive four outputs which covered basic and widely used Operating systems namely Windows, Linux, Android, and MAC.

The calculated loss rate was close to 0.01% with near-perfect accuracy using test data. The Mmodels were tested repeatedly with test data to record fluctuations in output and corrected accordingly till the changes were insignificant

5 Methodology

This study compared the widely used Machine Learning algorithms suitable for our use case. They consist of Decision Trees, KNN, Random forest, Bayes, and Artificial Neural Network algorithms. Each of the algorithms is explained below, along with their advantages and disadvantages and the approach taken in their implementation.

5.1 *Decision Trees*

Decision trees initially learn, then form decisions for splitting, and finally output in a tree-like structure. It has the advantages of not converting data into the decimal, less data cleaning required, and it works fast when the tree's depth/height is specified. However, since depth has a significant role, results change sometimes. For our implementation, Depth = (5, 15) was used.

5.2 *Artificial Neural Networks*

ANNs are machine learning algorithms that are meant to learn from data patterns. It is separated into three layers input, hidden, and output layers. The input layer takes the data from the user/source. There is no limit to the number of layers. It offers the advantages of being fault-tolerant, great accuracy when there is a massive volume of data contrary to other machine learning algorithms. Overfitting is a problem when the dataset is not too large.

5.3 *K-Nearest Neighbours*

KNN is a machine learning method that uses the closest neighbours' info by measuring separation to previous data when novel data is entered. The distance is calculated using the Euclidean calculation method. The KNN algorithm shines when it requires fast processing speed, and comparison data is not significant because, unlike others, learning is not necessary. Still, when the dataset is on the smaller side, performance takes a hit. For the KNN model, three separate learning models, with K = 5, 40, 100, were implemented.

5.4 *Random Forest*

Random Forest is based upon ensemble learning, which combines multiple classifiers to solve a complex problem and slightly improves performance. More specifically, Random Forest is a classifier that employs many decision trees and takes its average to improve the accuracy of the dataset. Although it combines the Decision trees, it takes a considerably big time and is not particularly good for OS that is rare to be seen. For the Random Forest model, three separate learning models, with the number of trees = 10, 50, and 120 variations, were implemented.

5.5 *Naive Bayes*

The Naive Bayes algorithm is loosely based on the famous Bayes Theorem of probability and statistics. It is simple and yet one of the powerful ML algorithms today and is categorized as a probabilistic classifier. One of the reasons for this is that it assumes one feature in a class does not affect the other. The drawback of Bayes is that it does not relate all the parameters together and treats everything independently, which can prove results to be unpredictable at times. Multiple runs were done using this model and the average was taken.

In Table 2, a summary comprising of the algorithms used and the different models implemented per algorithm is shown. Furthermore, a comprehensive list of limitations and overall accuracy of all the algorithms tested are discussed in Table 3.

Table 2 Accuracy for different parameter settings

Algorithm	Parameters	Accuracy (%)
Decision tree	Depth = 5	93.96
	Depth = 10	95.02
	Depth = 15	95.22
KNN	K = 5	91.26
	K = 40	96.17
	K = 100	96.25
Random forest	Trees = 10	92.46
	Trees = 50	94.18
	Trees = 120	95.88

Table 3 Comparison of algorithms

Model	Limitations	Accuracy (%)
Decision trees	Unstable and High training time	95.62
KNN	In the case of large data, the speed suffers	96.22
ANN (Artificial Neural Network)	For smaller quantity of data, accuracy rate declines and possible Overfitting occurs	75.22
Naive Bayes	All features are assumed to be independent, the relationship between features is not considered	79.88
Random forest	It is not suitable for rare outcomes and overfitting problems possible	95.88

6 Experimentation and Results

This section explains the various different experiments performed, metrics used, and the results obtained through each experiment by breaking them into sub-sections.

6.1 Comparing ML Algorithms

Here, in Fig 2, we can see different models compared using metrics such as precision and accuracy. Here, Y-axis denotes the percentage, and X-axis the algorithm used. F1-score is scaled to 100 for percentage depiction purposes. KNN appeared as the best ML model for OS identification. Random Forest Classifier also performed well and overtook KNN when the dataset was huge.

6.2 Comparison Between Parameters

A study was also done on the parameters involved to recognize the parameter that had the most effect on identification. For this, we took an approach of removing parameters one at a time and retraining the proposed model without the removed part. The parameter whose absence showed the greatest significant decline was termed the most influencing parameter. As shown in Table 4, it is evident that TCP SYN TTL [shows a decline of 23.65%] is the most influencing parameter of the bunch.

Table 4 Influence of parameters used

Parameter	Description	Decline (%)
SYN size	Request sent before a connection	16.44
TCP SYN TTL	The lifecycle of a single TCP SYN packet	23.65
TCP win	The window size of TCP connection	17.63
HTTP UA OS	HTTP User-agent Operating System	19.28
HTTP UA OS MAJ	HTTP user-agent Major Version of the OS	5.88
HTTP UA OS MIN	HTTP user-agent Minor Version of the OS	4.16
HTTP UA OS BLD	HTTP user-agent Build Version of the OS	9.39

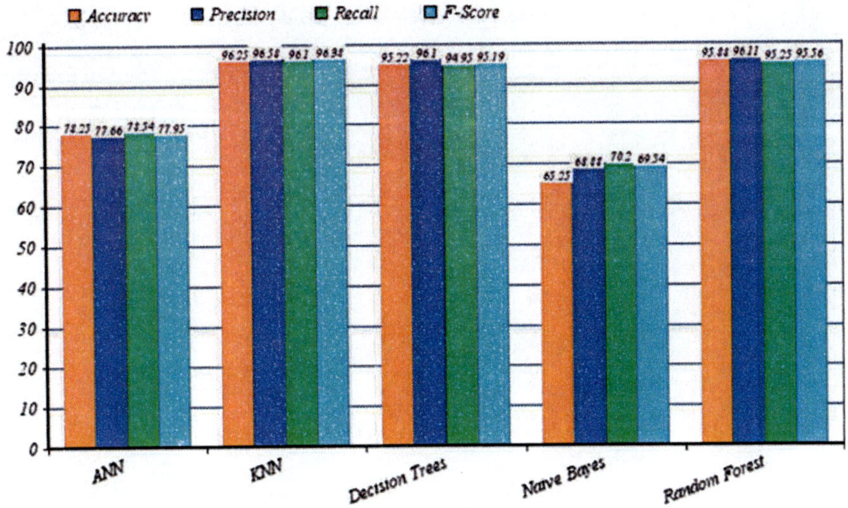

Fig. 2 Comparison between different Models

6.3 *Comparing Ease of Prediction Across OS*

We also took time to compare the accuracy of prediction across different operating systems. We can see the chart comparing the accuracy of predicting different operating systems in Fig. 3 with the Operating System category on the y-axis and percentage accuracy obtained for each operating system in the x-axis.

Comparing the results with OS identification solutions that use traditional Rule-based matching methods, an overall significant rise of 20% accuracy was observed over recognized operating systems and the 5% improvement for unknown data samples.

6.4 *Comparison with Existing Tools*

We compared the KNN, which has the best accuracy, to some of the most popular alternatives present in the market. In our case, p0f [15, 18] was used for Operating System identification. WireShark is a network packet analyzer that was used to capture essential data from the data obtained in p0f for comparison. The data from four different types of Operating systems (Mac, Linux, Windows, and Android) was available, disregarding the different versions and distributions available for the same OS.

A total of 1956 known and 490 unknown data samples from p0f were used for this study. Network Miner, another OS fingerprinting tool that uses only packet sniffing for identification, was used to obtain the anonymous data.

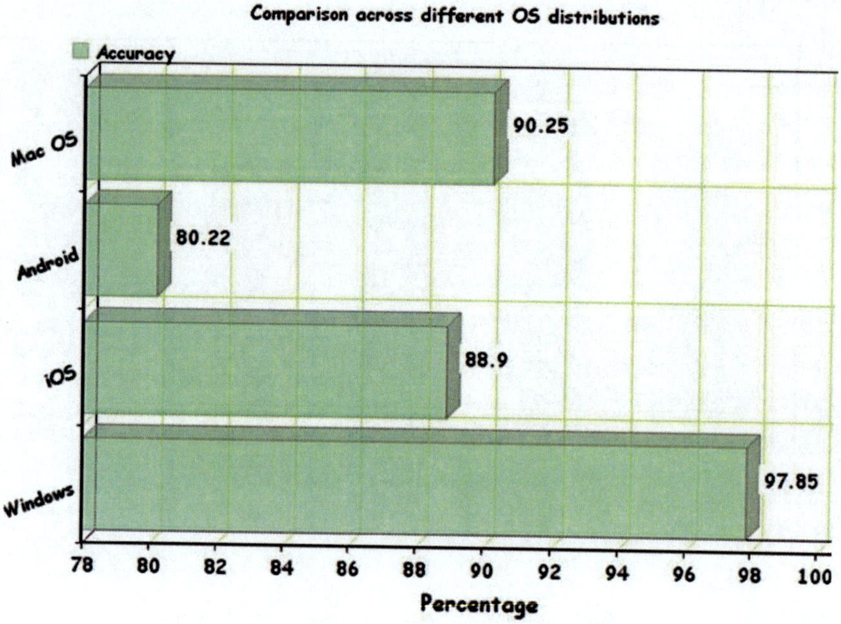

Fig. 3 Comparison between the different OS accuracy

In our results, we found that our best performing model KNN gave an accuracy of 95% (Fig. 2), with a 72% probability for the unknown OS. Among the 1956 datasets used, p0f had a precision of 52%, and the Artificial neural network model correctly identified the OS with a chance of 79%.

7 Conclusion and Future Scope

This proposed model using machine learning and Operating system attributes (SYN size, TCP win, TCP SYN TTL, HTTP UA OS, HTTP UA OS MAJ,

HTTP UA OS MIN, HTTP UA OS BLD) achieved probability of accurate determination of OS more than 96%, much higher than traditional methods. On the other hand, individual OS versions could not be precisely categorized due to them having similar attribute values and generally little implementation changes between them. Moreover, recently launched Operating Systems could not be identified in many cases in the rule-based strategy as the information about them is scarce.

Comparing the parameter's influence for identification of the Operating system, one can infer that the TTL(Time to Live) of a SYN Packet differs across different operating systems. Using TLS features with HTTP parameters enhanced the machine learning model's efficiency, and as a result, we found that the proposed model, using the Machine Learning approach in tandem with the conventional rule-based matching method, can yield better results than the tools we use now.

References

1. Al-Shehari, Taher., Shahzad, Farrukh.: Improving operating system fingerprint- ing using machine learning techniques. Int. J. Comput. Theory. Eng. 6. King Fahd university of petroleum and minerals, Saudi Arabia
2. Song, Jinho., Cho, ChaeHo., Won, Yoojae.: Computers and Electrical Engineering 78. Chungnam National University, Korea (2019)
3. Anderson, Blake., McGrew, David.: OS Fingerprinting: New Techniques and a Study of Information Gain and Obfuscation:2017 IEEE Conference on Communica- tions and Network Security (CNS), Cisco Systems
4. Tyagi, R., Paul, T., Manoj Bs., Thanudas B.: Packet Inspection for Unauthorized OS Detection in Enterprises. IEEE Security Privacy. 13. 60–65 (2015)
5. Lippmann, R., Fried, D., Piwowarski, K., Streilein W.: Passive Operating System Identification from TCP/IP Packet Headers :IEEE Workshop on Data Mining for Computer Security (DMSEC), pp. 40–49 (2003)
6. Spitzner, L.: Passive Fingerprinting **3**, 1–4 (May 2003)
7. Gu., Yufei, Fu., Yangchun, Prakash, A., Lin, Z., Yin, H.: OS- SOMMELIER: Memory-Only Operating System Fingerprinting in the Cloud : SOCC'12, October 14–17. CA USA, San Jose (2012)
8. Dierks, Tim., Rescorla, Eric.: The Transport Layer Security (TLS) Protocol: Version 1.2. RFC 5246 (Proposed Standard) (2008)
9. Durumeric, Zakir., Ma, Zane., Springall, Drew., Barnes, Richard., Sullivan, Nick., Bursztein, Elie., Bailey, Michael., Alex Halderman, J., Paxson, Vern.: The security impact of https interception. In: Network and distributed systems symposium (NDSS17) (2017)
10. Elkan, Charles.: The foundations of cost-sensitive learning: In International Joint Conference on Artificial Intelligence,(IJCAI), pp. 973–978 (2001)
11. Friedl, Stephan., Popov, Andrei., Langley, Adam., Stephan, Emile.:Transport Layer Security (TLS) Application-Layer Protocol Negotiation Extension :RFC 7301 (Pro- posed Standard) (2014)
12. Lastovicka, Martin., Spacek, Stanislav., Velan, Petr., Celeda, Pavel.: Using TLS Fingerprints for OS Identification in Encrypted Traffic: NOMS 2020–2020 IEEE,pages 1–6 04/2020
13. Greenwald, Lloyd., Tavaris Thomas, T.: Toward undetected operating system fingerprinting. : In USENIX Workshop on Offensive Technologies (WOOT), pp. 1–10 (2007)
14. Husak, Martin., Cˇerm´ak, Milan., Jirs´ık, Tom´aˇs., Cˇeleda, Pavel.: HTTPS traffic analysis and client identification using passive SSL/TLS fingerprinting. EURASIP Journal on Information Security volume 2016, Article number: 6 (2016)
15. Majkowski, M.: SSL fingerprinting for p0f. https://idea.popcount.org/2012-06-17- ssl-fingerprinting-for-p0f/
16. Matsunaka, T., Yamada, A., Kubota, A.: Passive OS fingerprinting using DNS traffic analysis. Advanced Information Networking and Applications (2013)
17. Allen, Jon Mark.: OS and application fingerprinting techniques. SANS.edu Grad- uate Student Research
18. Michal, Zalewski.: p0f v3 (version 3.09b). https://lcamtuf.coredump.cx/p0f3/README

Comparing HDD to SSD from a Digital Forensic Perspective

Mahmoud Jazzar and Mousab Hamad

1 Introduction

Due to the rapid and huge growth of the hardware and software industry, it is not tangible as to what opportunity HDD and SSD may contribute to forensic investigators. The wide spread of software and hardware within industrial and technological firms determines the characteristics of such properties to become very common. As such, SSD is faster in reading and writing files and folders, and the wattage of HDD is much more than SSD [1]. HDD has been used for a long time to store data files and folders. The updates in SSD technology include features such as trimming, garbage collection, and wear-leveling [2, 3]. With the new and efficient high-speed SSD technology, forensic examiners and investigators need to face the new technology updates. As such, SSD can alter and even eliminate deleted data files, and has the ability to clean up itself without any instruction [2, 4]. This problem is not evident in old technology which on the other hand introduces new challenges for forensic investigators [4].

Typically, there are two types of data acquisition to understand the evidence at the storage level. Static and live data acquisition can be used to obtain a bit-stream or image copy of the drive. The type of data question from a forensic perspective is always arguable particularly in case there are data alterations that exist when migrating data from old to new hard drives. Similarly, if we try to compare similar data files stored on different hard drives, expectation over time is indispensable. According to the existing literature [5–7], it is evident that SSD updates itself and

M. Jazzar (✉) · M. Hamad
Faculty of Graduate Studies, Palestine Technical University – Kadoorie, P.O. Box 7, Tulkarm, Palestine
e-mail: mjazzar@ptuk.edu.ps

M. Hamad
e-mail: m.a.hamad4@ptuk.edu.ps

B. Agarwal et al. (eds.), *Proceedings of International Conference on Intelligent Cyber-Physical Systems*, Algorithms for Intelligent Systems,
https://doi.org/10.1007/978-981-16-7136-4_14

alters the stored data or even deletes it. On the other hand, some features like data shuffling and encryption applied on HDD may not be useful for SSD [4]. As such, the updated SSD technology applies data compression and data encryption operations on data before storing data on disks [8]. Such operations make SSD much flexible and faster when dealing with a huge number of data files, i.e., improved security and increase of SSD lifetime [3].

SSD has a transformation in the mimic compared to HDD data storage standards. In addition, the SSD controller overcasts drive management from the operating system. As such, SSD uses an operation called the garbage collection to maintain the drive performance [2, 8]. Furthermore, the behaviors of SSD might be unexpected and cannot be easily predicted due to the technical diversification of technology [3]. This means that conducting a forensic investigation on SSD will always be a challenge for forensic investigators as we cannot identify if we can find the desired data or not very quickly. Although investigators have possession of typical hard drive technology, the huge technology change influences the need for further development in investigation tools, resources, skills, and methods.

The purpose of this paper is to bring the potential difference to data alteration when using similar data files and folders on HDD and SDD from a forensic perspective. This paper examines the practical digital forensic differences between HDD and SSD. In addition, it supports the recommendation of further forensic considerations such as anti-forensics using different tools for comparison between the drives similar in size, content, and under different conditions. The rest of the paper is organized as per the following. Section 2 introduces the background and physical differences between HDD and SSD. Section 3 represents digital forensic analysis differences and conflicts between HDD and SSD. Section 4 contains the experimentation setup, Sect. 5 discusses the results and findings, and finally conclusion remarks are presented in Sect. 6.

2 Background

Generally, binary data is stored on HDD. To arrange and manage these binary data, two types of addressing are used by the operating system to locate data forms. Cluster-head-sector (CHS) and logical block addressing (LBA). The hard drive contains magnetic disks and heads designed to read and write data. HDD can manage data for years without distress or devaluation. In addition, they can endure millions of rewrites over a lifetime [7, 9]. NAND flash memory is the main component in SSD [10]. The controller inside the SSD has a mix of hardware and software properties. The controller has an interpreter which can translate the CHS and LBA which are the same technologies used in HDD in the NAND flash array [9].

Normally, the operating system cannot exactly decide the original location of data inside SSD. The smallest object within the SSD is the block and it is the only object that can be deleted. When a block is deleted, then the memory must be erased before being rewritten again. Using NAND flash heavily would lead to bad blocks due to

Fig. 1 NAND cell [5]

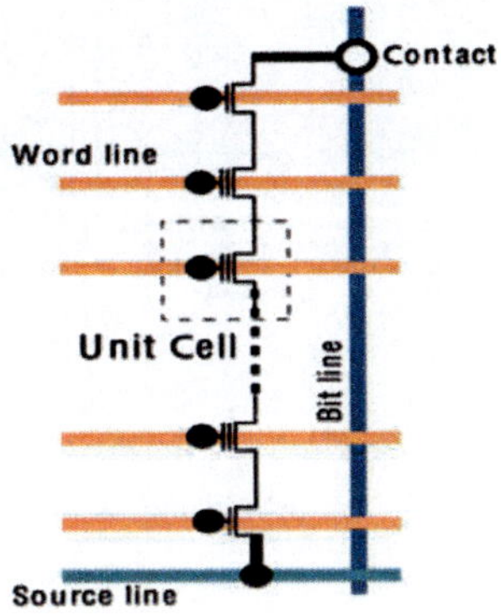

limited rewrite operations. Wear-leveling is a special feature inside SSD, as such, the write operations spread across all memory blocks in an effort to reduce the load to single block [11]. Wear-leveling is an important feature that increases the lifecycle of the SSD because it checks on how many times the flash memory chip in SSD has been written. Therefore, it will store the new data into the memory space which has not been used or used less than other spaces. Plus, the NAND cell is very similar to the HDD. The cell is structured as pages and can store sequential data [5]. Therefore, the cell might have blocks and influence error corrections. Figure 1 below illustrates NAND cell architecture.

A few research works focus on comparing different versions of hard drives from a forensic perspective. In [12], SSDs are considered as change of rules for forensic investigation. As described earlier, SSDs store data in random places within sectors, meaning that the memory allocation process diverges to be described. Forensic tools are custom-made mainly to conduct investigations on HDD and do not guarantee recovery processes for SSD technology.

SSDs do not rely on moving parts [5]. The term solid state refers to the fact that the data is stored in fixed layers of electronic transistors. The transistors have a fast read speed which reaches tens of microseconds and the writing speed reaches hundreds of microseconds; this speed is 10 × compared to hard disk drive speed in both read and write operations. The latency for reads and writes is usually 3–10 ms, i.e., 30–3000 times slower [13]. The whole operation of quick format to HDD and the overwrite operation are explained in [14, 15]. As such, deletion of specific data does not mean physical deletion through formatting. The operation just marks the data inside to overwrite blocks. The hard drive then waits for the same area to be used again before completely destroying the data. When the overwriting process ends, the point of no return is reached and the data cannot be recovered permanently. However, SSD is completely different as data is stored to represent the blocks in grid fashion; blocks are made up of individual rows called pages. According to [15], when data is deleted then it is erased physically using a background such as garbage collection process rather than being marked for overwriting. Garbage collection monitors the file allocation table to decide which blocks are no longer in use and performs the

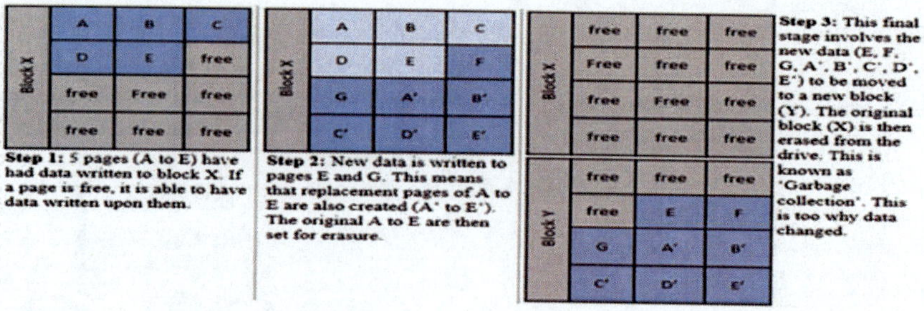

Fig. 2 SSD memory allocation process [15]

operation so that it speeds up the rewrite process. Correspondingly, trimming for garbage collection is applied to notify the SSD about the deleted data blocks inside the memory. Once the SSD gets the information from the operating system about the free block, the memory controller delivers instructions to wipe out the deleted blocks [15]. Figure 2 illustrates the SSD memory allocation method.

The disk drive is supplied with electric current, and the dish rotates in the form of circles, which causes the production of magnetic fields. These magnetic fields pass through and move the player, which moves the head and records data. Methods have been applied to modify the recording and become vertical, which causes an increase in the capacity of the drive [11]. As illustrated in Fig. 3, a new layer has been added to improve the efficiency of recording and storing data. The binary system is represented by the movement of magnetic fields from top to bottom and from bottom to top.

Forensic procedures and analysis of an SSD are clarified in [16]. Experimental investigation in [5, 13] explained the challenges for digital forensic investigators on one occasion of SSD analysis. Nisbet in [14] performed an analysis of SSD on three trim enabled file systems were as [17] proposed a method to find out if acquisition from the SSD is expensive or not. Key features for SSD and behavior scenarios are defined in [18]. However, those features and methods are still challenging for investigators to acquire clean and non-altered data.

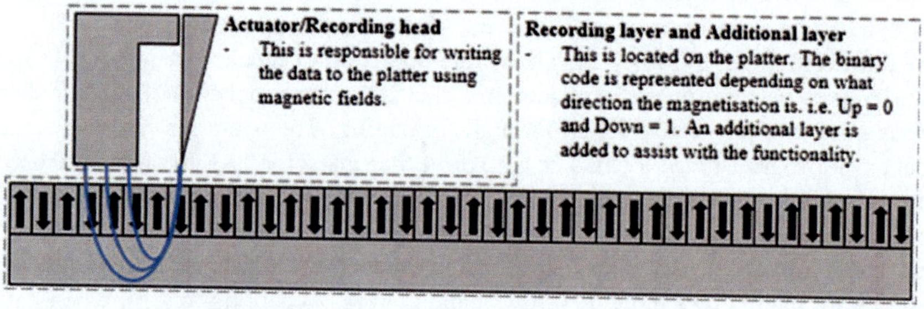

Fig. 3 Modern HDD technology [15]

3 Forensic Analysis Differences

HDD has been used for a long time and decades. Therefore, forensic investigators practice traditional and advanced methods in analyzing this type of drive [9, 13, 16]. It can be said that most of the digital investigators have an advanced level of examination on HDD [13]. Even with the use of anti-forensic techniques such as data hiding, steganography, and VPN, examiners and investigator can still find an entrance to restore and recover data due to the CHS and LBA addresses. Those addresses point directly to the physical location of the data on the drive [9, 13]. When data is deleted by the operating system, data stays clean on the drive but the entry in the file allocation table or address is no longer used. Forensic tools such as FTK, EnCase, and Autopsy take the advantage of all these features of HDD to locate files and anomalous artifacts within any investigation process [9, 13].

There are two main problems when analyzing SSD during a forensic investigation. First is the combined controller of software and hardware in SSD. As known, the operating system issues commands to SSD during file management. In the case of SSD, the mimic controller blocks what is actually happening to the data stored in the NAND flash array [10]. This process of mimicry blocking is also known as the flash translation layer (FTL). The FTL is responsible for many tasks which must be done inside the flash memory to increase its lifespan [9, 13].

The second main problem is caused by the controller's FTL which is the garbage collection or the trimming. If TRIM is enabled, it will be a huge problem for forensic analysis because it gives an instruction to the SSD controller to transfer the healthy data out of the block that is going to be deleted and then gives permission to wipe out the entire block. This is done by setting all transistors to the same value. Therefore, recovering data after this operation is hard or even not possible [13, 19]. However, not all SSD manufacturers implement the trim drivers. This means there is still a chance which can find the deleted artifacts. In addition, in some cases, the data blocks that have been set to erasure using trimming before blocks are deleted; they return with zero values when the controller uses a protocol called Deterministic Zeros After Trim (DZAT) [9, 19].

The SSD controller also manages wear-leveling. This is because FTL will start transferring data at the same time of writing to all the flash blocks by SSD controller and this to support wear-leveling. However, the operating system is confused now and access to physical blocks is almost impossible without ATA commands [6, 9]. Bad blocks are vulnerable for forensics analysis. The block is marked bad when the SSD controller cannot write to it. Therefore, a bad block cannot be overwritten but can still be read [9]. According to [14], some SSDs implement data compression, and the controller's proprietary FTL does not know about it. Therefore, when an investigator tries to access the data directly, they might not get any useful data.

4 Experimentation Method and Setup

An image of 10 GB partition is established for both HDD and SSD drives using imaging tools such as FTK Imager, FeX Imager, EnCase Imager, ProDiscover Basic, OSForensic, and Autopsy. Experimentation using these tools is compared for similar and different hard drives. The test partitions run with Windows 10 using Dell Inspiron 5558 with 8 GB RAM and Intel Core i7 5500u processor. The hard drives have full storage of 250 GB storage. Both drives are connected to a laptop through SATA to a USB cable and to a USB 3.0 port. Table 1 illustrates the specification of the test hard drives.

US Department of Defense's DOD-5220.22-M (E) (3-passes) technique used as disk wipe program to ensure that all the files are deleted and cannot be recovered to start the forensic procedure. Disk Wipe is a program that works well on hard drives and supports all Windows standard file systems such as NTFS, FAT, and FAT32. Figure 4 illustrates the drive information and the erasing pattern using the Disk Wipe utility.

After hours of operation to complete, the hard drives are eventually verified by the hash. To continue the process, we added an exact copy of test files and media to both drives, and perform partial deletion to check if the acquired image could carve the files. Table 2 illustrates the hash time for different forensic tools. The established image was then used for three different scenarios to forensically measure the consistency and alteration of data using different hard drives. A separate test folder is established for each scenario as illustrated in Fig. 5.

Three test folders for normal deletion, quick and full format tests were created. Once an image is all set, some considerations like image type such as logical, physical, and destination type raw (dd) or E01 standard, and time factor need to be maintained.

Acquiring image data files is essential for any kind of forensic investigation. The image is bit by bit or copy of the original hard drive which was generated using different forensic imaging tools for experimentation such as FTK Imager, Fex Imager, and Encase imager.

4.1 Normal Deletion Scenario

The arranged test folders were used for testing normal deletion on the hard drives and a physical image with E01 format was created. Figure 6 illustrates sample caption for FTK-acquired hard drive image properties of the HDD and SSD once the creating of the disk image is complete for the normal deletion scenario. HDD image size is 6.72 GB, and the time stamp elapsed for taking the HDD image is 1 h and 30 min including the verification operation. The SSD image of size 420 MB is used for this scenario. The time elapsed for taking an SSD image is 42 min including the verification operation.

Table 1 Test drive specification

Specifications		
WD BLUE HDD 250 GB Model: WD2500BEVT Source: Market-based (used)		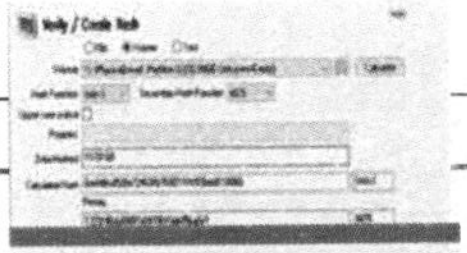
Kingston 240 GB A400 SATA 32.5" Internal SSD model: SA400S37/240G Source: Brand new (Market based)		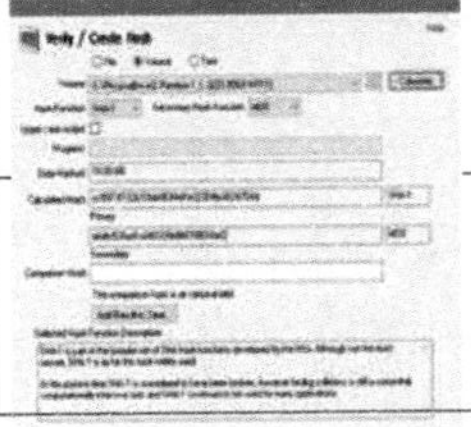

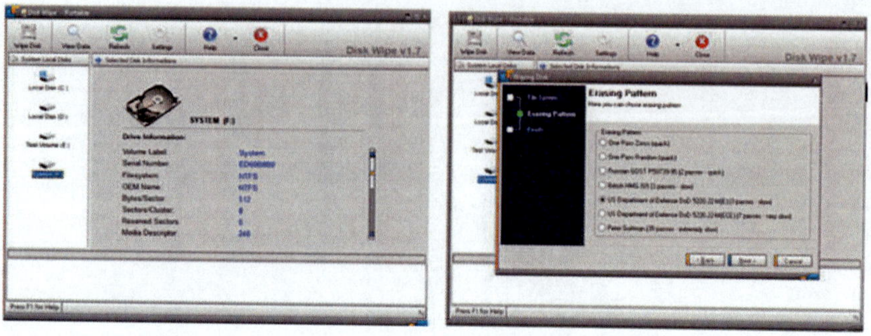

Fig. 4 HDD disk wipe

Table 2 Hash timing

Tool	HDD (Min.)	SSD (Min.)
FTK imager	47	16
FEX imager	50	14
OSForensics	48	16.5

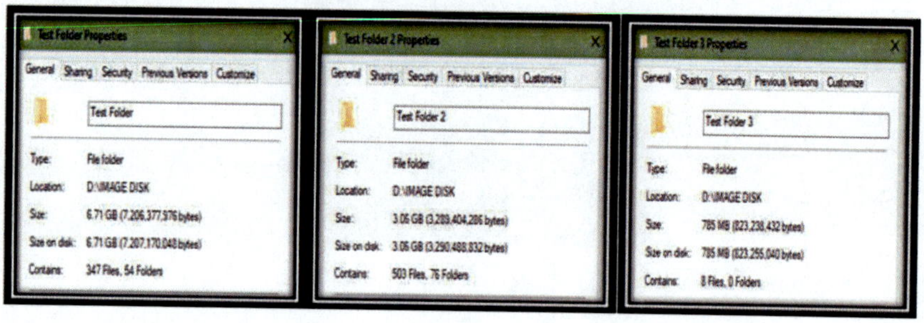

Fig. 5 Test folder properties

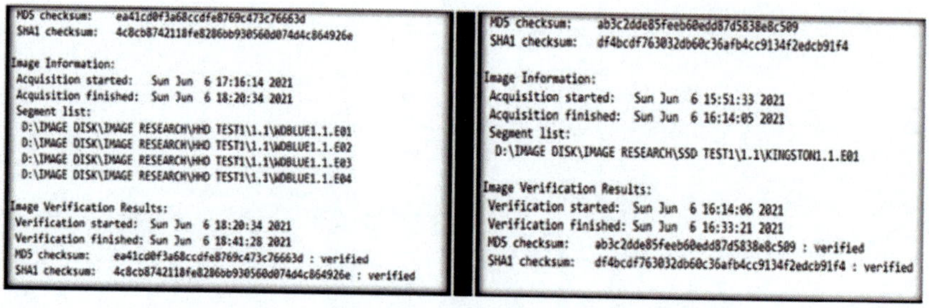

Fig. 6 Sample FTK-acquired images properties for HDD (left) and SSD (right)

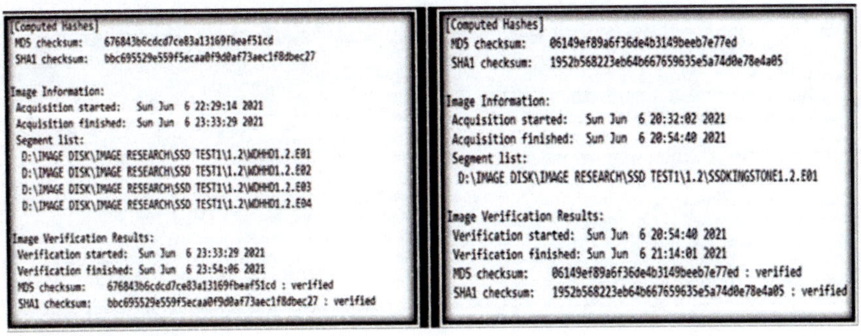

Fig. 7 Sample FTK-acquired images properties for HDD (left) and SSD (right)

4.2 *Quick Format Scenario*

The arranged test folder was used for testing the quick format on hard drives and a physical image with E01 format was created. Figure 7 illustrates sample caption for FTK-acquired hard drive images properties of the HDD and SSD once the creating of the disk image is complete for the quick format scenario. HDD image size is 6.39 GB, and the time stamp elapsed for taking the HDD image is 1 h and 22 min including the verification operation. The time elapsed for taking the SSD image is 42 min including the verification operation and the image size is 228.94 MB. The image properties demonstrate the computed hash value, image information such as the acquisition process start and finish times, and the verification results.

4.3 *Full Format Scenario*

The arranged test folder was used for testing full format on hard drives and a physical image with E01 format was created. Figure 8 illustrates sample caption for FTK-acquired hard drives image properties of the HDD and SSD once the creating of the disk image is complete for the full format scenario. The HDD image size is 437 MB, and the time stamp elapsed for taking the HDD image is 1 h and 14 min including the verification operation. The time elapsed for taking the SSD image is 45 min including the verification operation and the image size is 419 MB.

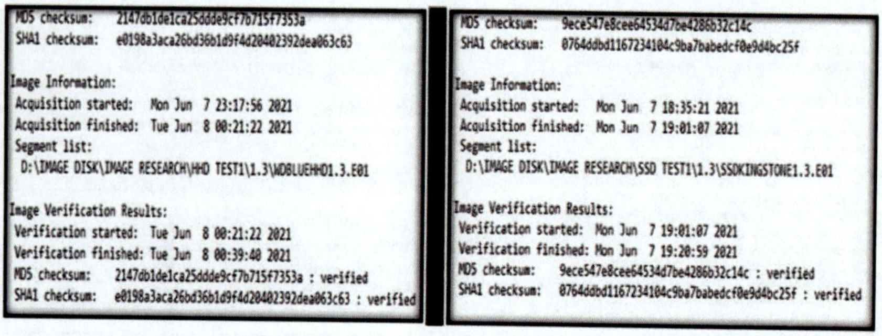

Fig. 8 Sample FTK-acquired images properties for HDD (left) and SSD (right) (full format scenario)

5 Results and Discussion

The hard drive contents for forensic preprocessing such as the creation test folders and images permit the postmortem analysis on the hard drives. Specialized forensic tools such as Autopsy and Encase were used to check on the curved and deleted items' recovery. It was noticed that the tools can poke on unallocated files, however, could not completely analyze those files.

During the image verification and validation process of original and similar data files from HDD and SSD drives, it can be clearly noticed that the hard drives epitomize slightly different results. This confirms that alteration on original files occurs during the evidence processing.

The conducted forensic analysis over the same image on HDD and SSD using Autopsy, FTK, and Encase forensic tools represents differences in the analysis time, percentage of the recovered data files, and more. Although the same image was used on HDD and SSD, the results conclude that data alteration occurs when using different hard drives and the anticipated evidence cannot be recovered completely. Figure 9 illustrates differences from two processed samples of data files recovered

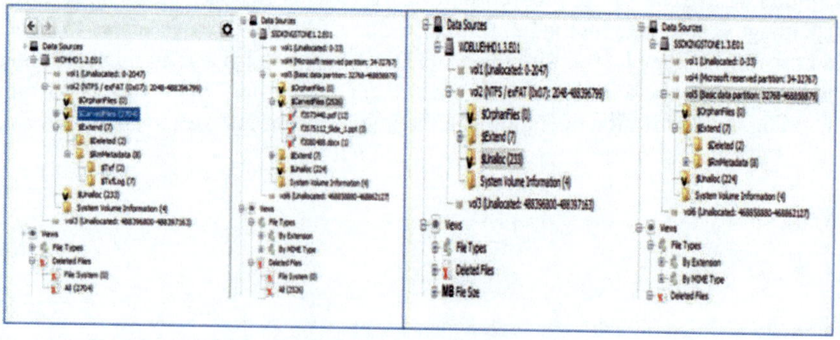

Fig. 9 Samples of recovered data files from quick and full format scenarios using autopsy

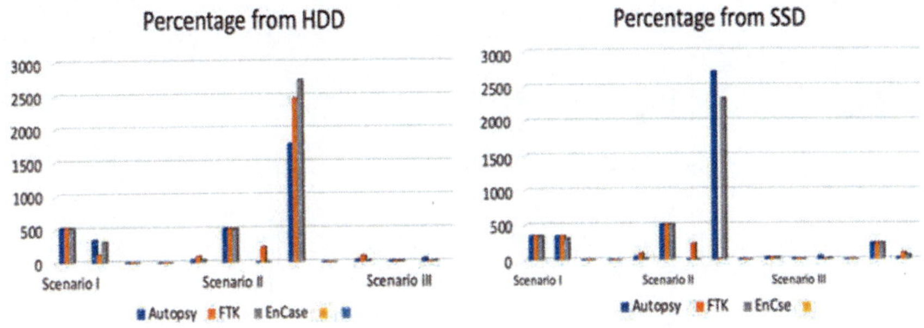

Fig. 10 SSD and HDD furnish different results once exposed to different forensic tools

from HDD and SSD (left and right) drives from one scenario to another. Similarly, Fig. 10 demonstrates differences in the percentage of data files recovered using different forensic tools.

Table 3 illustrates detailed sample results on each scenario from the Autopsy podium.

6 Conclusion Remarks

Despite the technical differences between HDD and SSD, they achieved a significant role for data and storage media. However, data recovery and analysis from a forensic perspective may not always disclose precise conclusions for similar data stored on such hard drives for different reasons. Deleted data on hard drives are not completely sponged out. Although operating systems have reference to each file on the hard drive, experimentation results disclosed incompetence to analyze scenarios with the full format on hard drives. As such, records from SSD deleted data cannot be recovered if the trim was enabled when the data was deleted, and significantly contribute to slow additional write operations to functional blocks. This makes the job more challenging for digital examiners and investigators giving cyber-criminals additional advantages.

Table 3 Autopsy results on each scenario

Disk type	Scenario	Number of deleted files	Number of recovered files	Number of carved files	Number unallocated files	Size of deleted files	Size of image used	Time required for image (Min.)	Analysis time (Min.)
HDD	Normal deletion	347	**347**	0	0	6.71 GB	6.72 GB	90	40
	Quick format	503	0	**2704**	0	3.06 GB	6.39 GB	82	30
	Full format	8	0	0	**233**	785 MB	437 MB	74	20
SSD	Normal deletion	347	**347**	0	0	6.71 GB	420 MB	42	20
	Quick format	503	0	**2526**	0	3.06 GB	229 MB	42	15
	Full format	8	0	0	**224**	785 MB	430 MB	45	12

Acknowledgements The authors wish to thank Palestine Technical University-Kadoorie (PTUK) for supporting this research work as part of the PTUK research fund.

References

1. Gibson, M., Medina, N., Nail, Z.: SSD forensics: evidence generation and analysis. Studies Big Data, pp. 203–218 (2019)
2. Geier, F.: The differences between SSD and HDD technology regarding forensic investigations. Linnaeus University (Sweden), pp. 1–67 (2015)
3. Bednar, P., Katos, V.: SSD: new challenges for digital forensics. lund university (Sweden), Oct., pp. 1–8 (2011)
4. Iqbal, M., Soewito, B.: Digital forensics on solid state drive (SSD) with TRIM feature enabled and deep freeze configuration using static forensic methods and ACPO framework. Int. J. Comput. Sci. Inf. Secur. (IJCSIS) **18**(11), (2020)
5. Bell, G., Boddington, R.: Solid state drives: the beginning of the end for current practice in digital forensic recovery? J. Digit. Forensics, Secur. Law **5**(3), (2010)
6. Vieyra, J., Scanlon, M.: Solid state drive forensics: where do we stand? University College Dublin (Ireland), Mar., pp. 1–16 (2018)
7. Gubanov, Y., Afonin, O.: Why SSD drives destroy court evidence, and what can be done about it. Belkasoft Ltd, pp. 1–10 (2013)
8. Yug, Y.G.: Why SSD drives destroy court evidence, and what can be done about it. Q 3 2012: State of the art in SSD forensics (2012)
9. Tom, B.: SSD vs. HDD: What's the difference? https://www.pcmag.com/news/ssd-vs-hdd-whats-the-difference
10. Micron: An introduction to NAND flash and how to design it in to your next product. TN -29–19: NAND Flash 101. https://user.eng.umd.edu/~blj/CS-590.26/micron-tn2919.pdf
11. Wiebe, J.: Forensic insight into solid stat drives. DFI News, CRU-DataPort/WeibeTech (2013)
12. Kumar, V.: Pros and cons of cassette player. Sooper Articles (2013)
13. Bonetti, G., Viglione, M., Frossi, A., Maggi, S., Zanero, S.: A comprehensive black-box methodology for testing the forensic characteristics of solid-state drives. In: Proceedings of the 29th Annual Computer Security Applications Conference (ACSAC '13). Association for Computing Machinery, New York, NY, USA, 269–278 (2013)
14. Nisbet, A., Lawrence, S., Ruff, M.: A forensic analysis and comparison of solid state drive data retention with trim enabled file systems. Edith Cowan University, Perth, Western Australia, SRI Security Research Institute (2013)
15. Cox, J., Bednar, P.: Potential difficulties during investigations due to solid state drive (SSD) technology. In: Cabitza F., Batini C., Magni M. (eds) Organizing for the Digital World. Lecture Notes in Information Systems and Organisation, vol 28. Springer, Cham. (2019) https://doi.org/10.1007/978-3-319-90503-7_7
16. Reddy, N.: Solid state device (SSD) forensics. In: Practical Cyber Forensics. Apress, Berkeley, CA. (2019) https://doi.org/10.1007/978-1-4842-4460-9_12
17. Shah, Z., Mahmood, A., Slay, J.: Forensic potentials of solid state drives, pp. 113–126. Lecture Notes of the Institute for Computer Sciences, Social Informatics and Telecommunications Engineering (2015)
18. King, C., Vidas, T.: Empirical analysis of solid state disk data retention when used with contemporary operating systems. Digit. Investig. **8**, S111–S117 (2011)
19. Afonin, O.: Life after trim: using factory access mode for imaging SSD drives (2019), https://blog.elcomsoft.com/2019/01/life-after-trim-using-factory-access-mode-for-imaging-ssd-drives/

An Efficient Novel for Soil Fertility Evaluation

Lokesh Surendra Jain, Bindu Garg, and Suraj Rasal

1 Introduction

For achieving gain selected objectives, fertility of the soil and nutrition of plant are very important and vital factor for plant development. Thus soil nutrition performs a critical role in plant systems. An element is taken into consideration fundamental if it's far needed for nutrition of plant and completing of the plant's way of life-cycle. Regularly, 17 elements are taken into consideration to satisfy the standards and they may be divided into macronutrients and micro-nutrients. The crucial macronutrients may be segregated into main macronutrients, consist of Phosphorus (P),Potassium (K), Nitrogen (N), and other macronutrients, Sulfur(S), Calcium (Ca), and Magnesium (Mg). The eight micro-nutrients are Nickel (Ni), Copper (Cu), Manganese (Mn), Iron (Fe), Chlorine (Cl), Boron (B), Zinc (Zn), and Molybdenum (Mo). Other sustenance variables might be significant for certain plants to profit crop quality or growth, even though no longer being important for metabolic techniques or crowning glory of the plant's lifestyles cycle frequently this factors are stated as useful factors. Beneficial elements consist of Sodium (Na), Cobalt (Co), Vanadium (V), Selenium (Se), and Silicon (Si). Silicon is available in fixations at the request for some of the large scale nutrients in bunch of plant tissues.

Soil is important parameter for the growth of plant. Soil includes the list of nutrition's which plays vital role in the plants life, health, and growth. Observing plants growth and analyzing its nutrition requirement is also key factor to maintain the plant health and growth. Sufficient and proper nutrients lead better growth of plant. But if plants did not receive sufficient nutritions, it may lead to severe plant problems like plant health issue, drying, and plant loss. This may lead for high level damage if we consider large area-based farm or agricultural area. Hence observing nutrition's

L. S. Jain (✉) · B. Garg · S. Rasal
Department of Computer Engineering, Bharati Vidyapeeth (Deemed University To Be) College of Engineering, Pune, Maharashtra, India

B. Agarwal et al. (eds.), *Proceedings of International Conference on Intelligent Cyber-Physical Systems*, Algorithms for Intelligent Systems,
https://doi.org/10.1007/978-981-16-7136-4_15

level of soil is very important parameter. Nowadays farming technology sick's the need of some technology which will help to grow the plant and farm with healthier environment.

This research paper emphasis on the maintaining the quality of soil and doing its analysis for better health and growth of the plant. Quality of soil is observed through the proposed architecture and technique. This research paper includes the list of proposed procedures through which soil-based information is fetched through smart way. After receiving soil based information, its analysis is done through proposed machine learning-based approach. Further, through proposed system, it predicts and suggests the required parameters to improve the health of soil. Another important aspect of this research paper is that same information exchange can be from any location. User can access or retrieve the information from any location across the globe using internet. Preferred technologies are Internet of Things and 5G network. The main focus of this research paper is proposing the technique to observe the quality of soil and suggesting the good parameters for plants healthier life. It consists of the list of proposed procedures through which proposed work is carried out. Initial step is retrieving soil nutrition's information using nutritions and PH sensor through smart way and without harming the soil and plant. Like traditional systems, hazardous chemicals are not used. This retrieved information is passed through the connected system to the sensor using internet to the server. Through the server, user can observe the retrieved information. Raspberry Pi is preferred server for circuitry connection and computing analysis which is acting as server also. Further, proposed machine learning approaches are applied for data analysis and recommendations. Important factor for this research paper is PH level of soil. By analysis of the PH values of soil, the weightage of micronutrients are observed and analyzed. More emphasis on some of the micronutrients is given like Nitrogen, Potassium, Phosphorous, Copper, Zinc, and Iron. Its detailed research is explained in the further points. Recommendations are given to the used in smart way to improve the quality of soil. Also, while recommending the nutrients parameters, the best qualitative nutritional products are suggested. This project can be accessed through the desktop-based systems or smartphone applications also. After applying the suggested changes for particular analyzed soil, its health improves in drastic manner which enhances the crops or plants health and production.

2 Literature Survey

Effectively overseeing soil supplements to give ideal plant sustenance has been a center act of farming creation all through mankind's history. For a long time, agrarian creation depended on the reusing of natural residuals, for example, compost and yield buildups. Industrialization and populace development of the nineteenth century requested expanded horticultural creation and corresponded with the advancement of financially delivered manures. Giving satisfactory treatment permitted the genuine capability of preparation to be acknowledged and mechanical creation of inorganic compost before long followed by advancement of the Haber–Bosch.

In a macrosense, the connection between the advancement of N and P composts reflects one of managing premises of soil richness unequivocally expressed as hypothesis of the base created by Sprengel and mainstream sized as 'Liebig's law' expresses that creation is constrained by the measure of the most restricted supplement comparative with the plant's required. In different words, regardless of how much amount N is included; yield could be constrained if P is inaccessible in enough amounts. When satisfactory P was given through preparation, after the improvement of P manures, at exactly that point could additional crop be cultivated by N fertilizer. The careful usage of the intelligent strategy, Sprengel and counterparts propelled the order of soil richness and plant nourishment to satisfy world developing need for nourishment. Most prominent commitment was understanding the mineral enhancements outside of plant that were needed by plant advancement. Perceiving fundamental segments and understanding that they ought to be offered external to plant was fundamental move in horticultural production. A few supplements were provided through reusing of harvest deposits and composts, and N could be given through N-fixing crops in turn in any case, most of the supplements were lost from the framework (because of framework wasteful aspects) and yield creation was constrained and farming terrains were frequently exhausted following a couple of long periods of the cropping. Daniel Webster comprehended the significance of protection of issue to this framework when he expressed it is upon this key thought of consistent creation without weariness, that the arrangement of development, and undoubtedly. The improvement of the ideas of soil ripeness and plant nourishment and the acknowledgment that supplements must be provided remotely through preparation to stay away from the 'fatigue' of the soil during the nineteenth century by another type of soil physicists and agronomists assisted with maintaining a strategic distance from the Malthusian Catastrophe. These standards under stuck the 'Green Revolution' of twentieth century. Somewhere in the range of 1960 and 2000, worldwide grain creation multiplied while worldwide N utilize expanded seven overlaps and P utilize expanded three and one-half overlay. In any case, this expansion underway and attendant increment in compost use have not come without cost. Thus, in spite of the fact that overseeing soil supplements for ideal plant sustenance and yield keeps on being a squeezing worry considering worldwide populace development, the administration of supplements in a productive way to ensure assets is a significant part of conversation of soil richness and plant nutrition.

During the 1980s, North American Author talk about and characterize another idea soil quality representing the numerous measurements (physical, compound, and natural) and capacities of soil. The principal meanings of soil quality were near those of manageable farming. Basically, preserving or improving soil quality is tied in with keeping up the drawn out components of soils, for instance, it is about reasonability. The ebb and flow most ordinary definition is 'Soil quality is the wellbeing of a specific sort of soil to work inside its current circumstance, support plant and creature efficiency, keep up or improve water and air quality, and backing human wellbeing and residence'. Complement is put on both inborn properties of soil ('a specific kind of soil') and dynamic intuitive cycles. Nowadays, a couple of creators in spite of everything fight that dirt nourishment and soil quality may be exchangeable; also, the wording stays comparative with the control or on the other hand the application segment.

3 Proposed System

Soil pH is an indication of acidity or alkalinity of soil and is estimated in pH units. pH is described as a bad logarithm of hydrogen particle concentration. The pH scale is going from zero to fourteen with pH 7 in view of the fair-minded point. As the quantity of hydrogen particles inside the soil increases the soil pH decreases thus becoming greater acidic. From pH 7 to 0 the soil is progressively more acidic and from pH 7 to 14 the soil is more and more alkaline or basic (Table 1).

Suitable soil is one that has excessive water keeping capacity but drains freely leaving air space. Water and vitamins in the sort of soil can be easily available to flora. A good soil might be barely acid (pH 6–6.8) at which stage the vitamins required by vegetation are maximum freely to be had. Topsoil will have all of the nutrients required for growth in the correct balance.

There are various types of soils are available that are as follows.

1. Sandy Soil

Sandy Soil is mild, heat, and tends to be in nature of acidic and less in vitamins. Sandy soil is fast to warm up in spring as compared to clay soils. Nerveless have a bent to dry get into the time of year and be afflicted by low nutrients which could be washed away by suggests that of rain. The addition of organic depends on will give plants an extra raise of nutrients with the help of enhancing the nutrient and water holding capability of the soil.

Mostly found in the western area of Rajasthan, Southern Haryana, the South-west part of Punjab, north-western parts of Gujarat, and along east and west coasts of India.

2. Clay Soil

Clay Soil may be a significant soil kind that edges from excessive value of nutrients. Preserve Associate in nursing extreme quantity of water. Channel of Clay soils generally take more time to warm up in summer. Also it blended in with drying out and converted into broken form in summer.

Table 1 pH range

pH range	Nature (Acidic/Alkaline)
0–4	Strongly acidic
4–6	Weakly acidic
6–7	Neutral
7–8.5	Weakly alkaline
8.5–14	Strongly alkaline

Usually found in Maharashtra, Madhya Pradesh, Chhattisgarh states of India.

3. Loam/Loamy Soil

Loam soil is combination of clay, Silt, and sand that area unit combined to avoid the adverse consequence of each kind of soils.

Nutrient deficiency and its effect

1. Calcium—Tender leaves gets dry and plant dies.
2. Magnesium—leaves Start drying from tip.
3. Sulfur/Sulfur—Brownish spots on leaf.
4. Iron—veins of leaves becomes green.
5. Manganese—Younger leaves shows chlorotic spot on veins.
6. Copper—leaves become yellow and fall off.
7. Boron—leaves become petiole and start folding.
8. Nitrogen—stems of your plants will turn purple or reddish.

Nutrient required for plant growth

The factors required for plant boom fall into three groups.

1. Major elements—required in enormous quantities—potassium (K), nitrogen (N), phosphorus (P)
2. Minor factors—sulfur (S), iron (Fe), magnesium (Mg), calcium (Ca)
3. Trace elements—manganese (Mn), copper (Cu), molybdenum (Mo), Selenium (Se), Aluminum (Al), boron (B), zinc (Zn).

While all those factors are normally gift in maximum soils, several are often gift in inadequate amount to enable a satisfactory yield.

Most contemporary compound/combined fertilizers include tiers of maximum or all of the above in stability that suits the boom of most plants. Hence those compound fertilizers are the favored nutrient source. There may be a few florae and a few situations or boom tiers where more amount of a selected element may want to be supplemented, e.g—extra potassium to provoke flowering or fruiting, or to provide disorder resistance, chelated aluminum sulfate to make certain blue hydrangeas.

Acid soils with a pH of under 6 typically have deficiencies in

- Potassium
- Phosphorus
- Calcium
- Molybdenum
- Magnesium

Acid soils with a pH of significantly less than 4 generally have harmful amounts of:

- Aluminum and Manganese

Alkaline soils with a pH of additional than 7 the following nutrients are regularly absurd:

- Zinc
- Boron
- Copper
- Iron
- Manganese.

Raising Soil pH of Soil

Soil pH can be raised in different types:

1. Limestone (Calcium Carbonate)

The most ordinarily used approach to elevate the soil pH scale is applying carbonate or agricultural lime. If the soil is simply too acidic, then agricultural rock (calcium carbonate) must be applied. The number needed can vary hoping on the pH scale and therefore the soil sort. The solubility of lime is genuinely low, in this manner if its miles applied handiest to the soil surface, it ordinarily influences exclusively the most noteworthy layer of the soil, not a scope of centimeters down.

The adequacy of a liming material is straightforwardly identified with its virtue. Immaculateness of a liming material could be a component of the carbonate equivalent (CCE). It's a degree to the measure of acid that the liming material can neutralize, when contrasted with regular carbonate ($CaCO_3$).

The higher the calcium carbonate identicalness, the more remarkable the item in neutralizing acidity (Table 2).

2. Potassium Carbonate

Potassium carbonate is especially dissolvable as are regularly applied by means of trickle water system. In light of its inordinate dissolvability, carbonate are often just distributed for the duration of the premise space put along with water system water and arrive at more profound soil. It will speedily have an effect on chemical reactions inside the foundation sector; as a result, elevate root sector pH.

Applying potassium carbonate frequently as part of the fertilization will forestall the pH drop. Potassium carbonate also contributes potassium to the nutrient content material of the irrigation water.

Table 2 Limestone (in Pounds) required increasing pH (per 1000 square Feet)

pH	Pounds number		
	Soil_ sandy	Soil_ loam	Soil_clay
4–6.5	60	161	230
4.5–6.5	50	130	190
5.0–6.5	40	100	150
5.5–6.5	30	80	100
6.0–6.5	15	40	60

Table 3 Fresh nutrient content of cow dung

Organic matter_value	Nitrogen_value	Phosphorus_value	Potassium_value	pH_value	Moisture Content_value
14.5%	0.30–0.45%	0.15–0.25%	0.10–0.15%	6.0–7.5	80–90%

Table 4 Nutrition content of solid cow dung

Organic matter_value	Nitrogen_value	Phosphorus_value	Potassium_value	Calcium_value	Moisture Content_value
20%	0.32%	0.14%	0.30%	0.40	77%

When making use of potassium carbonate through the irrigation water, it's far important to hold the pH below 7 in order to avoid emitter clogging.

Sometimes growers need to extend the buffer capacity of the irrigation water, at the same time as keeping pH degrees low sufficient. In this case, it's far possible to add potassium carbonate to water, and an equal time to acidify the water. The acid will neutralize a number of the carbonate ions, at the same time as the pH degree will still be low sufficient to save you emitter clogging.

3. Cow Dung

It is found that Cow dung is a best supply of natural fertilizer. Cattle fertilizer is essentially obtained from grain and grass. Cow Dung constituents of 1% potassium, 2% of Phosphorus, 3% nitrogen. Subsequent to convert into natural fertilizer it could act as an unique role in soil, that facilitates to increase fertility of the soil, natural count number, soil bodily, and environmental behavior of microbial (Tables 3 and 4).

Lowering Soil pH of Soil

1. Sulfur

Sulfur gets converted to sulfuric acid with the assistance of soil microscopic organisms yet it needs some time. The change charge of sulfur relies on the fineness of sulfur, measure of soil moisture, soil temperature, and furthermore the presence of the microorganisms. These elements, the transformation rates of sulfur, are often terribly gradual and take numerous months if the conditions don't seem to be ideal (Table 5).

2. Aluminum Sulfate

Aluminum sulfate can alternate the pH scale because aluminum produces the acidity because it soluble within soil. This compound immediately makes the soil extra acidic because of a chemical reaction related to aluminum. It alters the pH of the soil so quickly, it is able to be greater tough to control the soil acidity (Table 6).

Table 5 Sulfur (in Pounds) needed to lowering the Soil pH (per 10 square feet)

Present pH	Desired pH				
	6.5	6.0	5.5	5.0	4.5
8.0	0.3	0.4	0.5	0.6	0.7
7.5	0.2	0.3	0.4	0.5	0.6
7.0	0.1	0.2	0.3	0.4	0.5
6.5		0.1	0.2	0.3	0.4
6.0			0.1	0.2	0.3

Table 6 Pounds of aluminum sulfate needed to lowering the pH (per 10 square feet)

Present pH	Desired pH				
	6.5	6.0	5.5	5.0	4.5
8.0	1.8	2.4	3.3	4.2	4.8
7.5	1.2	2.1	2.7	3.6	4.2
7.0	0.6	1.2	2.1	3.0	3.6
6.5		0.6	1.5	2.4	2.7
6.0			0.6	1.5	2.1

3. Gypsum/Calcium Sulfate

Gypsum (Calcium Sulfate, $CaSO_4.\ 2H_2O$) is a sedimentary mineral. It is very crucial for the treatment of alkaline soil. It is a tremendous supply of Sulfur for plant nutrients and enhancing crop yield. It can lessen Aluminum toxicity specifically in the subsoil. It allows in reducing runoff and erosion by retaining Phosphorous and other nutrients from the soil. Gypsum replaces Sodium and drained descending and out of achieving of plant roots and it may be carried out a supply of $Ca++$ ions to replace the Sodium on the exchange complex within the soil. While using Gypsum, there has to be sufficient natural drainage to the underground, in any other case artificial subsurface drainage systems have to be present.

A. **Equipment Used**

1. Raspberry PI—It is used to connect all the Sensors.
2. Temperature Sensor (DHT22)—The probes are inserted in the farms to induce the temperature of the soil. It can measure temperature from –40 °C to 80 °C.
3. Moisture Sensor—Moisture sensor is applied to measure dampness present in the soil and will comprehend the water within area to regulate water deliver for plants.
4. Analog to Digital converter (ADS1115)—Converting the pH sensor analog signal to digital signal. It includes a programmable gain amplifier to boost up smaller single/differential signals to the full range.

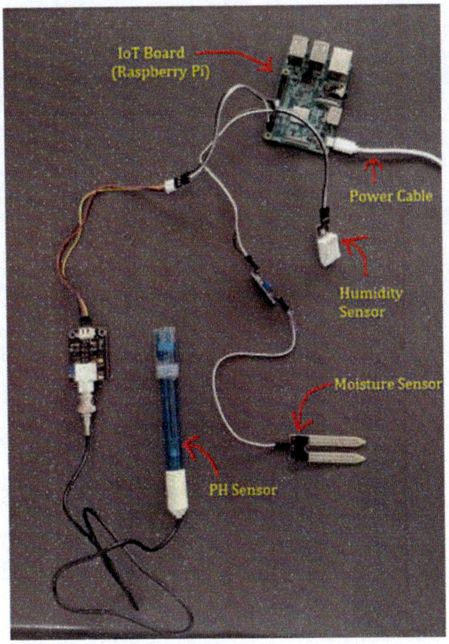

Fig. 1 Circuit diagram

5. pH Sensor—pH meter, electrical instrument, is utilized to degree hydrogen particle inside the soil. A pH meter depends on voltage take glance to determine hydrogen ion. It's accustomed decide the pH of soil.

B. **Development and Implementation**

Programming Languages Used (Fig. 1)

1. HTML
2. JSP
3. Python
4. MySQL
5. Android

Server Used

1. Apache Tomcat

The pH sensor will connect with ADS1115 and other sensors like Humidity and Temperature (DHT22). Moisture Sensor can be directly connected with the Raspberry PI as the output of this sensors is Digital signals and the output of pH sensor is analog signal so it converts this analog signal into digital signal using ADS1115. After getting this input from sensors it will be further processed by raspberry pi using some algorithms. The processed result can be viewed using website or Android Application. The result can be downloaded in PDF format which will have Recommendation and Suggestion regarding the Soil Quality and Fertility.

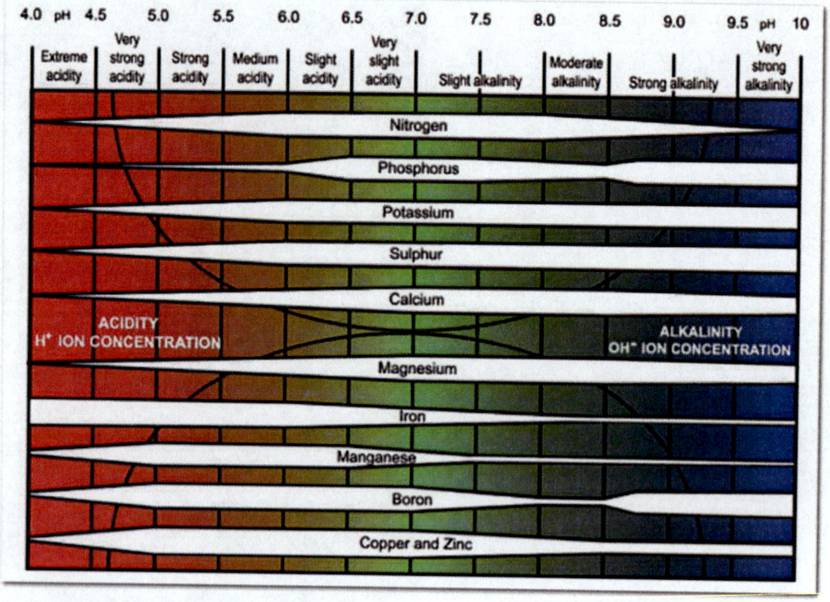

Fig. 2 Nutrition availability and Soil pH

4 Result

Nutritional content of soil pH

The presented concept tends to become acceptable manure to advise farmers to grow their yield in order so that there will be no cause of ruination the crop and land it also helps in enhance the nutrition of soil and crops production (Fig. 2).

Based on above Chart the approximate values are generated by using GetData Graph Digitizer software. The GetData Graph Digitizer software is used for digitizing graphs and plots. It's necessary to get original (x, y) data from graphs. It is used when data values are not available (Table 7).

5 Discussion

The study demonstrates Soil assumes in the field of agriculture, atmosphere, and yield developed during earlier years. According to the availability of nutrients, suggestions for developing the specific yield and appropriate fertilizer will be given. Using classification algorithm, prediction of appropriate crops supported the values we have a tendency to get from our device which we tend to additionally can provide acceptable fertilizers required for that land. The prediction of the crop can be done by

Table 7 Soil pH and its nutritional content

pH	Nitrogen	Phosphorus	Potassium	Sulfur	Calcium	Magnesium	Iron	Manganese	Boron	Zinc and copper	Molybdenum
4	15	36	29	29	32	25	100	50	36	30	23
4.5	30	36	33	52	39	40	100	71	64	64	44
5	53	40	52	71	54	52	100	83	80	87	53
5.5	71	47	77	83	68	66	100	92	91	100	64
6	85	65	92	94	79	78	91	100	95	100	100
6.5	85–90	90	100	100	89	85	91	100	84	83	100
7	100	100	100	100	100	100	87	94	76	70	100
7.5	100	89	100	100	100	100	82	87	60	60	100
8	100	75	100	100	91	89	78	83	44	50	100
8.5	89	53	100	100	82	81	69	75	40	43	100
9	78	95	100	100	68	66	60	63	80	38	100
9.5	57	95	100	100	53	52	55	54	80	25	100
10	21	95	100	100	39	40	44	38	80	5	100

Soil pH which is measured of the acidity and alkalinity in soils. The Macronutrients (Potassium, Nitrogen, and Phosphorous) and Micronutrients (Zinc, Iron, and Copper) are essentials for healthy plant development. Soil pH could be an important parameter for crop efficiency. Soil pH influences the soils chemical, physical, and natural properties and therefore plant growth will increase.

The objective of the experiment is to provide an Embedded-based soil analyzer that can be created with the brisk and dependable computerized framework which is used to analyze different soil nutrients with the assistance of pH value. Soil pH could be a proportion of Hydronium particle (H+) concentration historically tested in labs to determine what quantity chemical to use to the sector. Various Data of all types of crops are collected like pH value of crop, min–max temperature of the soil, and a database is formed. With the help of all the databases, predictions will be made of the required soil. The device which focuses on increasing the crop productivity at a lesser cost where the system can facilitate help in cultivating by giving important idea identifying with the yields and all fundamental data.

6 Conclusion

Fertility of soil and nutrition of plant cover the executives of soil condition to give the fundamental nutrition within necessary adds-up to plants for ideal performance. Basic nutrients are those elements that assume an important job in plant development, advancement, multiplication, or metabolic capacities. Every nutrient is needed into specific focus in plant tissue, beneath that typical plant capacities are restricted. Along these lines, soil nutrition analysis to specify the quantity of obtainable important nutrients present in fertility of soil and nutrition of plant management program. Various factors on the far side simply nutrient concentration within soil impact potential of soil to provide fertilizer to the plant and furthermore the ability of plant to need up and use those fertilizer. The substance, natural, and actual soil characteristics and procedures impact plant Nutrients use rather like other environmental factors like pressure, environment, and management of crop practices. Consequently, soil richness and plant sustenance that coordinates everything of soil and yields the executive's orders to give ideal nutrient offer to plants for a unique objective (e.g., plant food used for nourishment, vitality) while ensuring common assets and natural quality.

References

1. Epstein.: The anomaly of siliconin plant biology. In: Proceedings of the National Academy of Sciences of the United States of America vol. 91, pp. 11–17 (1994)
2. Epstein, E., Bloom, A. J.: Mineral nutrition of plants: principles and perspectives. Sunderland, MA: Sinauer Associates. (2005). https://doi.org/10.2307/2484208

3. Jungk, A. CarlSprengel.: The founder of agricultural chemistry: a reappraisal commemorating the 150th anniversary of his death. *Journal of Plant Nutrition and Soil Science*, **172**, 633–636 (2009)
4. Erisman, J. W., Sutton, M. A., Galloway, J., Klimont, Z., Winiwarter, W.: How a century of ammonia synthesis changed the world. *Nature Geoscience*, **1**, 636–639 (2008)
5. Dawson, C. J., Hilton, J.: Fertilizer availability in a resource–limited world: Production and recycling of nitrogen and phosphorus. *Food Policy*, **36**, S14–S22 (2011)
6. Webster, D., Everett, E.: The works of Daniel Webster biographical memoir [by Edward Everet] and speeches on various occasions. C.C. Little and J. Brown, Boston (1851)
7. Tilman, D., Cassman, K. G., Matson, P. A., Naylor, R., Polasky, S.: Agricultural sustainability and intensive production practices. *Nature*, **418**, 671–677 (2002)
8. Parkin T. B, Doran J.W.: Defining and assessing soil quality. Pp. 3–22 in 'Defining soil quality for a sustainable environment', ed. by J.W. Doran and A.J. Jones. SSSA Special Publication No. 35. Soil Science Society of America: Madison, WI (1994)
9. Patzel N., Sticher H. and Karlen D.L.: Soil fertility—phenomenon and concept. Journal of Plant Nutrition and Soil Science—Zeitschrift fur Pflanzenernahrungünd Bodenkunde 163(2), pp. 129–142 (2000)
10. Karlen, D.L., Ditzler, C.A., Andrews, S.S.: Soil quality: why and how? Geoderma **114**(3–4), 145–156 (2003)
11. Doran J.W., Parkin T.B., Jones A.: Quantitative indicators of soil quality: a minimum data set. Pp. 25–38 in 'Methods for assessing soil quality', ed. by J.W. Doran, D.C. Coleman, D.F. Bezdicek and B.A. Stewart. SSSA Special Publication No. 49. Soil Science Society of America: Madison, WI (1996)
12. Moebius B.N., van Es H.M., Schindelbeck R.R., Idowu O.J., Clune D.J. Thies J.E.: Evaluation of laboratory-measured soil properties as indicators of soil physical quality. Soil Science 172(11), pp. 895–912. https://doi.org/10.1097/ss.0b013e318154b520 (2007)
13. Moebius-Clune B., Idowu O., Schindelbeck R., van Es H., Wolfe D., Abawi G. et al.: Developing standard protocols for soil quality monitoring and assessment. Pg. 833–842 in 'Innovations as key to the green revolution in Africa: exploring the scientific facts', ed. by A. Botiono, B. Waswa, J.M. Okeyo, F. Maina and J.M. Kihara. Springer: Dordrecht, Netherlands (2011)
14. T. Venkat Narayana Rao.: Prediction Of Soil Quality Using Machine Learning Techniques. ISSN 2277–8616, pp. 2–5 (2019)
15. Jay Gholap, Anurag Ingole, Jayesh Gohil, Shailesh GarGade, Vahida Attar. Dept. of Computer Engineering And IT, College of Engineering, Pune, Maharashtra-411005, India. Soil Data Analysis Using Classification Techniques and Soil Attribute Prediction. arXiv: 1206.1557, pp. 2–4 (2012)
16. Ansif Arooj, Mohsin Riaz, Malik Naeem Akram.: Evaluation of predictive data mining algorithms in soil data classification for optimized crop recommendation. https://doi.org/10.1109/ICACS.2018.8333275. (2018)
17. Emrullah ACAR, Mehmet Sirac OZERDEM, Burak Berk USTUNDAG. Machine Learning based Regression Model for Prediction of Soil Surface Humidity over Moderately Vegetated Fields. https://doi.org/10.1109/Agro-Geoinformatics.2019.8820461, pp. 3–5 (2019)
18. Yang XiaoXia, Zhang Chengming.: A soil moisture prediction algorithm base on improved BP. https://doi.org/10.1109/Agro-Geoinformatics.2016.7577668 (2016)
19. Bindu Garg, Shubham Aggarwal, Jatin Sokhal.: Crop yield forecasting using fuzzy logic and regression model R. https://doi.org/10.1016/j.compeleceng.2017.11.015, pp. 384–401 (2018)
20. Bindu Garg, Tanya Sah.: Prediction of Crop Yield Using Fuzzy-Neural System. https://doi.org/10.1007/978-3-030-19562-5_21, pp. 213–220 (2020)
21. Lokesh Surendra Jain, Anirban Paul, Bindu Garg, Suraj Rasal. Embedded Model for predicting quality of Soil. https://doi.org/10.1007/978-981-33-4073-2_40, pp. 439 – 450 (2020). Available From: https://link.springer.com/chapter/10.1007/978-981-33-4073-2_40
22. McGrath, J.M., Spargo, J., Penn, C.J.: Soil Fertility and Plant Nutrition. Elsevier BV (2014). https://doi.org/10.1016/B978-0-444-52512-3.00249-7

Exploration of Demographic Factors that Proliferated COVID-19

Md Shadab Warsi, Imran Hussain, Aqeel Khalique, and Sherin Zafar

1 Introduction

Information visualization has continuously played a significant part within the logical examination and derivation of data. The explanatory visualization of information has continuously been given noteworthiness, and it can be followed from John Snow's examination of the cholera plague in 1854 [1]. By doing such analysis, the analyst was able to urge the center information to discover out from crude numbers. The later COVID-19 widespread postures modern challenges, for its exponential development and in like manner colossal financial effect [14]. This research study has utilized charts to speak to the continuous spread of the disease over different nations. This research work points to analyzing the information with the assistance of a direct relapse calculation and utilizing Python tools to portray and bring out a result by comparing the COVID-19 episode for different nations around the world. We have analyzed the dataset collected from WHO and different open Web assets. We visualized this dataset utilizing information visualization apparatuses to foresee the different parameters for the COVID-19 breakthrough. The upcoming sections of this paper will highlight upon related work in Sect. 3, methodology in Sect. 4, and Conclusion in Sect. 5. References are listed after that.

M. S. Warsi · I. Hussain (✉) · A. Khalique · S. Zafar
Department of CSE, SEST, Jamia Hamdard, New Delhi, India

S. Zafar
e-mail: sherin.zafar@jamiahamdard.ac.in

B. Agarwal et al. (eds.), *Proceedings of International Conference on Intelligent Cyber-Physical Systems*, Algorithms for Intelligent Systems,
https://doi.org/10.1007/978-981-16-7136-4_16

2 Related Work

Troublesome circumstances like a worldwide spread can be controlled with the assistance of strength and flexibility by the individuals. The exceptional proverb of this investigation educates us as the world inundates by a few unanticipated crises, appropriately, the mindfulness of unused methods and apparatuses take the shape to foil it. All things considered, the world has confronted numerous pandemics; in any case, the spread of the COVID-19 being phenomenal in history in this way makes it diverse from other ones. In the later situation, COVID-19 estimation and analysis are a few of the foremost sought-after topics that are being investigated in much detail. A few partners evaluated the harms and conditions within the chosen time outline through distinctive approaches. Kourba [1] compared the seriousness of the COVID-19 episode with other plagues, viz., Ebola 2014, MERS 2012, and SARS 2003. He watched that the patterns of other scourges were tall at the starting, but it begins appearing as a decay after 2 months.

To defeat the pandemic by utilizing the current innovation, numerous clarifications and arrangements have been given by the specialists to get reasonable results. Khadilkar [2], Arias [3], and Karim [4] have applied control measures and imaginative innovations like Artificial knowledge and AI to restrict the impacts of the COVID-19 pandemic. Khadilkar et al. [2] have expounded the ascent of contamination rates during the lockdown and without lockdown by utilizing instruments like support learning and Arias [3] has utilized AI model and bend fitting to assess COVID-19 disease-specific regions while Karim [4] considered COVID-19 patients with the assistance of Chest X-beam pictures and propose an AI-help application that contains Deep Neural Network (DNN) in view of the programmed discovery of COVID-19 side effects followed by featuring class-separating locales utilizing slope directing strategy. Nanning Zheng et al. [5] have proposed a mixture of man-made consciousness (AI) models for COVID-19 expectation.

Test discoveries on pestilence information from a few conventional Chinese territories and urban areas show that individuals with Covid have a higher pace of disease in the third to eighth day after contamination, which is more in accordance with the plague's real transmission rules. The model's forecast results are firmly steady with genuine plague occasions, which shows that the proposed half-breed model can investigate the proliferation law and improvement example of the infection even more precisely comparated with past models, and pertinent news can additionally support the expectation model's exactness. Moreover, they have found an effective device for gauging the law of transmission just as the pattern of advancement of likely general wellbeing. Likewise, Ghosh et al. [6] have created three development models to anticipate tainted individuals throughout the following 30 days. The models like the remarkable, strategic, and SIS are the pieces of their investigation alongside the everyday pace of disease (DIR). They have dissected the outcomes all things considered from all recreations as opposed to doing it separately. They have conjectured the DIR to be zero or negative to build up the way that the COVID-19 spread has been choked. Indeed, even a little certain DIR (say 0.01) demonstrates that the infection multiplies locally. As far as having the option to pronounce a finish to the pandemic, DIR should get zero or negative for 14 days straight.

Muhammad et al. [7] have broken down the connection between the seriousness of COVID-19 and region-explicit climatic boundaries. Considering their discoveries, they inferred that the nations, which are in the low-temperature zone, have gotten a quick expansion in the COVID-19 cases in contrast with the nations situated in generally hotter climatic areas regardless of their better financial conditions. A relationship between meteorological boundaries and COVID-19 cases was likewise investigated. Normal sunlight is related to the all-out COVID-19 cases with a coefficient of assurance of 0.42, while normal high-temperature shows a connection of 0.59 and 0.42 with complete COVID-19 cases and passing cases individually. The normal temperature and sunlight hours have shown a positive connection toward the spread pace of COVID-19. Thus, territorial meteorological boundaries (mist concentrates, greatest and least temperature, day length, and so forth) are among the supporters of the quick spread of Covid in many nations of the world. Their investigation gives proof that the savagery of COVID-19 cases becomes weak when the nearby climate conditions become hotter. A large portion of the Asian nations, for example, Vietnam, Laos, Pakistan, India, and so forth, which fall under the heat and humidity, have a high temperature during May–June, bringing about a lesser number of influenced cases per million populaces. Their examination focuses on the following estimates, for example, social removing, hand washing, and long summer that help to break the chain of the COVID-19 pandemic. This investigation is valuable for policymakers and wellbeing offices in breaking the chain and understanding the quick spread of the pandemic. The outcome will likewise help the global wellbeing associations and neighborhood organizations to battle against the spread of COVID-19.

3 Methodology

For this examination, we have gathered information from 214 nations to investigate all potential reasons which can affect the spread of the COVID-19 episode around the world. We have gathered information from different open-source stages like WHO, World Population Review, WorldMeters.Info, World Climate Guide, Population Pyramid of World [8], and so forth; the information hotspots for the various boundaries utilized in our investigation are referenced in Table 1.

3.1 *Demographic Parameters*

The different important terminologies and parameters used in this work are explained below. Control parameters are factors that affect the COVID-19 outbreak worldwide. We have included 6 parameters for our study and analysis.

Table 1 Data sources for parameters used in the present study

Parameter	Data source
Median age	https://en.wikipedia.org/wiki/List_of_countries_by_median_age
Average temperature	https://www.climatestotravel.com/
Population density	https://worldpopulationreview.com/countries
BMI	https://en.wikipedia.org/wiki/List_of_countries_by_body_mass_index
Total Covid and death cases	https://www.worldometers.info/coronavirus/countries-where-coronavirus-has-spread/
Total population	https://www.worldometers.info/world-population/population-by-country/

1. **Body Mass Index (BMI)**

   The Body Mass Index of a person is measured as the weight of the person in kilograms divided by the square of height in meters. A high BMI is an indication of high body fat. BMI is an important tool to analyze body weight, and accordingly health issues [9]. The BMI is a simple and easy-to-understand tool to know a person's underweight, normal weight, overweight, or obeseness based on tissue mass (muscle, fat, and bone) and height. The general accepted BMI ranges are underweight (under 18.5 kg/m2), normal weight (18.5–25), overweight (25–30), and obeseness (over 30) [10]. BMIs under 20 and over 25 are associated with severe health complications and mortality too [10].

2. **Population Density**
   Population density is one of the important geographical tools, measured as the number of people residing in an area per square Kilometer. It is a measurement of population density per unit area. It is generally applied to living beings, especially human beings [11].

3. **Death Percentage**

$$Death\ Percentage = \left(\frac{No.\ of\ Death(s)}{Total\ No.\ of\ Infected\ Cases}\right) * 100\% \quad (1)$$

   In simple terms, it is the probability of death after getting infected.

4. **Infection Percentage**

$$Infection\ Percentage = \left(\frac{Total\ Infected\ Cases}{Total\ Population}\right) * 100\% \quad (2)$$

In simple terms, it is the percentage of people who got infected out of the total population.

5. **Median Age**
Middle age is characterized as the age that isolates a specific populace into two mathematically similarly measured gatherings. It could be identified as a large portion of individuals are more youthful in this age and half are more seasoned. It is the single record that features the age circulation of a populace [12].

6. **Average Temperature**
The normal yearly temperature is estimated by taking the normal of the base and greatest everyday temperatures in a country during the period 1961–1990 [13]. It depends on gridded climatology, done by the Climatic Research Unit in 2011.

4 Results and Discussion

In the current examination, results have been gotten through the careful cycle directly from the start as far as possible. A dataset has been built by gathering information identified with different boundaries. This dataset has been exposed to the thorough interaction of information cleaning. We have cleaned the information utilizing NumPy. Subsequent to acquiring the cleaned dataset, we eliminated the exceptions and afterward standardized the information, to get more exact outcomes out of the dataset. The standardized dataset is then exposed to Linear Regression. At last, we have applied Linear Regression to break down our outcomes and plotted the diagram for distinct examination. The investigation is completed utilizing Python language. Its libraries like NumPy end up being valuable in information cleaning (Fig. 1).

In this examination, information of 214 nations with a sum of 71,503,614 affirmed cases and 1,612,833 passing cases has been investigated as announced until December 2020. The USA, India, and Brazil were the nations that announced the greatest number of tainted and passing cases. Our perceptions and inductions are talked about in the accompanying sub-headings.

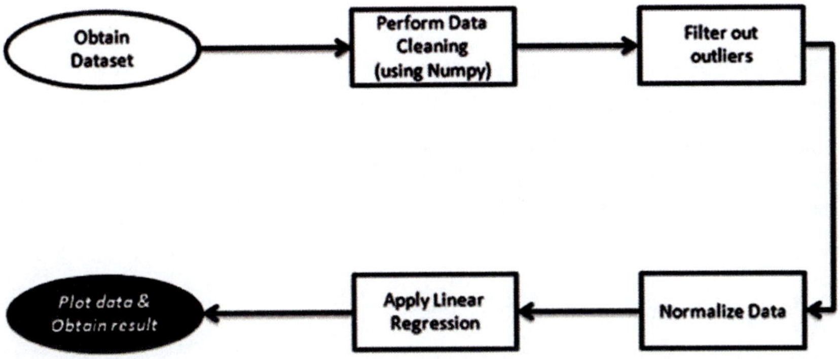

Fig. 1 Overview of methodology

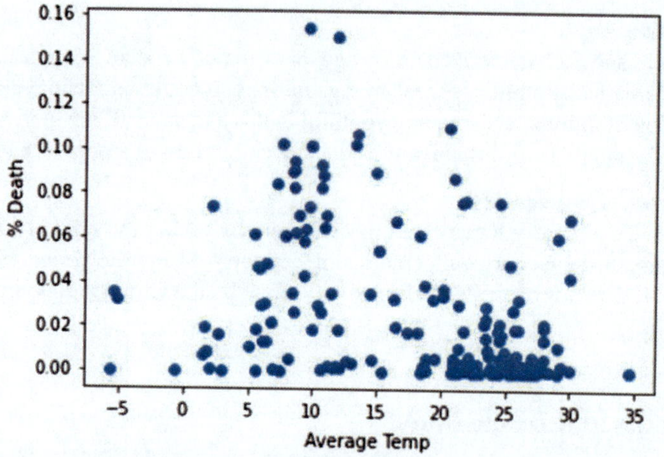

Fig. 2 Graph showing deaths versus average temperature

4.1 Deaths Versus Average Temperature

Figure 2 shows that Burkina Faso's infected percentage is (0.0189) with an average temperature (28.29 °C). Benin's infected percentage is (0.0248) with an average temperature (27.55 °C). Andorra has (9.486) infected percent with average temperature (7.6 °C) while in the case of Georgia, it is (4.8%) with an average temperature (5.8 °C).

These results suggest that an increase in COVID-19 deaths is being experienced more where the average temperature is lower, whereas its incidence is lower wherein the average temperature is higher.

The initial conclusion drawn from these observations is that the higher the temperature the lower the infection rate (lesser deaths), and the lower the temperature the higher the infection rate (higher deaths).

4.2 Death Versus Median Age

Figure 3 observed that the countries with higher median age, witnessing higher infection rates (more deaths) as compared to the countries, have lower median age. This assumption will help us to strategize better to fight against COVID-19.

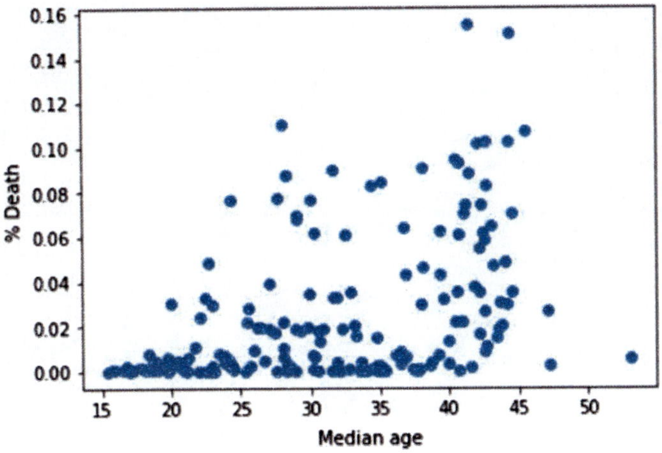

Fig. 3 Graph showing deaths versus median age

4.3 *Death Versus Population Density*

Figure 4 shows that Indonesia's Population Density is 151 people per km^2 while Mauritania has a Population Density of 5 persons per km^2. These two countries have vastly different population densities but still, their infection rates are almost similar at 0.224 and 0.226%, respectively. Similarly, Nicaragua's Population Density is 54 persons per km^2 whereas Saint Vincent and the Grenadines have a Population

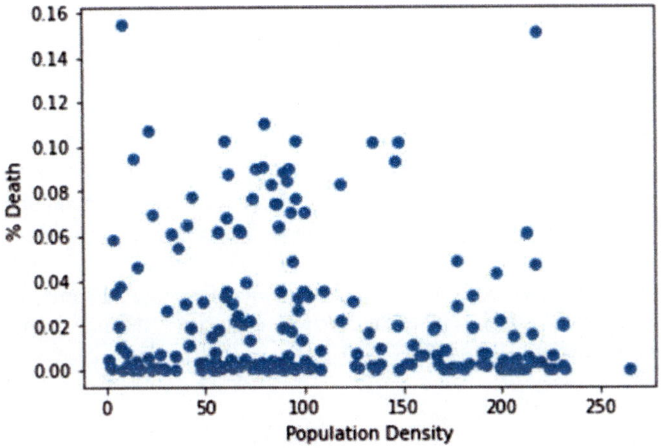

Fig. 4 Graph showing deaths versus population density

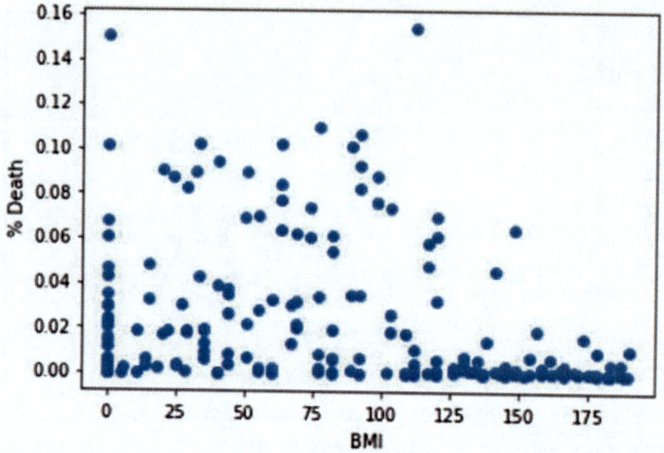

Fig. 5 Graph showing deaths versus body mass index

Density of 284 persons per km^2. These two countries have large differences in population densities but still, their infection rates are almost similar at 0.087 and 0.088%, respectively.

From the above result, it is observed that the spread of COVID-19 is independent of population density. It has also been enumerated through the graph.

4.4 Death Versus Body Mass Index

Figure 5 shows that Vanuatu has an infection percentage (0.003) with an observed Body Mass Index (BMI) (82), whereas in the case of Laos the infection percentage is (0.005) with BMI (166). To investigate further, countries like French Polynesia is having an infection percentage (5.55) with BMI (120), and Malaysia is having an infection percentage (0.25) with Body Mass Index (120). Another conclusion has been drawn with countries like Saint Lucia wherein infection percentage is (0.14) with BMI (6) and for Venezuela infection percentage is (0.37) in respect of BMI (14).

From the above observations, it has been found that there is ambiguity in the results. Thus, we can conclude that the Body Mass Index (BMI) of a person has no bearing on his Infection from COVID-19.

5 Conclusion

Results in the above section revealed that Burkina Faso's infected percentage is (0.0189) with an average temperature (28.29 °C). Benin's infected percentage is (0.0248) with an average temperature (27.55 °C). Andorra has (9.486) infected percent with average temperature (7.6 °C) while in the case of Georgia, it is (4.8) percent with an average temperature (5.8 °C). These results suggest that an increase in COVID-19 deaths is being experienced more where the average temperature is lower, whereas its incidence is lower wherein the average temperature is higher. The initial conclusion drawn from these observations is that the higher the temperature the lower the infection rate (lesser deaths), and the lower the temperature the higher the infection rate (higher deaths). Also, Indonesia's Population Density is 151 people per sq. km while Mauritania has a Population Density of 5 persons per km^2. These two countries have vastly different population densities but still, their infection rates are almost similar at 0.224 and 0.226%, respectively. Similarly, Nicaragua's Population Density is 54 persons per km^2 whereas Saint Vincent and the Grenadines have a Population Density of 284 persons per km^2. These two countries have large differences in population densities but still, their infection rates are almost similar at 0.087 and 0.088%, respectively. From the above result, it is observed that the spread of COVID-19 is independent of population density. It has also been enumerated through the graph. Vanuatu has an infection percentage (0.003) with an observed Body Mass Index (BMI) (82), whereas in the case of Laos the infection percentage is (0.005) with BMI (166). To investigate further, countries like French Polynesia is having an infection percentage (5.55) with BMI (120), and Malaysia is having an infection percentage (0.25) with Body Mass Index (120).

Another conclusion has been drawn with countries like Saint Lucia wherein infection percentage is (0.14) with BMI (6) and Venezuela whose infection percentage is (0.37) in respect of BMI (14). From the above observations, it has been found that there is ambiguity in the results. Thus, we can conclude that the Body Mass Index (BMI) of a person has no bearing on his infection from COVID-19.

This study explored the effect of temperature, median age, population density, and Body Mass Index on the fast worldwide outbreak of the COVID-19 pandemic (deaths in this pandemic). There are persuasive results from those countries with higher median age experiencing higher infection rates (more deaths) as compared to the countries having lower median age. The novel Coronavirus is convincingly lower (lesser deaths) in high-temperature zone and appears to spread faster in cooler climates. The spread of COVID-19 has no relationship in respect of Body Mass Index and population density. As future course of works, more reliable data is required to understand the behavior of COVID-19.

References

1. Anis Koubaa, et al.: Understanding the COVID19 outbreak: a comparative data analytics and study. Mar (2020) https://www.riotu-lab.org/resources/covid19-analytics.pdf
2. Harshad Khadilkar, et al.: Optimising lockdown policies for epidemic control using reinforcement learning. US National Library of Medicine, National Institutes of Health, PMCID: PMC7311597, June (2020)
3. Villalobos Arias, et al.: Using generalized logistic regression to forecast population infected by COVID-19. April (2020) https://arxiv.org/abs/2004.02406
4. Md. Rezaul Karim, et al.: Deep COVID explainer: explainable COVID-19 predictions based on chest X-ray images. April (2020) https://creativecommons.org
5. Nanning Zheng, et al.: Predicting COVID-19 in China using hybrid AI model. IEEE Trans. Cybern. **50**(7), July (2020)
6. Palash Ghosh, et al.: COVID-19 in India state-wise analysis and prediction. JMIR Public Health and Surveillance, April (2020) https://doi.org/10.2196/20341
7. Muhammad Mazhar Iqbal et al.: The effects of regional climatic condition on the spread of COVID-19 at global scale. Sci. Total Environ. Oct (2020)
8. Population pyramids of the world, by PopulationPyramid.net Dec. (2019)
9. National center for chronic disease prevention and health promotion, 17 Sep. (2020), https://www.cdc.gov/healthyweight/assessing/bmi/index.html0
10. Wikipedia contributors. (17 Mar 2021). Body mass index. In Wikipedia, the free encyclopedia. Retrieved 11:05, 19 Mar (2021). https://en.wikipedia.org/w/index.php?title=Body_mass_index&oldid=1012617044
11. Wikipedia contributors. (16 Mar 2021). Population density. In Wikipedia, the free encyclopedia. Retrieved 11:06, 19 Mar (2021), https://en.wikipedia.org/w/index.php?title=Population_density&oldid=1012533226
12. Wikipedia contributors. (2 Mar 2021). List of countries by median age. In Wikipedia, the free encyclopedia. Retrieved 11:07, 19 Mar (2021), https://en.wikipedia.org/w/index.php?title=List_of_countries_by_median_age&oldid=1009823802
13. Wikipedia contributors. (10 Mar 2021). List of countries by average yearly temperature. In Wikipedia, the free encyclopedia. Retrieved 11:07, 19 Mar (2021), https://en.wikipedia.org/w/index.php?title=List_of_countries_by_average_yearly_temperature&oldid=1011434571.
14. Harjule, P., Rahman, A., Agarwal, B.: A cross-sectional study of anxiety, stress, perception and mental health towards online learning of school children in India during COVID-19. Journal of Interdisciplinary Mathematics **24**(2), 411–424 (2021). https://doi.org/10.1080/09720502.2021.1889780

AI Approach for Autonomous Vehicles to Defend from Adversarial Attacks

Kritika Dhawale, Prachee Gupta, and Tapan Kumar Jain

1 Introduction

Neural networks have proved their capabilities and have outperformed most of the ML-related problems. Its increasing popularity is no daze. It is a widely accepted model for image-related tasks like detection, recognition, etc. But when it comes to the espousal of these models for security reasons, it failed miserably.

In 2014, a group of Google and NYU researchers [2] discovered that fooling Convolution Networks with an imperceptible but carefully designed nudge in the input was all too simple. They added some carefully constructed noise to the input image, and the same neural network now predicted the image incorrectly, as shown in Fig. 1 [2]. Neural Networks can be easily fooled by crafting small patches using generative models and placing them over the input image. These small perturbations intentionally performed on images are popularly known as adversarial attacks, and the images formed by these attacks are known as adversarial samples or attacked images.

The image above shows that an ML model predicts the image of a panda as a "Panda" with a confidence score of 57.7%. But when the same panda image is incorporated with some noise and then resulting image is passed to the same ML model, it now predicts the image as "Gibbon" with a confidence score of 99.3%.

These adversarial attacks can be printed, placed in any scene, photographed, and then introduced to image classifiers. They trigger the classifiers to disregard the rest of the scene's objects and report only the target class. These intrusions may not look like a hitch for humans in normal conditions, but is a serious complication when it comes to automated vehicles. The stratification of road signs and further trajectory planning is a crucial feature for any of these vehicles. If these vehicles fail to classify the traffic signs correctly, it can become very malicious for us. We can imagine a situation of misclassification for stop signs with speed 100.

K. Dhawale (✉) · P. Gupta · T. K. Jain
Department of Electronics and Communication Engineering, Indian Institute of Information Technology Nagpur, Nagpur, India
URL: http://iiitn.ac.in/

B. Agarwal et al. (eds.), *Proceedings of International Conference on Intelligent Cyber-Physical Systems*, Algorithms for Intelligent Systems,
https://doi.org/10.1007/978-981-16-7136-4_17

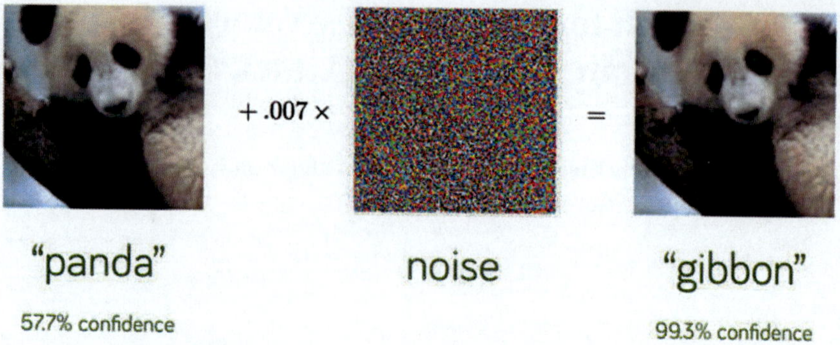

Fig. 1 Example of adversarial attack

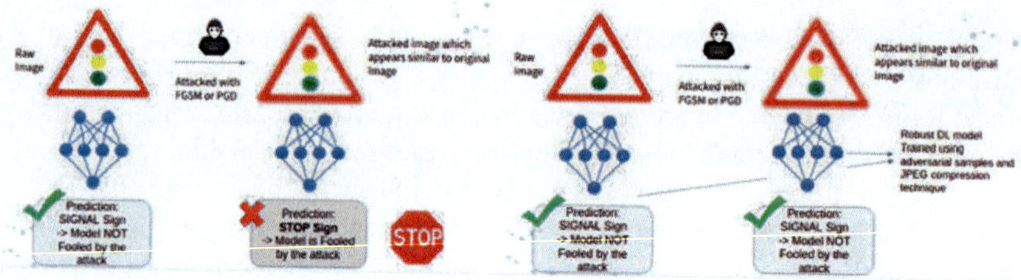

Fig. 2 Thought process

Figure 2 explains the thought process behind our objective. On the left-hand side of the figure, the Signal sign when passed to a traditional Traffic sign classifier, it correctly predicts the sign. Now if the same “Signal” sign is attacked, the traditional model predicts it as a “Stop” Sign, which is just fooling the ML model. Our motivation comes from this and aims to build a model which will not get fooled by any of the attacked images and still correctly classify the Traffic Sign as shown in the right-hand side of the above Fig. 2.

In this paper, a new approach for the safety of automated vehicles is proposed. Diverse research has been done for adversarial attacks available and their effects. Also tried to create a buckler to avoid these attacks up to the maximum extent. In this work, a complete pipeline is introduced. Out of all the available attacking techniques like Fast Gradient Sign Method, Projected Gradient Descent (PGD), Target Class Gradient Method, DeepFool Method, Carlini and Wagner Method, etc. [1], we have employed FGSM attacking technique. And the accuracy with these adversarial samples reached 10.39%. Further, it raised up to 93.10% after applying JPEG compression as defensive measure.

In Sect. 2, some of the previous work on adversarial examples is discussed. Section 3 explains the pipeline to construct more robust neural networks that are resistant to adversarial samples and the dataset used. Part 4 includes the implementation and experimental findings. Section 6 is about the discussion over the results from the experiments. Finally, Sect. 6 covers the consideration of the work's shortcomings and potential future extensions.

2 Related Work

This chapter includes a detailed study of the need for traffic signs detection facility in any automated vehicles and advancement in the approaches. The chapter also embraces the attacks that are available to mislead these automobiles, with some defensive techniques.

In a work done by Tabernik and Skocaj [3], the problems that can appear in the recognition and detection of traffic signs in massive traffic systems for automated vehicles are mentioned. The paper also covers approaches like CNN, R-CNN to direct the complete pipeline of detection and recognition with automatic end-to-end learning.

There are sundry papers available with different approaches for different types of problems related to traffic signs, like Hierarchical Deep Architecture and Mini-Batch Selection Method For Joint Traffic Sign and Light Detection [4], Traffic Sign Detection under Challenging Conditions: A Deeper Look Into Performance Variations and Spectral Characteristics [5].

The main aim that this paper focuses on problems faced by automated vehicles while dealing with traffic signs from adversarial attacks and defensive measures. The most common adversarial attacks are Fast Gradient Sign Method (FGSM), Projected Gradient Descent (PGD), CarliniWagnerL2Attack, DeepFool Attack, etc. [3]. The Fast method is the simplest method for the generation of adversarial attacks on images derived from linearizing the cost function [14].

Along with attack techniques, there are several counters available also. Image compression techniques are popular among them. JPEG-based defensive compression framework proved very functional for rectification of adversarial attacks without compromising the accuracy of classification [15]. After the instigation of localized attacks, Localized and Visible Adversarial Noise (LaVAN), and Adversarial patch, the challenges in security also increased. To overcome this problem, "Local Gradients Smoothing" proved very effective. The work includes regulation of gradients in the noisy zone before passing it to the DNN model [12].

Nguyen et al. [16] pioneered the use of state-of-the-art neural networks to deceive them. In this paper, it is demonstrated that deep neural networks can be easily tricked into reliably classifying images that are not recognized by humans as belonging to specific groups. Reference [17] demonstrated the vulnerability of the machine learning model and neural network model for adversarial examples. Reference [18] delves further into the reasons for adversarial attacks and shows how flaws in neural network-based training can make them vulnerable to adversarial samples. It also presents a new class of algorithms for crafting adversarial samples and formalizes the space of adversaries against deep neural networks.

Reference [19] discussed the dilemma of black-box adversarial threats, which is similar to the problem space that our research aims to solve. Adversarial samples are not only created by perturbing pixels in an image; [8, 14] investigated how adversarial examples can be transferred to the real world, such as adversarial printed road signs with graffiti-like art on top. These road signs appear to be vandalized, but the graffiti overlays lead neural networks' classification away from the right forecast. Standard detectors such as FasterRCNN [20] and YOLO [21] are not fooled by physical adversarial stop signals, according to recent work by [22, 23]. We can see a summary of related work in Table 1.

The main objective of this paper is to design a model that promises a better experience against adversarial attacks in automated vehicles.

Table 1 Summary table for related work

S. No.	Author	Year	Work done
1	Tabernik et al. [3]	2019	Deep learning method for the detection of traffic signs with large-intra-category appearance variation
2	Pon et al. [4]	2018	The deep hierarchical architecture that allows a network to detect both traffic lights and signs from training on the separate traffic light and sign datasets
3	Temel et al. [5]	2019	A Deeper Look Into Performance Variations and Spectral Characteristics
4	Morgulis et al. [6]	2019	How to utilize adversarial attacks to attack real-life systems in the physical world
5	Sitawarin et al. [7]	2018	Generation of adversarial samples which are robust to the environmental conditions and noisy image transformations
6	Sitawarin et al. [8]	2018	Out-of-Distribution attacks, Lenticular Printing attack
7	Aung et al. [9]	2017	A complete model including attacks and defensive approaches used defensive distillation with 91.46% accuracy
8	Shaham et al. [10]	2018	Experiment with low-pass filtering, PCA, JPEG compression, low resolution wavelet approximation, and soft-thresholding
9	Liu et al. [11]	2019	JPEG-based defensive compression framework
10	Naseer et al. [12]	2018	Focuses on frequency changes in the attacked images. Further adoption of Local Gradients Smoothing (LGS) scheme
11	Mustafa et al. [13]	2019	Proposes a computationally efficient image enhancement approach that provides a strong defense mechanism

3 Proposed Methodology

3.1 Pipeline

The pipeline can be dived into four main parts, viz., generating adversarial samples, performing JPEG compression on those samples in order to remove the noise in it, mixing all the data together and lastly training a robust model which is resilient to adversarial attacks. The technique flowchart in Fig. 3. It illustrates the scope of designing a powerful neural network based on the goals.

So, to first start with crafting the adversarial samples, we needed to build a classifier which has near state-of-the-art performance on test set of GTSRB dataset. We trained and tested a Convolutional Neural Network Model—43 class sign board classifier trained on GTSRB dataset. With the help of this model, adversarial samples are generated on the original dataset using Fast Gradient Sign Method (Sect. 4.2) and Projected Gradient Descent Method (Sect. 4.2). These adversarial samples are then experimented to attack the state-of-the-art model, which resulted into failing miserably with an accuracy of merely 10%.

The next step is to perform JPEG compression on these adversarial attacked images as a defense technique against adversarial attacks methods. This defensive technique can be used to make a more robust model against adversarial attacks. Further, we intend to mix the original data, the adversarial attacked images and JPEG compressed images accordingly into 43 classes in order to train a new robust model. The final step in our pipeline is to train a model with the mixed data that we generated. This model will be resilient to attacked images and will not get fooled by any of the attack method. This simulates a more practical situation, in which this robustly trained neural network is likely to be more complex than a replica network designed by a criminal to attack it.

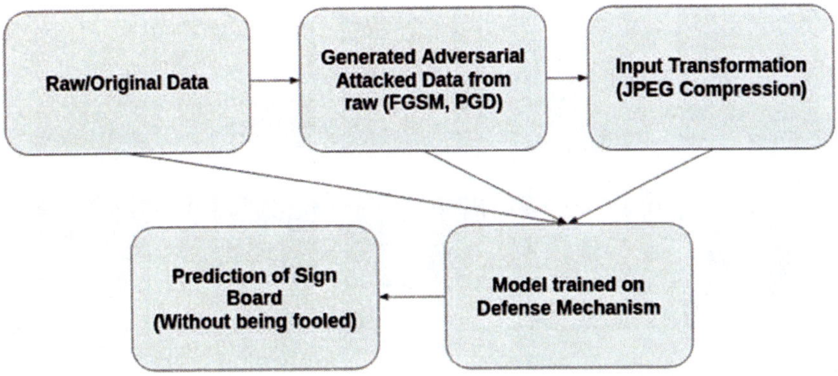

Fig. 3 Flowchart of the process

3.2 Dataset

We have used German Traffic Sign Recognition Benchmark Dataset (GTSRB). There are total 43 categories of different traffic sign boards. It is a Single-image, multi-class classification problem with an average image size 30*30 pixels. The resolution of the images are too small to deal with. It contains approximately 40,000 images for training in which we used 6535 images after cleaning the dataset. The number of samples per class was plotted, and random images from each class were plotted in an image-grid for exploratory data visualization.

Figure 4 shows the samples from the original training set of the GTSRB dataset, whereas Fig. 5 shows the distribution of images per class in the training set. The dataset is not at all balanced, therefore we perform data preprocessing in Sect. 3.

3.3 Data Preprocessing

The GTSRB dataset contains approximately 40,000 images in the training set which is already a lot. Further, we also had to generate adversarial samples and jpeg compressed images, so we needed to take a smaller portion of this training dataset. For each image in the training directory of the original GTSRB datatset, there are 30 different augmentations, from which we only took 5 augmentations.

Fig. 4 Sample dataset

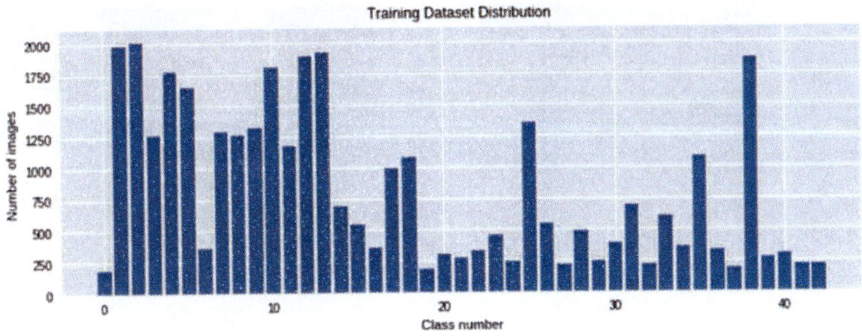

Fig. 5 Dataset distribution per class

- The training dataset was scaled down to 6535 images in total.
- Generated 6535 adversarial samples using FGSM and PGD attack methods.
- Performed JPEG compression on former generated adversarial samples to get 6535 compressed images.
- Mixed the new constructed samples accordingly into the classes. So, now we have a total of 32,675 training images.
- Images were reshaped to 32*32 pixels and rescaled by 255. Training data was split into training, validation, and Testing set in the ratio of 70:20:10.

Here, Fig. 6 shows attacked images with different intensity values known as epsilon value (ep). With ep = 0, that means no attack is performed and the original image and attack image will be exactly same. For ep = 0.01, we can see small perturbations embedded in the image. As we increase the epsilon value, the intensity of the perturbations also increase and the capability of the attack to fool the model also increases. Figure 7 is the image on which JPEG compression was applied and image was converted into grey scale as part of feature engineering.

Fig. 6 Adversarial samples with different epsilon values

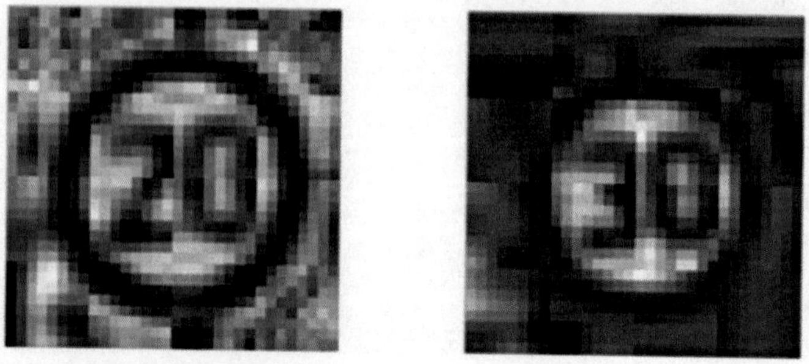

Fig. 7 JPEG compressed samples

4 Implementation and Experiment

Convolutional neural networks (CNNs) are used for feature extraction and have gained a lot of popularity due to the various architectures available in some frameworks. Convolution layers and Pooling Layers are the two key components of the CNN's Convolution section. By computing the output of neurons linked to local layers, the convolution layers add depth. Down-sampling is accomplished by pooling layers. The feature vectors are extracted from the input images using CNN. Based on the feature vector given by CNN, an ANN model classifies the data.

4.1 Network Architecture

The detailed architecture for our 43 class classifier is described in Fig. 8. The network takes in a 3-channel RGB image of width 32 pixels and height 32 pixels.

The model is an inspiration of mini VGGnet, containing 7 layers, hence the name is VGG7. The layers consist of 4 convolution layers with ReLu activation function and 2 max pooling layers with dropout in between. The last layers are dense layers with ReLu and Softmax functions.

4.2 Crafting Adversarial Samples

The main aim of adversarial inputs is to tamper with the performance integrity of deep learning algorithms. These inputs may be created with the intent of causing the model to misclassify the input image into a different class than it should be, or at the very least lowering the output trust. There are copious attacking techniques available as, discussed in previous sections. In this paper, we have preferred Fast Gradient Sign (FGSM) method and Projected Gradient Descent (PGD) method.

```
Layer (type)                 Output Shape
=================================================================
conv2d (Conv2D)              (None, 28, 28, 32)
_________________________________________________________________
conv2d_1 (Conv2D)            (None, 24, 24, 32)
_________________________________________________________________
max_pooling2d (MaxPooling2D) (None, 12, 12, 32)
_________________________________________________________________
dropout (Dropout)            (None, 12, 12, 32)
_________________________________________________________________
conv2d_2 (Conv2D)            (None, 10, 10, 64)
_________________________________________________________________
conv2d_3 (Conv2D)            (None, 8, 8, 64)
_________________________________________________________________
max_pooling2d_1 (MaxPooling2 (None, 4, 4, 64)
_________________________________________________________________
dropout_1 (Dropout)          (None, 4, 4, 64)
_________________________________________________________________
flatten (Flatten)            (None, 1024)
_________________________________________________________________
dense (Dense)                (None, 256)
_________________________________________________________________
dropout_2 (Dropout)          (None, 256)
_________________________________________________________________
dense_1 (Dense)              (None, 43)
=================================================================
Total params: 356,939
Trainable params: 356,939
Non-trainable params: 0
_________________________________________________________________
```

Fig. 8 Model architecture

Fast Gradient Sign Method (FGSM) The fast gradient sign method generated adversarial samples using the parameters of the neural network. The fast gradient sign method generated adversarial samples using the parameters of the neural network [2]. In particular, FGSM involves the addition of predeclared noise in the same direction of the gradient of the cost function with respect to the data. It updates the initial sample x in a single step in the direction of the gradient of a loss function $J(x, y; \theta)$ [17].

$$x_{adv} = clip_{[0,1]}\{x + \epsilon.sign(\nabla_x J(x, y;))\} \tag{1}$$

where epsilon regulates the maximal l perturbation of the adversarial samples and the clip[a,b] (x) function constrains x to the set [a, b] [24].

Projected Gradient Descent (PGD) The PGD attack is a white-box attack, which ensures the attacker has a copy of your model's weights and has access to the model gradients. The iterative version of FGSM is Projected Gradient Descent (PGD). PGD greedily solves the problem of optimizing the loss function of each iteration: [24]

$$x_{adv}^{(t+1)} = clip_{[0,1]}\{x_{adv}^{t} + \epsilon.sign(\nabla_x J(x, y;))\} \tag{2}$$

$$x_{adv}^{0} = x_{raw} \tag{3}$$

The clip here is a projection function that can be exchanged with other functions like tanh to keep each iteration's contribution within the operational range.

4.3 Defensive Input Transformation

After exploring various defensive techniques, we concluded that JPEG compression is a supreme approach for a condition for an attack that aims to perturb in ways that are visually unapparent to the naked eye. The JPEG compression technique can effectively reduce pixel "noise," mainly because JPEG is designed to bring down image information indiscernible to humans [25]. There are mainly six steps involved in getting a compressed (JPEG) image from an input image-

1 **SPLITTING:** In this step, the image gets divided into blocks of 8 × 8 blocks, with each block referred to as 1 pixel.
2 **COLOR TRANSFORMATION:** In this step, we convert the R, G, B models to the Y, Cb, Cr models. The letters Y, Cb, and Cr stand for brightness, blue color, and red color. We translate them to chromium colors, which are less sensitive to human vision and can therefore be omitted.
3 **DCT:** Each block is then subjected to the Direct Cosine Transform. The Discrete Cosine Transform (DCT) represents an image as a sum of sinusoids of varying magnitudes and frequencies.
4 **QUANTIZATION:** Here quantization of data takes place.
5 **SERIALIZATION:** Here serialization refers to the zig-zag scanning pattern for redundancy exploitation.
6 **ENCODING:** In the last stage, we apply to encode either run-length encoding or Huffman encoding. The main aim is to convert the image into text and, by using any encoding technique, convert it into binary form (0, 1) to compress the data.

There can be an addition of extra steps in between serialization and encoding and, i.e., vectoring. Refer Fig. 9. for a complete flowchart of the process.

4.4 Adversarial Training

We divided the training and testing adversarial sets using adversarial examples produced with Fast Gradient Sign Method and Projected Gradient Descent Method. The neural network feeds back the training adversarial samples and defensive input transformed images as new training samples. Adversarial training is equivalent to using

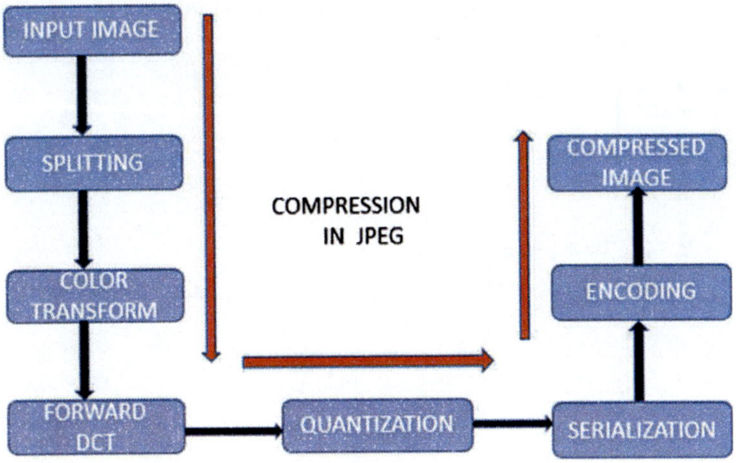

Fig. 9 JPEG compression: flow chart including all the steps of JPEG compression

brute force to solve more complex neural networks, but creating more adversarial samples is costly and not scalable.

5 Results and Discussions

The GTSRB dataset has 43 different traffic sign board classes. The original/raw GTRSB dataset (we call it as legit samples) was trained on mini VGGnet, i.e., VGG7 and was able to achieve a whopping 99.04.% training accuracy and 98.76% testing accuracy which is near the state-of-the-art. We used this model to test the adversarial samples (Adv samples) that we generated from FGSM and PGD attack methods, and found out that they fooled the model very easily with an accuracy of merely 10.39% as shown in Table 2.

We also calculated F1 score so as to take into account the false positives and false negatives the model is predicting. Our dataset was also quite imbalanced and therefore F1 score metrics plays a better role in evaluating our model.

Table 2 Results without defense

	Without defense mechanism	
	On legit samples	On Adv samples
Training accuracy	99.04%	NA
Testing accuracy	98.76%	10.39%
F1 score	0.9876	0.1039

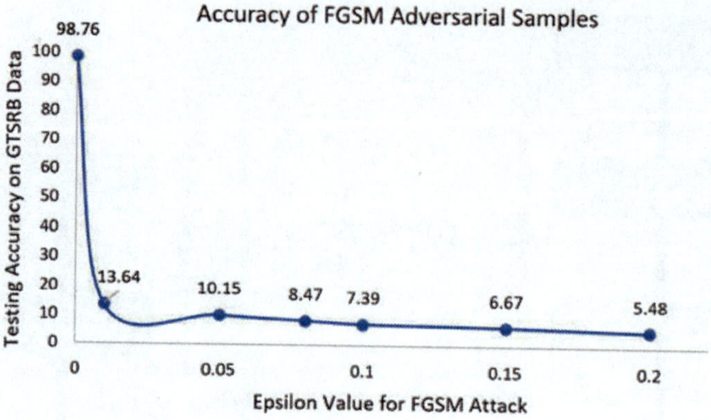

Fig. 10 Accuracy of FGSM adversarial samples

Table 3 Results with defense

	With defense mechanism	
	On legit samples	On mixed samples
Training accuracy	NA	93.56%
Testing accuracy	NA	93.10%
F1 score	NA	0.931

Another finding to note is Fig. 10, which plots the model accuracy (trained on original GTSRB) as epsilon increases. The number of images correctly identified from the total number of images tested is used to determine accuracy. With increasing epsilon, we can see an exponential decline in accuracy.

Further, we took into consideration the adversarial samples, jpeg compressed images and the original data as a whole new training set (mixed samples). When we trained the same model architecture as discussed above with this new dataset (we call it as a model with defense mechanism), we got a training accuracy of 93.56%, whereas the testing accuracy was 93.10%. The F1 score was also remarkable, i.e., 0.931 as shown in Table 3.

Considering the images generated by performing JPEG compression (as defensive method) in the training set, a smoother classifier model is developed by lowering their sensitivity to input perturbations. These smoother classifiers are found to be more resistant to adversarial samples and have better class generalizability. By lowering their sensitivity to input perturbations, a smoother classifier model is developed. These smoother classifiers have greater class generalizability and are more immune to adversarial samples.

The plot shows the variation of Accuracy per epoch on Training as well as Validation data (Fig. 11). This shows that the model came to a point of equilibrium after 44 epoch.

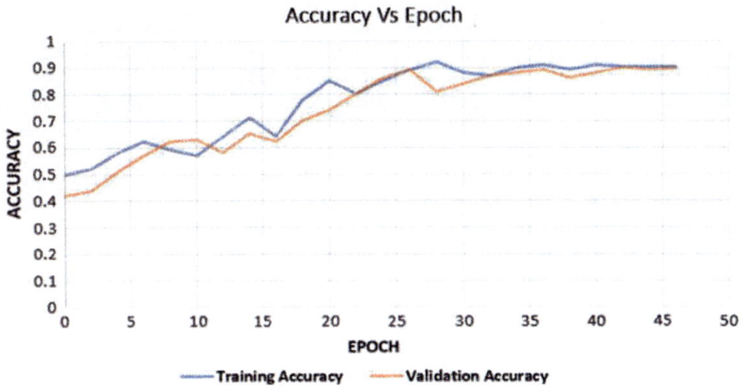

Fig. 11 Accuracy versus Epoch graph for adversarial training

6 Conclusion and Future Work

In this new era of technological automation, it becomes a necessity to look after safety measures. Especially when these are linked with our lives like in automated vehicles. One wrong prediction can cost our lives in these cases. In this paper, we have designed a robust model that focuses on avoiding adversarial attacks. We have concentrated on training a model with a defensive mechanism to eliminate adversarial noise. We have generated the attacked images using FGSM and PGD techniques. And for defense, we have employed JPEG compression.

Our model reached up to an accuracy of 98.76% without training the attacked images. After including attacked images generated from FGSM and PGD methods, the accuracy dropped to 10.39%. Following by defensive measure, i.e., JPEG compression the accuracy gets elevated to 93.10%.

In future work, we can think of multiple defensive layers. We can go for some preprocessing techniques as well.

References

1. Xu, H., Ma, Y., Liu, H.-C., Deb, D., Liu, H., Tang, J.-L., Jain, A.K.: Adversarial attacks and defenses in images, graphs and text: a review. Int. J. Autom. Comput. **17**(2), 151–178 (2020)
2. Goodfellow, I.J., Shlens, J., Szegedy, C.: Explaining and harnessing adversarial examples, arXiv preprint arXiv:1412.6572
3. Tabernik, D., Skočaj, D.: Deep learning for large-scale traffic-sign detection and recognition. IEEE Trans. Intell. Transp. Syst. **21**(4), 1427–1440 (2019)

4. Pon, A., Adrienko, O., Harakeh, A., Waslander, S.L.: A hierarchical deep architecture and mini-batch selection method for joint traffic sign and light detection. In: 2018 15th Conference on Computer and Robot Vision (CRV), pp. 102–109. IEEE (2018)
5. Temel, D., Chen, M.-H., AlRegib, G.: Traffic sign detection under challenging conditions: A deeper look into performance variations and spectral characteristics. IEEE Trans. Intell. Transp. Syst. **21**(9), 3663–3673 (2019)
6. Morgulis, N., Kreines, A., Mendelowitz, S., Weisglass, Y.: Fooling a real car with adversarial traffic signs, arXiv preprint arXiv:1907.00374
7. Sitawarin, C., Bhagoji, A.N., Mosenia, A., Mittal, P., Chiang, M.: Rogue signs: deceiving traffic sign recognition with malicious ads and logos, arXiv preprint arXiv:1801.02780
8. Sitawarin, C., Bhagoji, A.N., Mosenia, A., Chiang, M., Mittal, P.: Darts: deceiving autonomous cars with toxic signs, arXiv preprint arXiv:1802.06430
9. Aung, A.M., Fadila, Y., Gondokaryono, R., Gonzalez, L.: Building robust deep neural networks for road sign detection, arXiv preprint arXiv:1712.09327
10. Shaham, U., Garritano, J., Yamada, Y., Weinberger, E., Cloninger, A., Cheng, X., Stanton, K., Kluger, Y.: Defending against adversarial images using basis functions transformations, arXiv preprint arXiv:1803.10840
11. Liu, Z., Liu, Q., Liu, T., Xu, N., Lin, N., Wang, Y., Wen, W.: Feature distillation: Dnn-oriented jpeg compression against adversarial examples. In: 2019 IEEE/CVF Conference on Computer Vision and Pattern Recognition (CVPR), pp. 860–868. IEEE (2019)
12. Naseer, M., Khan, S., Porikli, F.: Local gradients smoothing: defense against localized adversarial attacks. In: 2019 IEEE Winter Conference on Applications of Computer Vision (WACV), pp. 1300–1307. IEEE (2019)
13. Mustafa, A., Khan, S.H., Hayat, M., Shen, J., Shao, L.: Image super-resolution as a defense against adversarial attacks. IEEE Trans. Image Process. **29**, 1711–1724 (2019)
14. Kurakin, A., Goodfellow, I., Bengio, S. et al.: Adversarial examples in the physical world (2016)
15. Liu, Z., Liu, Q., Liu, T., Xu, N., Lin, X., Wang, Y., Wen, W.: Feature distillation: Dnn-oriented jpeg compression against adversarial examples. in 2019 ieee. In: CVF Conference on Computer Vision and Pattern Recognition (CVPR), pp. 860–868
16. Nguyen, A., Yosinski, J., Clune, J.: Deep neural networks are easily fooled: High confidence predictions for unrecognizable images. In: Proceedings of the IEEE Conference on Computer Vision and Pattern Recognition, pp. 427–436 (2015)
17. Szegedy, C., Zaremba, W., Sutskever, I., Bruna, J., Erhan, D., Goodfellow, I., Fergus, R.: Intriguing properties of neural networks, arXiv preprint arXiv:1312.6199
18. Papernot, N., McDaniel, P., Jha, S., Fredrikson, M., Celik, Z.B., Swami, A., The limitations of deep learning in adversarial settings. In: IEEE European Symposium on Security and Privacy (EuroS&P), pp. 372–387. IEEE (2016)
19. Papernot, N., McDaniel, P., Goodfellow, I., Jha, S., Celik, Z.B., Swami, A.: Practical black-box attacks against machine learning. In: Proceedings of the 2017 ACM on Asia Conference on Computer and Communications Security, pp. 506–519 (2017)
20. Ren, S., He, K., Girshick, R., Sun, J.: Faster r-cnn: towards real-time object detection with region proposal networks, arXiv preprint arXiv:1506.01497
21. Redmon, J., Farhadi, A.: Yolo9000: better, faster, stronger. In: Proceedings of the IEEE Conference on Computer Vision and Pattern Recognition, pp. 7263–7271 (2017)
22. Lu, J., Sibai, H., Fabry, E., Forsyth, D.: No need to worry about adversarial examples in object detection in autonomous vehicles, arXiv preprint arXiv:1707.03501
23. Lu, J., Sibai, H., Fabry, E., Forsyth, D.: Standard detectors aren't (currently) fooled by physical adversarial stop signs, arXiv preprint arXiv:1710.03337
24. Wu, F., Gazo, R., Haviarova, E., Benes, B.: Efficient project gradient descent for ensemble adversarial attack, arXiv preprint arXiv:1906.03333
25. Das, N., Shanbhogue, M., Chen, S.-T., Hohman, F., Li, S., Chen, L., Kounavis, M.E., Chau, D.H.: Shield: Fast, practical defense and vaccination for deep learning using jpeg compression. In: Proceedings of the 24th ACM SIGKDD International Conference on Knowledge Discovery and Data Mining, pp. 196–204 (2018)

26. Goodman, D.: Transferability of adversarial examples to attack cloud-based image classifier service, arXiv preprint arXiv:2001.03460
27. Kumar, R. Sharma, M., Dhawale, K., Singal, G.: Identification of dog breeds using deep learning. In: 2019 IEEE 9th International Conference on Advanced Computing (IACC), pp. 193–198. IEEE (2019)

Quantum Layer-Inspired Deep Learning for Mechanical Parts Classification

Vikas Khullar, Raj Gaurang Tiwari, and Ambuj Kumar Agarwal

1 Introduction

With more and more innovations and inventions happening in the digital world, it has paved way for more growth opportunities in almost every field. Bridging the gap between man and machine is the most challenging task that researchers are facing now. Supervised learning methods provide known images (i.e., labeled data) to the machine during classification training [1, 2]. These techniques extract features from pictures and then search for the same set of features in new images. Convolutional Neural Networks (CNNs) are one form of deep learning used predominantly in the field of computer visualization in the past decade in applications such as image classification, object detection, image captioning, handwritten digit recognition, face recognition, image segmentation, action recognition, pedestrian detection, and so on [3, 4].

With the advent of deep learning methods, most notably CNN, all previous state-of-the-art machine learning techniques in the area of computer vision were rendered obsolete [5]. Though deep learning is nowadays being widely used in many applications for analyzing text data, voice data, data from sensors, it has found major advances in analyzing image data with more and more Convolutional layers [5–7].

Quantum Neural Network (QNN) layers now working based on conceptual quantum mechanics. Initially quantum neural computation was coined by Kak and Chrisley. Most of the QNN is dependent on feed-forward network theory. Quantum computers are developing as a novel way to solve glitches that conventional computers cannot solve. Quantum computers operate in a separate computing environment from conventional computers [8, 9]. Quantum computers can take advantage of superposition and entanglement, which are not possible in traditional computing

V. Khullar (✉) · R. G. Tiwari · A. K. Agarwal
Chitkara University Institute of Engineering and Technology, Chitkara University, Chandigarh, Punjab, India

B. Agarwal et al. (eds.), *Proceedings of International Conference on Intelligent Cyber-Physical Systems*, Algorithms for Intelligent Systems,
https://doi.org/10.1007/978-981-16-7136-4_18

settings, and achieve high efficiency through the use of parallelism between qubits. According to these benefits, the quantum computer is regarded as a novel approach to difficult algorithmic problems.

In this paper, we aim to move one step further by adding a quantum layer to the CNN to achieve higher metrics. We intend to find its applications in the domain of Mechanical Engineering. We develop a classification system for mechanical parts (i.e., Nuts, Bolts, Washers, and Locating Pins) by exploiting the conception of Quantum Layer-inspired Convolutional Neural Networks (QCNNs).

2 Related Studies

Hybrid quantum–classical systems allow the maximum use of current quantum computers. Under this context, parameterized quantum circuits can be viewed as highly articulate machine learning models [8, 9]. Kerenidis et al., [10] suggested a Quantum algorithm for implementing and training deep convolutional neural networks that could significantly accelerate the method. They demonstrated a novel quantum tomography algorithm with $\ell\infty$ norm guarantees, as well as novel implementations of probabilistic sampling in information processing. Additionally, they provide numerical models of the MNIST dataset's classification to demonstrate the QCNN's performance in practice. Yang et al. [11] used quantum Behaved Particle Swarm Optimization (BQPSO) with binary encoding to simplify the search procedure for the optimum architecture and minimize human involvement. To accomplish this, they suggested an innovative as well as stable binary encoding technique that does not involve domain information about CNNs from the users. Then, to guarantee the efficacy of advanced CNN architectures, a new quantum behaving emerging strategy was proposed. Their algorithm's success was evaluated using the classification accuracy of many standard datasets often utilized in deep learning. The investigational outcomes have shown that their suggested approach outperforms and is more stable than the conventional method.

Li et al. [12] proposed an explicit quantum model capable of effectively reproducing the conventional DCNN with a particular edifice. It fully exploits the parallelism inherent in the quantum model, both in storage as well as computation, resulting in an exponential speeding up over the conventional equivalent. Additionally, a competitive finding from numerical experimentation demonstrated the QDCNN model's viability and validity in multi-class image recognition. Zhou et al. [13] presented a new form of deep learning approach for classification called Deep Quantum Network (DQN). DQN inherits the ability to use fuzzy sets to model the configuration of function space. To begin, they proposed the DQN architecture, which is composed of quantum neurons and sigmoid neurons and is capable of guiding the entrenching of divisible samples in novel Euclidean space. The greedy layer-wise unsupervised learning technique was used to configure the DQN parameter. Then, using supervised learning and the global gradient-descent technique, the deep architecture's and quantum representation's parameter space is optimized. This article

Table 1 Mechanical parts dataset description [15]

Training data		Testing data	
Object name	Number of images	Object name	Number of images
Bolt	1523	Bolt	381
Locating pin	1523	Locating pin	381
Nut	1523	Nut	381
Washer	1523	Washer	381

introduced an exponential loss function to direct the supervised learning process. In tests using standard datasets, DQN performs admirably other feed-forward neural networks and neuro-fuzzy classifiers. Henderson et al. [14] conducted an analytical evaluation of the possible value of these quantum transformations by evaluating three categories of models based on the MNIST dataset: CNNs, quantum layer-inspired convolutional neural networks and CNNs with additional non-linearitys. Their findings indicated that QNN models outperformed strictly classical CNNs in terms of test set precision and training time.

3 Materials and Methods

3.1 Data-Set

We collected 1904 elements images for each of the four groups, i.e., Bolts, Nuts, Locating Pins, and Washers, from an online component repository available at https://www.kaggle.com/krishna8338/mechanical-parts-data [15]. Each image has a resolution of 28 × 28 pixels. The details of utilized data and its training–testing partitioning are mentioned in Table 1.

3.2 Methodology

We leverage the CNN and CNN with quantum layer concepts to allow our model to classify images into four categories: Bolts, Nuts, Locating Pins, and Washers. The model is trained by "observation" of a series of training photographs. The whole procedure is illustrated in Fig. 1.

The flow of methodology is depicted in Fig. 2. Initially, mechanical parts data was collected from mentioned resource [15]. Further data pre-processing and cleaning were conducted by reducing the size of images to 28 × 28 pixels and then convert into grayscale images. Then, the training of CNN with quantum Layer and without

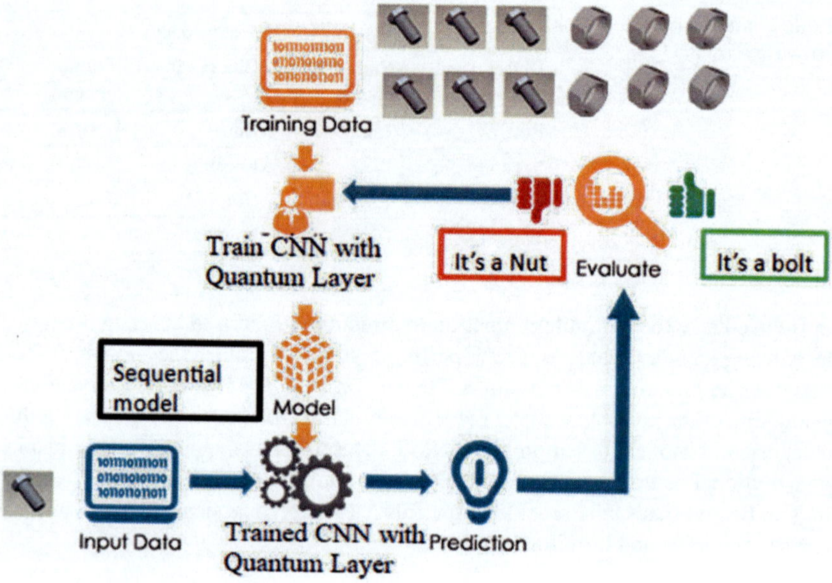

Fig. 1 Procedure of mechanical part classification

quantum layer was conducted for comparative analysis and to find the better approach. The detail of CNN with quantum layer is mentioned in the next section.

3.3 CNN with Quantum Layer

Deep learning is a kind of machine learning technique in which feature extraction is automated. With deep learning, the features are learned on their own with simpler concepts building the complex ones. Deep learning had been here since the 1940s, but with different nomenclatures [2]. Convolutional Neural Networks are one form of deep learning used predominantly in the field of computer visualization in the past decade in applications such as image classification, object detection, image captioning, handwritten digit recognition, face recognition, image segmentation, action recognition, pedestrian detection, and so on. CNNs are neural networks consisting of multiple layers which identifies the features layer by layer and construct the feature representations. Early layers detect simple patterns which collectively generate complex patterns or abstractions in the latter layers. The training images are fed to the network which in every layer convolutes with some set of weights and propagates through all the layers (in the forward direction). At the end of the network, the difference is computed using the loss function (errors). Based on these errors the weights are adjusted in every layer while propagating in the backward direction. The weights are adjusted by some optimization function. A complete cycle of forward

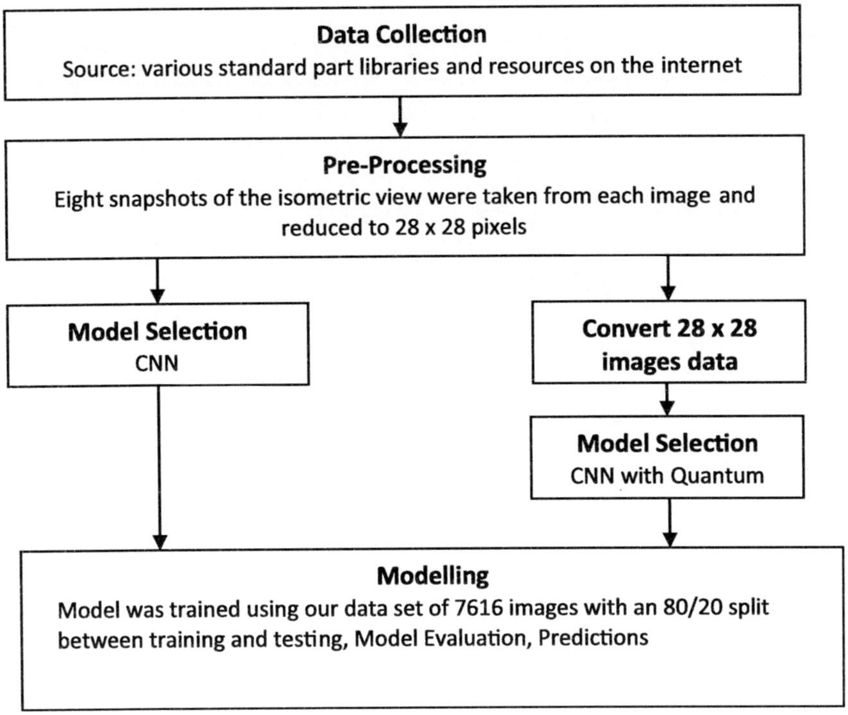

Fig. 2 Flow of methodology

and backward propagation corresponds to a single iteration during which the weights are updated and this continues until convergence [5]. In end, before applying conventional CNN, a layer of quantum concepts has added. The circuit diagram of added quantum layer is reflected in Fig. 3 which is included with the CNN model. A 4-bit quantum layer was implemented as a hidden layer between Convolutional layers. Here, the quantum circuit was implemented with the help of the Qiskit Python library. Further in Fig. 4 created quantum layer was added in between the Convolutional layer 1 (dense_9) and flatten layer (flatten_3).

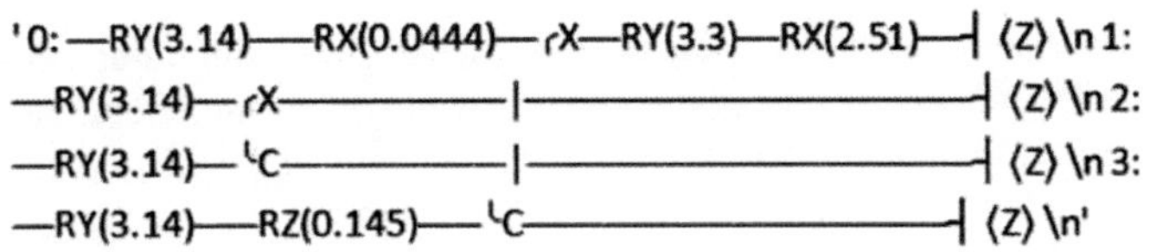

Fig. 3 Quantum layer circuit

Layer (type)	Output Shape	Param #
dense_9 (Dense)	(4, 14, 14, 32)	160
dense_10 (Dense)	(4, 14, 14, 64)	2112
flatten_3 (Flatten)	(4, 12544)	0
dense_11 (Dense)	(4, 4)	50180

Total params: 52,452

Trainable params: 52,452

Non-trainable params: 0

Fig. 4 CNN structure implemented with quantum layer

4 Results

The model is constantly adjusting its weight to reduce cost (loss) and thereby have the highest accuracy. Cost is a metric for the model's inaccuracy in estimating the image's class. Cost functions are used to quantify the poor performance of models. If the algorithm makes an erroneous prediction, the cost increases; if the algorithm makes an accurate prediction, the cost reduces. After 30 epochs of preparation, the effects of loss and accuracy are seen in Figs. 5 and 6. The loss rose in proportion to the number of times the model was conditioned. It improves for each epoch at classifying pictures.

The model significantly enhances efficiency on the validation package. More detailed epoch-wise comparative validation results of CNN with quantum layer and CNN without quantum layer are presented in Table 2.

The accuracy improved with each epoch's model training. It improves at classifying objects. The validity set's accuracy is smaller than the training set's because it has not been trained specifically on it. It had identified that from the very first training epoch CNN with quantum layer resulted in higher accuracy and lesser loss in comparison to without quantum layer model. Further, the efficiency of implemented algorithms was analyzed using various metrics viz., precision, recall, and F1-score as shown in Fig. 7 and Table 3. The absolute results of precision, recall, and F1-score with quantum layer model are 85.5, 82.2, and 82.25%, which is better in comparison to CNN model without quantum layer.

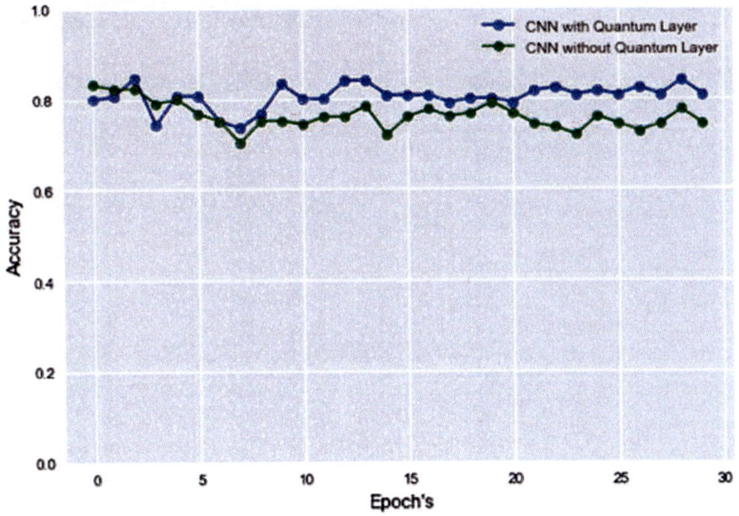

Fig. 5 Accuracy comparison between CNN with and without quantum layer

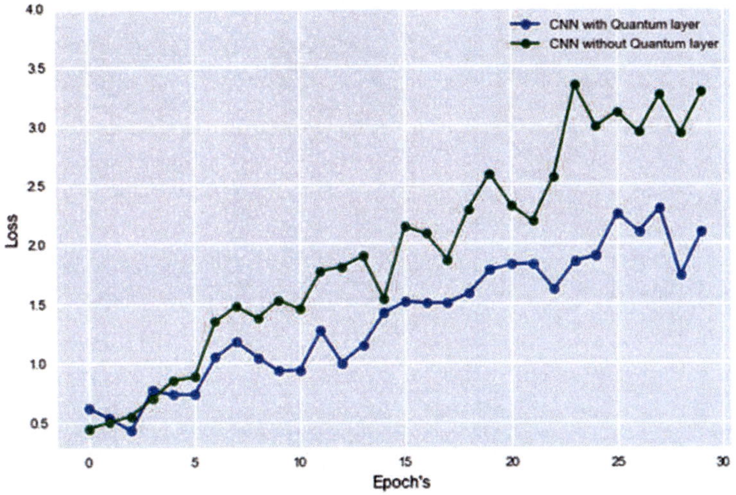

Fig. 6 Loss comparison between CNN with and without quantum layer

5 Conclusion

An earnest effort has been made to develop a mechanical part classification system that facilitates identification and prediction of Nuts, Bolts, Washers, and Locating Pins. In this paper, burly classifiers were constructed by exploiting the concept of

Table 2 Validation accuracy and loss comparative analysis for CNN model with quantum layer versus CNN model without quantum layer

Epoch's	CNN with quantum layer		CNN without quantum layer	
	Accuracy	Loss	Accuracy	Loss
1	0.798387	0.615146	0.830645	0.444206
5	0.806452	0.731584	0.766129	0.877152
10	0.798387	0.930611	0.741935	1.453034
15	0.806452	1.512678	0.758065	2.139245
20	0.790323	1.823704	0.766129	2.317274
25	0.806452	2.251515	0.741935	3.113592
30	0.806452	2.092529	0.741935	3.281013

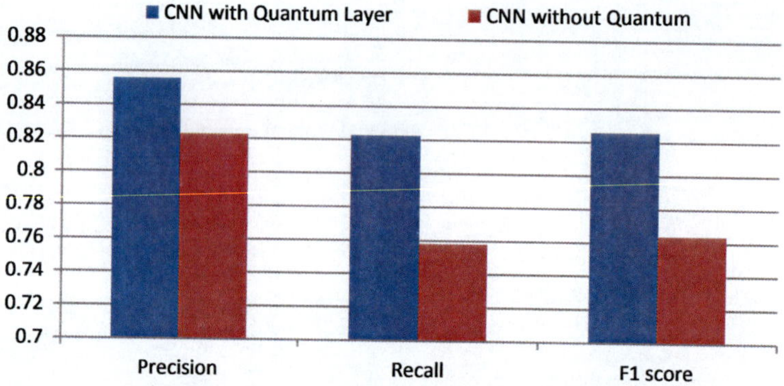

Fig. 7 Precision, Recall, and F1-score of CNN with and without Quantum Layer

Table 3 Comparison of Precision, Recall, and F1-score of CNN with and without Quantum Layer

Parameter	CNN with quantum layer	CNN without quantum
Precision	0.85577	0.82255
Recall	0.82258	0.75807
F1-score	0.82569	0.7641

CNN with and without quantum layers and with the intention that we get elevated accuracy. After comparing the accuracy of the classifier with a quantum layer with the classifier without the quantum layer, we conclude that the accuracy, as well as other parameters (i.e., precision, recall, and F1-score) of the proposed quantum layer-based model, is better.

References

1. Du, P., Bai, X., Tan, K., Xue, Z., Samat, A., Xia, J., Li, E., Su, H., Liu, W.: Advances of four machine learning methods for spatial data handling: A review. J. Geovisualiz. Spatial Anal. **4**, 1–25 (2020)
2. Tian, C., Lunke, F., Wenxian, Z., Yong, X., Wangmeng, Z., Chia-Wen, L.: Deep learning on image denoising: an overview. Neural Netw. **131**, 251–275 (2020)
3. Krizhevsky, A., Sutskever, I., Hinton, G.E.: Imagenet classification with deep convolutional neural networks. Adv. Neural. Inf. Process. Syst. **25**, 1097–1105 (2012)
4. Najafabadi, M.M., Villanustre, F., Khoshgoftaar, T.M., Seliya, N., Wald, R., Muharemagic, E.: Deep learning applications and challenges in big data analytics. J. Big Data **2**(1), 1–21 (2015)
5. Gu, J., Wang, Z., Kuen, J., Ma, L., Shahroudy, A., Shuai, B., Liu, T., Wang, X., Wang, G., Cai, J., Chen, T.: Recent advances in convolutional neural networks. Pattern Recogn. **77**, 354–377 (2018)
6. Khullar, V., Salgotra, K., Singh, H.P., Sharma, D.P.: Deep learning-based binary classification of ADHD using resting state MR images. Augment. Human Res. **6**(5) (2021)
7. Khullar, V., Singh, H.P., Bala, M.: Deep neural network-based handheld diagnosis system for autism spectrum disorder **69**(1), 66–77 (2021)
8. Oh, S., Choi, J., Kim, J.: A tutorial on quantum convolutional neural networks (QCNN). In: International Conference on Information and Communication Technology Convergence, pp. 236–239 (2020)
9. Benedetti, M., Lloyd, E., Sack, S., Fiorentini, M.: Parameterized quantum circuits as machine learning models. Quantum Sci. Technol. **4**(4), 043001 (2019)
10. Kerenidis, I., Landman, J., Prakash, A.: Quantum algorithms for deep convolutional neural networks. In: Eighth International Conference on Learning Representations, pp. 1–36 (2019)
11. Li, Y., Xiao, J., Chen, Y., Jiao, L.: Evolving deep convolutional neural networks by quantum behaved particle swarm optimization with binary encoding for image classification. Neurocomputing **362**, 156–165 (2019)
12. Li, Y., Zhou, R.G., Xu, R., Luo, J., Hu, W.: A quantum deep convolutional neural network for image recognition. Quantum Sci. Technol. **5**(4) (2020)
13. Zhou, S., Chen, Q., Wang, X.: Deep quantum networks for classification. In: 20th International Conference on Pattern Recognition, Istanbul, Turkey, pp. 2885–2888 (2010)
14. Henderson, M., Shakya, S., Pradhan, S., Cook, T.: Quanvolutional neural networks: powering image recognition with quantum circuits. Quantum Mach. Intell. **2**(1), 1–9 (2020)
15. Mitra, K.: Mechanical Parts Data, https://www.kaggle.com/krishna8338/mechanical-parts-data/activity, Accessed 13-04-2021

Handwritten Signature Verification Using Transfer Learning and Data Augmentation

Yash Gupta, Ankit, Sanchit Kulkarni, and Pooja Jain

1 Introduction

Signature is one of the most commonly accepted methods for personal verification and identification. Signature verification is important for banking, legal documents and still an important area of research in the field of machine learning and deep learning. Typically, signatures are of two types: (1) handwritten and (2) digital. Capturing handwritten signature needs a paper with pen or electronic pad with stylus. Apart from the ink impression on the paper, signature verification also requires to consider writing speed, pressure, etc.

In this paper, we focus on feature extraction and classification on the image dataset of handwritten signature stored in PNG format. We make several contributions for each feature extraction and classification. First, to obtain feature extraction for each algorithm and CNN architectures independently. Second, we present algorithms that are more suitable for classification famous as supervised learning algorithm. Data augmentation is another aspect of our paper where even the small dataset can be used to increase the dataset for performing the feature extraction task (Figs. 1 and 2).

The remainder of this paper is organized as follows. In Sect. 2, we have discussed the problem and literature survey. Section 3 introduces algorithms and architectures to efficiently extract features and discusses algorithms and methods of classification using those extracted features. In Sect. 4, perform a set of experiments on various architectures using concepts of transfer learning and data augmentation. Finally, we have a conclusion for our work and suggestions for future work in Sect. 5.

Y. Gupta · Ankit · S. Kulkarni · P. Jain (✉)
Computer Science and Engineering, Indian Institute of Information Technology, Nagpur, India
e-mail: pooja.jain@cse.iiitn.ac.in

B. Agarwal et al. (eds.), *Proceedings of International Conference on Intelligent Cyber-Physical Systems*, Algorithms for Intelligent Systems,
https://doi.org/10.1007/978-981-16-7136-4_19

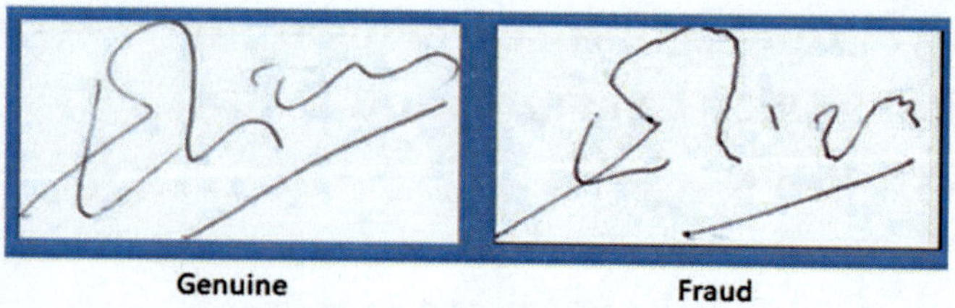

Fig. 1 Signature of the same subject left: real right: morphed

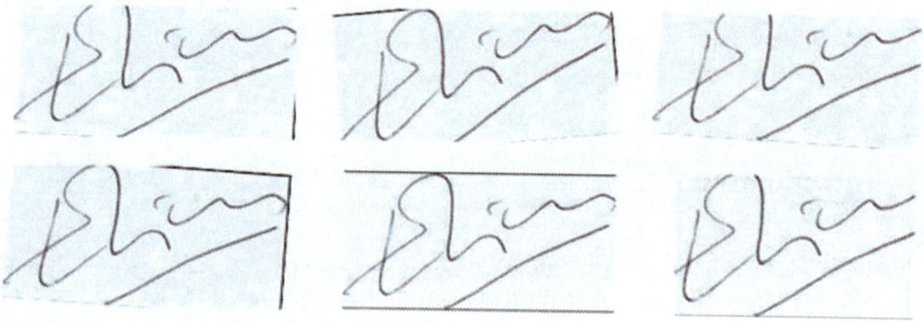

Fig. 2 Augmentation on same image rotation and zoom

2 Literature Survey

The objective is to develop the handwritten verification system using the latest advancement in deep learning. Input parameter to this system is paired of two signatures in portable network graphics images (PNG) format and outputs the Boolean value (1 or 0). This paper focuses on the experiment of convolution neural network architectures and different classification techniques. The deep learning-based method has emerged as successful tool for computer vision and pattern recognition-related applications [17]. It is a lot easier to verify the digital signature as compared with handwritten signature verification, this counts in the most challenging areas of pattern recognition. Although signature verification is a well-researched problem and there are many contributors of the same.

SVC2004 [1] "The first international signature certification competition". The competition has two competitions, competing first with 13 teams with ERR 2.84% and second with 8 teams with ERR 2.89%.

Many ways to get the right limit from the reference are being investigated. A positive result yields a false negative rejection rate of 2.8% and a false acceptance rate over 1.5%. A database test containing a total of signatures over 1200 of people greater than 100 shows that author-based thresholds provide better results than using the same limit [3]. The "Siamese" process of the neural network is used.

The authenticity of the test signature is established by aligning it with the reference user's reference signature, using a dynamic time. The authors [4] compared the test signature with the corresponding mean values found in the reference set, forming a three dimensional vector. This feature vector is then divided into one of two categories (real or fraud). The key component analysis received an error rate of 1.4% of the 619 test signatures and 94 people.

From Ref. [5], an online signature verification methodology has been introduced. The system uses a timeline set with Hidden Markov Models (HMMs). Development tests and tests are reported in the subcorpus of the MCYT bimodal biometric database containing more than 6500 signatures from a total of 145 studies [6]. The developers were familiar with the verification process and did their best to defraud the system. The acceptance rate for random forgeries, i.e., the accidental similarity of two different signatures, was 0.16%.

Classifiers based on neural net feeds are used. The factors used to distinguish are guessing times and symbols based on the upper and lower envelope. The output of the three separators is integrated using a connecting system. The integration of these separators based on signature verification is a distinctive feature of this work. Test results show that the combination of classifiers increases the reliability of visual results.

In Ref. [7], Datasets selected by CEDAR, MCYT and GPDS were performed. The performance of the algorithm proposed is based on three precision steps such as FAR, FRR and AER [9]. Compared with the standard system, the findings were found to be 20% error. The database SVC2004 was selected to verify the signature [10]. We tested our approach to GPSSynthetic, MCYT, SigComp11 and CEDAR databases that demonstrate the generality of our suggestion. The review [11] includes the implementation of state-of-the-art programs in selected subjects in five public databases.

The authors [8] used signal processing for the signature verification task. Vector representation of words is used for the analysis of sentiments [16]. The authors [12–15] used the Google Net, Inception-v1, Inception-v3, DAG-CNN and other architectures model for signature verification. We have in-depth studied in this paper about the VGG16, ResNet-50, Inception-V3 and Xception architectures.

3 The Proposed Methodology

A. Dataset

Data used are ICDAR 2011 Signature Dataset, which consists signature of 69 subjects and multiple genuine and forged signatures.

B. Proposed system

This paper is divided into two major steps (1) Feature extraction and (2) Classification. Following toward the tasks, CNN models are involved for feature extraction and supervised models for classification.

C. Feature extraction

Convolution neural network (CNN) is a popular neural network architecture for working on image dataset. It consists of certain number of layers such that output of previous layer is fed to next layer as input. These images are feed as 3D array than 1D or flattened array because CNN architectures are designed to treat images as human visual cortex. The architecture of CNN determines the function of each layer and connections in layers. There are four architectures of CNN used in the paper for experiment, VGG16, Inception-v3, ResNet-50 and Xception. Choosing a suitable architecture for the dataset is crucial to complete the first step, i.e. feature extraction of our handwritten signature verification system.

Convolution Layers: The convolution layer is the building block of the convolutional network that does a lot of complex computer-based lifting. CONV layer parameters contain a set of readable filters. Each filter is small in area (in terms of length and width) but expands to the full depth of input volume. During the progression, we slide (accurately, convince) each filter into the width and height of the input volume and calculate the dot products between the filter input. As we load the filter over the dimensions of the input, we will produce a map that provides the feedback for that filter to all areas below.

Max-pooling Layers: High integration, or greater cohesion, is a merging function that determines the maximum, or largest value in each section of the map for each feature. The results are sample or combined feature maps highlighting the feature that is most present in the piece, not the central presence of the feature in the case of a moderate combination. This has been found to be more effective in performance than standard integration of computer viewing functions such as image splitting. We can make concrete of the composite work by re-inserting it into the feature map of the active metal detector and manually calculating the first line of the composite map.

All the architectures used in the paper are modified by removing the fully connected layers with the output layer to fine-tune the already trained model to extract features.

Architecture 1: VGG16

VGGNet-16 has 16 layers of resolution and is very attractive due to its similar Architecture, similar to AlexNet, but has many filters. It is currently the most widely used way to remove elements from images. VGG16 weight loss is publicly available and used in many other programs and challenges as a first feature release. However, VGG16 has 138 million parameters, which is a challenge to train. When the model is specified in the database and the parameters are changed and updated for increased accuracy, we can use the parameter values.

Architecture 2: Inception-v3

In 2014, Google researchers introduced the first network that stood first in the competition, which has ImageNet dataset for discovery challenges.

The model is made up of a basic unit called the "Inception Cell" in which we perform a series of interaction. Inception-v3 has 24 M parameters.

Architecture 3: ResNet50

ResNet50 is a variety of ResNet model with 48 layers. It is the most popular and used ResNet model, and we have the design of ResNet50 in depth. ResNets was originally included in the image recognition function but as stated in the paper that the framework can be used for non-computer activities and for better accuracy. ResNet50 has 23 M parameters.

Architecture 4: Xception

Xception is known as Extreme Inception based entirely on divisively divisive structures. The construction of Xception has 36 disclosure layers that form the basis of network outsourcing. The 36 layers of convolutional are organized into 14 modules, all with residual connections around it, with the exception of the first and last modules. The Xception architecture is a series of deep dividing layers with remaining connections. Xception is an adaptation from Inception model and has 23 M parameters.

Optimizers used are (i) Stochastic gradient descent (SGD), (ii) Root Mean Square Propagation (RMSprop) (iii) Adaptive Gradient Algorithm (Adagrad), (iv) Active Design and Analysis Modeling (Adam).

D. Classification

After the feature extraction, they are stored in the comma separated value (CSV). We have obtained a different number of features as described in Table 1. Explaining back our dataset, we have prepared the pairwise data for each subject. The first column is genuine signature, and the second column is either genuine signature or fraud signature associated with the same subject. Labels used are 0 if both signatures are genuine and 1 otherwise.

File format used for genuine signature is PNG and naming schema used is XXX/YY_XXX for Genuine signature and XXX_forg/YY_ZZZXXX for forged signature. XXX denotes the person ID, YY denotes the signature number, ZZZ is the person ID signed the signature.

In the classification, we have used (i) Euclidian Distance, (ii) Cosine Similarity, (iii) Linear SVM, (iv) RBF SVM, (v) Sigmoid SVM, (vi) Poly SVM, (vii) Logistic

Table 1 Basic info about architecture

Architecture	VGG16	Inception-v3	ResNet-50	Xception
Parameters	138 M	24 M	23 M	23 M
Features Extracted	512	2048	2048	2048

Table 2 The random sample of five points

Column 1	Column 2	Target
065/06_065	065_forg/01_0118065	1
042/09_042	042_forg/03_0118042	1
029/05_029	029/02_029	0
001/001_12	001/001_15	0
065/01_065	065/05_065	0

Regression and (viii) Random Forest. We first created pairwise similarity between column 1 and column 2 using Euclidian distance and Cosine similarity. Pairs with the similarity greater than trainable hyper-parameter are not forged.

$$\textbf{Euclidian Distance} : d(p, q) = \sqrt{\sum_{i-1}^{n} (q_i - p_i)}$$

$$\textbf{Cosine Similarity} : sim(p, q) = \frac{p \cdot q}{\|p\| * \|q\|} = \frac{\sum_{i=1}^{n} (p_i * q_i)}{\sqrt{\sum_{i-1}^{n} pi^2} * \sqrt{\sum_{i-1}^{n} qi^2}}$$

∴ p, q are Euclidean points; p_i, q_i are feature vectors; n is dimension of vector.

Support Vector Machine, Logistic Regression, Random Forest: K(Xi, Xj) = (Xi,Xj + 1)d.

Features of image 1 and image 2 are concatenated to make total features of 2 * features of 1 image. The reason to choose SVM is that our previous experience and its results on high dimensional data (Table 2).

4 Results

The results obtained with feature extraction are presented in Tables 3, 4, 5, 6 and Graphs 1 and 2. The results obtained with classification are presented in Tables 7, 8, 9, 10. Bold value represents the best results.

Table 3 Training accuracy (3-fold)

	Optimizers			
	SGD	RMSprop	Adagrad	Adam
VGG16	0.8648	**0.9645**	0.8821	**0.9584**
Inception-v3	0.8042	**0.9827**	**0.9567**	**0.9922**
ResNet50	**0.9515**	**0.9991**	**0.9991**	**0.9974**
Xception	0.7730	**0.9835**	0.8215	**0.9939**

Table 4 Training loss (3-fold)

	Optimizers			
	SGD	RMSprop	Adagrad	Adam
VGG16	**0.4497**	**0.0918**	0.3716	**0.1069**
Inception-v3	0.4485	**0.0448**	0.2218	**0.0176**
ResNet50	0.1561	**0.0050**	**0.0324**	**0.0084**
Xception	0.5424	**0.0642**	0.4889	**0.0221**

Table 5 Validation accuracy (3-fold)

	Optimizers			
	SGD	RMSprop	Adagrad	Adam
VGG16	**0.7091**	**0.9717**	0.5111	**0.9556**
Inception-v3	0.5818	0.4202	**0.6020**	**0.6323**
ResNet50	0.4182	0.5879	0.5818	0.4182
Xception	0.5697	0.5818	0.5657	0.5899

Table 6 Validation loss (three-fold)

	Optimizers			
	SGD	RMSprop	Adagrad	Adam
VGG16	**0.5971**	**0.0793**	**0.9206**	**0.1127**
Inception-v3	**0.7371**	8.5688	**0.7872**	2.3959
ResNet50	1.2646	**0.6738**	1.4782	**0.7494**
Xception	**0.7339**	7.0186	**0.7754**	3.2455

Table 7 Model i—VGG16-Adam

	Accuracy	Precision	Recall	Time
Logistic Regression	0.9852	0.9789	0.9880	7.2 s
Random Forest	**0.9900**	**0.9851**	**0.9926**	**12.5 s**
Linear SVM	**0.9902**	**0.9909**	**0.9956**	**66 s**
RBF SVM	0.9353	0.9361	0.9146	75 s
Sigmoid SVM	0.5017	0.4322	0.4349	82 s
Poly SVM	0.8624	0.8977	0.7962	67 s

The first observation from the above tables VGG16 architecture outperformed all other architectures and features from the models that can be used for classification are with at least 95% training accuracy and 60% validation accuracy. Four models that we choose to test our classification algorithms are (i) VGG16—Adam, (ii) VGG16—RMSprop, (iii) Inception-v3—Adam and (iv) Inception-v3—Adagrad.

Table 8 Model ii—VGG16-RMSprop

	Accuracy	Precision	Recall	Time
Logistic Regression	0.9887	0.9836	0.9911	7.22 s
Random Forest	**0.9911**	**0.9868**	**0.9934**	**12.1 s**
Linear SVM	**0.9915**	**0.9860**	**0.9932**	**38.1 s**
RBF SVM	0.9625	0.9558	0.9534	56.8 s
Sigmoid SVM	0.5608	0.4982	0.4982	80 s
poly SVM	0.8699	0.9040	0.8060	62 s

Table 9 Model iii—Inception-v3—Adam

	Accuracy	Precision	Recall	Time
Logistic Regression	0.8897	0.8698	0.893	8.48 s
Random Forest	**0.9914**	**0.9872**	**0.993**	**11.8 s**
Linear SVM	0.9347	0.9100	0.946	406 s
RBF SVM	0.8955	0.8790	0.894	93 s
Sigmoid SVM	0.6680	0.6267	0.625	64 s
poly SVM	0.9066	0.8907	0.907	74 s

Table 10 Model iv—Inception-v3—Adagrad

	Accuracy	Precision	Recall	Time
Logistic Regression	**0.9920**	**0.9883**	**0.9939**	**7.39 s**
Random Forest	0.9891	0.9839	0.9919	14.8 s
Linear SVM	0.9894	0.9857	0.9923	19.2 s
RBF SVM	**0.9915**	**0.9882**	**0.9942**	**32.3 s**
Sigmoid SVM	0.9178	0.9101	0.9139	31.4 s
poly SVM	**0.9928**	**0.9892**	**0.9945**	**19.9 s**

We now, for the classification, refer to the selected architecture for feature extraction with the assigned roman number above. i, ii, iii and iv for VGG16—Adam, VGG16—RMSprop, Inception-v3—Adam and Inception-v3—Adagrad, respectively.

For the Classification Tasks, models selected for feature extraction are used with supervised learning algorithms and supporting performance metrics such as Accuracy for each model, Confusion Matrix whenever necessary. All the models are trained on CPU i5-7200U with 8 GB of RAM. We ran the tests for three cross-fold validation.

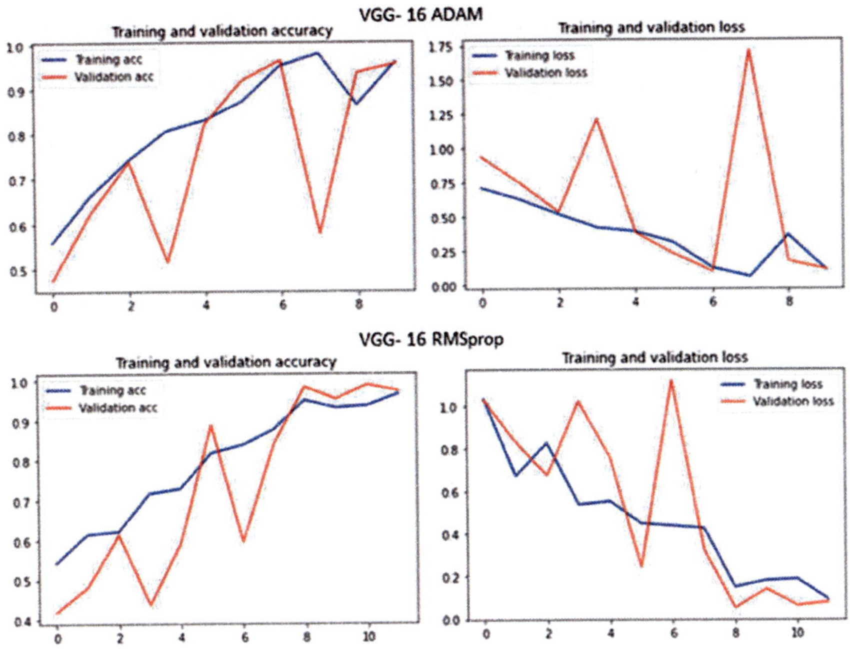

Graph 1 Feature extraction results for a selected model of VGG16

From our observation, Euclidean distance and Cosine similarity didn't perform well with our model while SVM outperformed all the models. Euclidean distance and Cosine similarity-based classification methods tend to overfit with our features. Average training time for Inception-v3-Adagrad architecture is significantly lower than other architectures. Bold marked model is with accuracy greater than 99% in Tables 7, 8, 9, 10. Best performing model is poly SVM with an accuracy of 99.28%. Other metrics to evaluate the model are also in Fig. 3. Figures 4, 5, 6, 7 are sample examples of our final handwritten signature verification system.

5 Conclusion

In this paper, we have presented methods for feature extraction and classification on signature dataset. This paper does not focus on manual crafted features, inspire feature extraction is done using CNN architectures. Experiment conducted on the ICDAR 2011 Signature Dataset showed features extracted from VGG16 outperformed all other architectures with small margins. Other architectures may perform better than other datasets because experiments are random for feature extraction. In addition, classification best results are obtained with poly SVM on feature extracted from Inception v3trained on Adagrad optimizer.

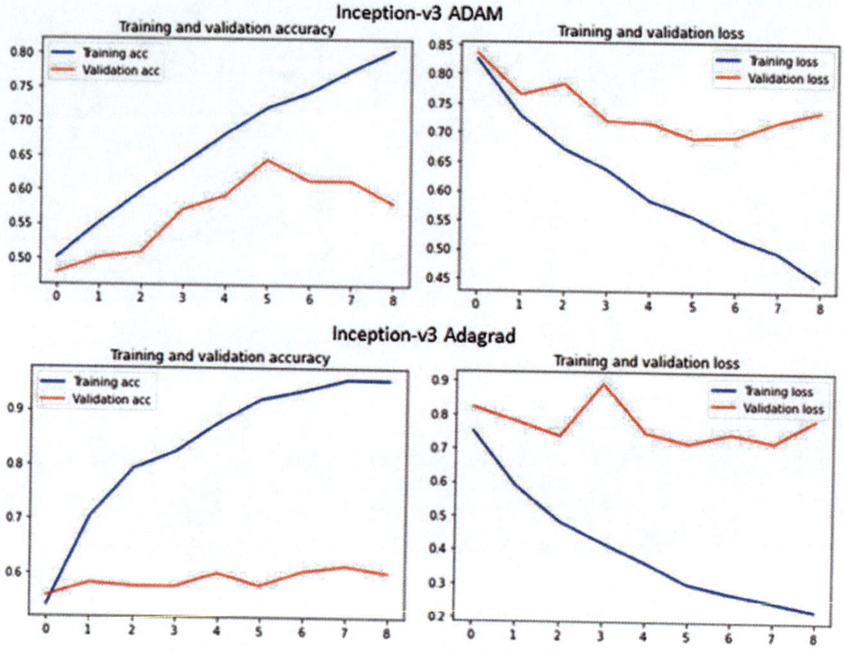

Graph 2 Feature extraction results for a selected model for Inception-v3

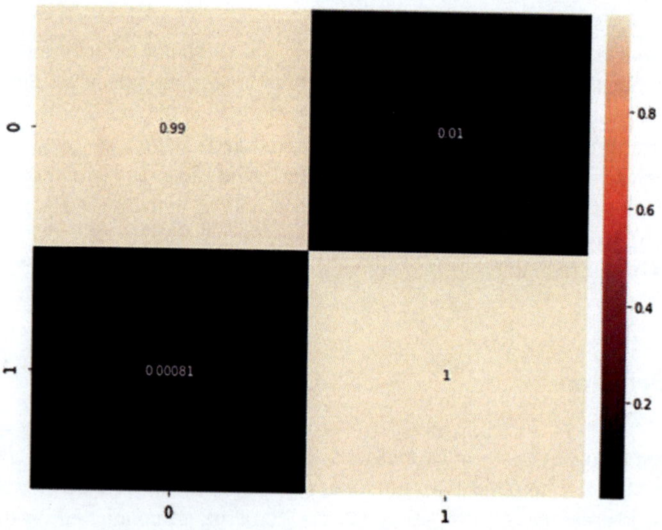

Fig. 3 Confusion matrix for best performing model

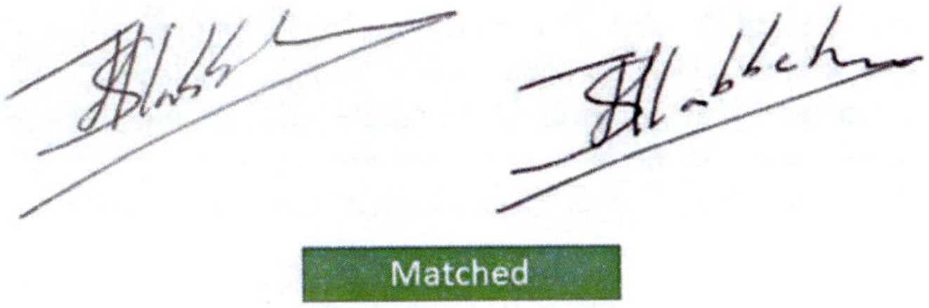

Fig. 4 Sample Example 1 after classification

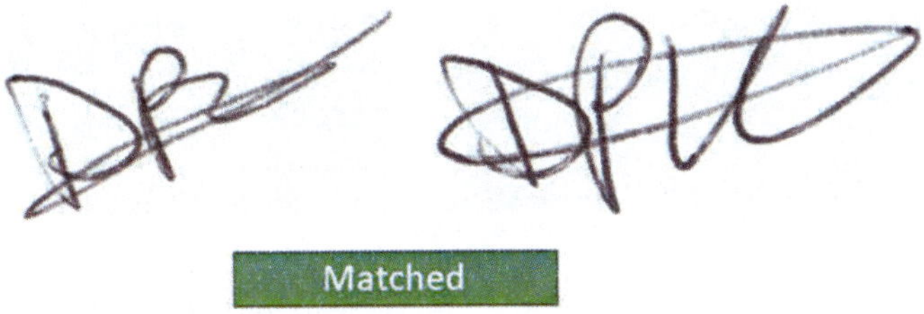

Fig. 5 Sample Example 2 after classification

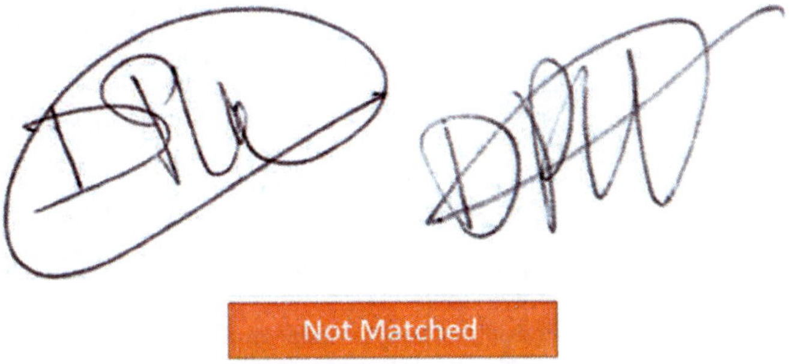

Fig. 6 Sample Example 3 after classification

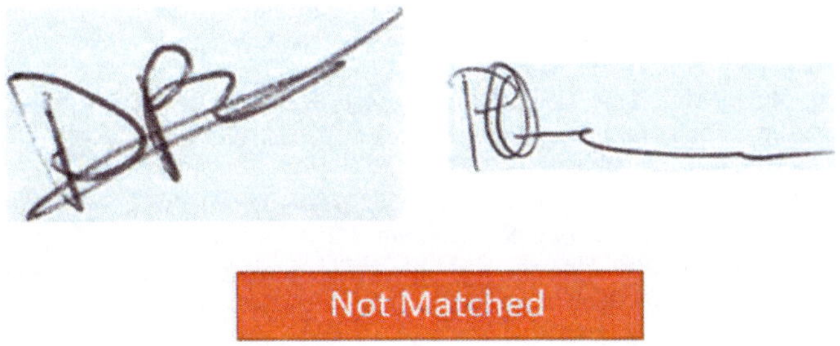

Fig. 7 Sample Example 4 after classification

Though studies till now have proved its best results on the recognition of handwritten digits (MNIST), its performance is not significant in the verification of signatures. As future work, we will focus on the development of a more purpose-specific neural network model. Furthermore, other different classification techniques specific to signature data can be explored.

Acknowledgements We thank the Indian Institute of Information Technology for constant support and providing opportunity.

References

1. Yeung, D.Y., Chang, H., Xiong, Y., George, S., Kashi, R., Matsumoto, T., Rigoll, G.: SVC2004: First international signature verification competition. In International conference on biometric authentication, pp. 16–22. Springer, Berlin (2004)
2. Jain, A.K., Griess, F.D., Connell, S.D.: On-line signature verification. Pattern Recogn. **35**(12), 2963–2972 (2002)
3. Bromley, J., Guyon, I., LeCun, Y., Säckinger, E., Shah, R.: Signature verification using a" siamese" time delay neural network. Adv. Neural. Inf. Process. Syst. **6**, 737–744 (1993)
4. Yildiz, M., Yanikoğlu, B., Kholmatov, A., Kanak, A., Uludağ, U., Erdoğan, H.: Biometric layering with fingerprints: template security and privacy through multi-biometric template fusion. Comput. J. **60**(4), 573–587 (2017)
5. Fierrez, J., Galbally, J., Ortega-Garcia, J., Freire, M.R., Alonso-Fernandez, F., Ramos, D., Gracia-Roche, J.J.: BiosecurID: a multimodal biometric database. Pattern Anal. Appl. **13**(2), 235–246 (2010)
6. Praino, A.P., Treinish, L.A.: IBM Thomas J Watson Research Center, Yorktown Heights, New York
7. Sharif, M., Khan, M.A., Faisal, M., Yasmin, M., Fernandes, S.L.: A framework for offline signature verification system: Best features selection approach. Pattern Recogn. Lett. (2018)
8. Ghosh, R.: A recurrent neural network based deep learning model for offline signature verification and recognition system. Expert Syst. Appl. **168**, 114249 (2021)
9. Tamilarasi, K.: Design and implementation of deep learning strategy based smart signature verification system. Microprocess. Microsyst. **77**, 103119 (2020)
10. Ruiz, V., Linares, I., Sanchez, A., Velez, J.F.: Off-line handwritten signature verification using compositional synthetic generation of signatures and Siamese neural networks. Neurocomputing **374**, 30–41 (2020)
11. Hameed, M.M., Ahmad, R., Kiah, M.L.M., Murtaza, G.: Machine learning-based offline signature verification systems: a systematic review. Signal Process.: Image Commun. **116139** (2021)
12. Khalajzadeh, H., Mansouri, M., Teshnehlab, M.: Persian signature verification using convolutional neural networks. Int. J. Eng. Res. Technol. **1**(2), 7–12 (2012)
13. Jagtap, A.B., Hegadi, R.S., Santosh, K.C.: Feature learning for offline handwritten signature verification using convolutional neural network. Int. J. Technol. Human Interact. (IJTHI) **15**(4), 54–62 (2019)
14. Alajrami, E., Ashqar, B.A., Abu-Nasser, B.S., Khalil, A.J., Musleh, M.M., Barhoom, A.M., Abu-Naser, S.S.: Handwritten signature verification using deep learning (2020).
15. Shabbir, S., Malik, M.I., Siddiqi, I.: Offline Signature Verification Using Feature Learning and One-Class Classification. Pattern Recogn Artif Intell **1322**, 242 (2021)

16. Sharma, Y., Agrawal, G., Jain, P., Kumar, T.: Vector representation of words for sentiment analysis using GloVe. In: 2017 International Conference on Intelligent Communication and Computational Techniques (ICCT), pp. 279–284. IEEE (2017)
17. Ram, S., Gupta, S., Agarwal, B.: Devanagri character recognition model using deep convolution neural network. J. Stat. Manag. Syst. **21**(4), 593–599 (2018). https://doi.org/10.1080/09720510.2018.1471264

Comprehensive Review on Machine Learning for Plant Disease Identification and Classification with Image Processing

Shital Jadhav and Bindu Garg

1 Introduction

Increasing population and decreasing manpower is a major challenge in Agriculture. Changing global environment, crop diseases and pests are factors with which farmers and agronomists are combating. The use of Artificial Intelligence for the detection of crop problems would be beneficial because of the remote nature of Farming and the unavailability of Agriculture experts on the field. Increasing use of technology such as the Internet of Things, Image Processing and mobile technology would provide a solution to this lack of expertise. Mobile phones are being used widely by Farmers that can act as tools for smart farming. These devices have different sensors like motion, environment, position and camera that can be used for making farming easier [27, 34].

Pests and diseases affect different parts of plants. Early identification can be done by the diagnosis of the problem and recommendation of a solution for it. Bacteria, fungi and viruses cause different diseases in plants along with it. There are other reasons like water stress or nutrient stress. Efficient analysis of different parts of plants with images acquired by a camera can help in the diagnosis of problems [25].

The multiple Image Processing problems are well addressed by Machine Learning (ML) methods. Plant leaves are a peculiar category of images and visual examination is used for the diagnosis of diseases. The problem with image processing is training time required for images and accuracy percentage varies with the image acquisition method. Real field image processing is challenging due to the complex nature of images.

S. Jadhav (✉) · B. Garg
Department of Computer Engineering, Bharati Vidyapeeth's College of Engineering, Pune 411043, India

B. Garg
e-mail: brgarg@bvucoep.edu.in

B. Agarwal et al. (eds.), *Proceedings of International Conference on Intelligent Cyber-Physical Systems*, Algorithms for Intelligent Systems,
https://doi.org/10.1007/978-981-16-7136-4_20

The research questions identified for conducting this survey are as follows [30]:

i. What are the standard practices of collecting crop image data?
ii. What are different machine learning techniques for crop disease identification and classification?
iii. How to solve challenges in real field image segmentation?
iv. Which Convolutional Neural Network (CNN) model would be effective for disease classification?
v. How to improve present CNN architecture for disease identification and classification?
vi. How to reduce parameters of the CNN algorithm to run on mobile devices?
vii. How to improve accuracy for mobile-based CNN?

This paper represents answers to these research questions divided into sections, background, literature survey, analysis and future challenges.

2 Background

2.1 Use of Image Processing in Crop Disease Identification

Agriculture Image Processing is a new horizon for developers and researchers. A visible range of images can be captured with mobile cameras at a low cost. Image processing applies various algorithms in stages to understand images. An intelligent system can be developed with image processing for early detection of plant diseases. There is a requirement for new algorithms and models specific to crop disease identification problems [35].

Segmentation of real field images in agriculture is the critical step. It is a process that divides images into different segments by the identification of objects based on regions. The quality of the identification of healthy and unhealthy parts of a crop depends on the segmentation algorithm and classification method used [24] (Fig. 1).

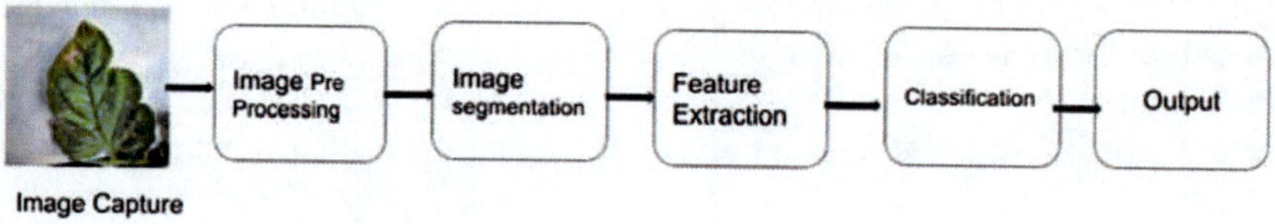

Fig. 1 Image classification stages

2.2 *Machine Learning Approaches*

The majority of work on image processing before 2015 used traditional approaches like Color Transforms, Color Co-occurrence, Gray Co-occurrence, Wavelet Transform, Clustering and Deep Learning methods for image preprocessing and classification. The techniques are used in three categories like Traditional, Deep Learning and Hybrid [29].

2.2.1 Traditional Perspectives

Identification of disease is carried out traditionally with methods that detect a change in texture, shape and size of the leaf. Segmentation techniques are used to identify areas of interest with handcrafted feature extraction [15]. Study of various segmentation techniques are useful for the identification of infected leaves from real field images. Algorithms like K-Nearest Neighbor, Decision Tree, Random Forest, Support Vector Machines (SVM), etc. are used to classify images.

SVM with one or more hyper-planes in multidimensional space is used to find distance with respect to nearest training data points belonging to different classes. Kernel is a mathematical function provided for manipulating data. It provides inner multiplication between two standard feature points. Types of kernels are Gaussian, Gaussian Kernel Radial Basis and Sigmoid Function [16] (Table 1).

2.2.2 Deep Learning Perspectives

Convolution Neural Networks (CNN) are mainly used for image classification and feature extraction. This is a fully automated way of learning features instead of handcrafted feature extraction. It requires a large amount of training images to obtain accuracy [13]. CNN has three layer categories: Input image layer, Output layer and Intermediate hidden layer obtained by convolution of the filter. Depending on the type of feature to be extracted, it has categories like Edge Detector CNN, Region-based CNN (RCNN) and various models developed for classification of images. Chronological successful CNNs for classification are in year 2012, AlexNet (8 layers) and ZFNet (8 layers); in year 2014, VGGNet (19 layers) and GoogLeNet (22 layers); in year 2015, Residual Net (18,34,50,101,152 layers) and Inception (27 layers); and in year 2017, MobileNet (28 layers). Another factor determining the accuracy and computational efficiency of the model is the Activation function. Hyperbolic Tangent (TanH), Logistic (Sigmoid), Rectified Linear Unit (ReLU) and Softmax are commonly used activation functions in CNN [23, 28] (Table 2).

Table 1 Comparison of various segmentation techniques

Segmentation methods	Popular models	Description	Pros	Cons	Citation
Edge-based segmentation	Canny Edge Detector **Sobel** Edge Detector Laplace Edge Detector	Detected edges in an image are assumed to represent object boundaries, and used to identify these objects	Very important local image preprocessing method	Image noise and unsuitable information have negative influence on segmentation	[40]
Region-based segmentation	Region Growing	Starting in the middle of an object and then growing outward until it meets the object boundaries	Region growing techniques are better in noisy images	The computation is time-consuming, Noise or variation of intensity may result in holes or over segmentation. This method may not distinguish the shading of the real images	[40]
Thresholding Technique	Ostu Methods Iterative Optimal Thresholding Multispectral Thresholding	Transform input image to output binary image based on threshold value	Computationally inexpensive and fast	Don't Perform well when there is noise in image, non-uniform lighting conditions and object is small compared to background	[3, 5, 8, 11, 22]
Clustering techniques	K Means Fuzzy C Mean	Group together patterns that are similar in some sense	Fuzzy methods based on partial membership useful for real problems	Complexity of Algorithm, Result varies due to change in image capture condition	[6, 22]
Contour-based Method	Bounding Box-based Snake and Active Contour	Draw shape around object iterate to fit	Kernel based on canny edge and thresholding betters with iteration	Active user involvement	[14]

Table 2 Comparison of various CNN architectures

CNN models	Year	Dataset	Input size	Number of layers	Number of parameters	Characteristics
LeNet LeNet5	1998	Handwritten numbers	32*32 60*60	7	60 k	Gradient-Based Learning applied to document. Recognition [7]
VGG	2014	Imagenet	224*224*3	19	7.3 m	Small convolution kernels and modularized design [1, 6, 9, 10, 18, 26]
GoogLeNet	2014	ImageNet			6.8 m	Inception Module is designed and dimension reduction using 1*1 kernel [12, 18]
AlexNet	2012	ImageNet		8	60 m	Use of ReLU and Dropout [1, 6, 9, 10, 18, 19]
Inception V3	2015	ImageNet	224*224*3	27	23.6 m	The BN layer added and 3*3 kernel replaces previous 5*5 [19, 26]
ResNet	2015	ImageNet CIFAR 10	224*224*3	110 50	6.61 m 25.5 m	Design of residual learning block is added [14, 19, 33]
SENet	2017	ImageNet	224*224*3	–	35.7 m	Channel relationship of feature maps reformed [26]
DenseNet 40, 100	2017	CIFAR 10	above 32*32*3		1 m, 27.2 m	Densely Connected Network Structure [19]
MobileNets	2017	CIFAR 10	224*224*3	28	4.2 m	Uses depthwise and pointwise convolutions [14, 36, 37]
ShuffleNets	2017	ImageNet	224*224*3	50		Channel shuffling group convolutions [38]
Xception	2017	JFT Google dataset	299*299*3	36	22.8 m	Data parallelism in model with synchronous gradient [39]

2.2.3 Hybrid Perspectives

In some cases, results are observed to have more accuracy with traditional techniques than using resource-consuming deep learning techniques. When restricted data and resources are available, it is suggested to use traditional techniques. With the availability of processor, storage and data at a low cost, a new automatic approach would be beneficial. Depending on the complexity of the problem, a combination of traditional and deep learning techniques is carried out by researchers [4, 29, 31, 32].

3 Literature Survey

Crop leaves have different morphological features. Various techniques have been used till now for the detection and classification of crop leaves. In the literature survey, various papers were analyzed based on keywords Smart Farming or Agriculture and Plant Leaf Disease and Image Processing and ML or CNN. Around 60 papers were evaluated based on their research contribution and citations.

In 2016, Mohanty et al. [1] focused on evaluating transfer learning and training from scratch for the publically available PlantVillage dataset for 14 crop types and 26 diseases. All the images were collected in a controlled environment. Performance of standard models GoogleNet and AlexNet achieved high accuracy of around 99% for plant diseases. This approach proved to be better compared to previous hand engineering methods for feature extraction and enhancement.

In the same year, Sladojevic et al. [2] carried out plant disease detection for peach, apple and grape. Original leaf image dataset of 4483 augmented by using Image transformations such as affine, perspective and rotation. Automated feature extraction is carried out by using the CaffeNet deep convolutional neural network model. In every layer, it refines the extracted feature of the previous layer. Graphical processing units were used for training images that took around 40 h. The model is used on a single computer for identification and verifying the performance of the model. Fine-tuning accuracy of 96.3% was obtained.

In 2017, Petrellis [3] uses color area, number of spots to determine the extent of disease early and first stages. Detection of disease signature is based on rules related to the color shape of spots and historical weather data. Format of disease signature helps agriculturists to act as end-users of the developed application. Agriculturists as end-users can customize and extend the plant disease mobile application. The work focuses on two plant grapevines and citrus obtaining an accuracy of 90%. To reduce the cost of toxic treatment, crop disease needs to be identified at an early stage. The author developed a mobile application for the classification of plant disease by evaluating various features extracted by photographs using disease signature. This window-based system analysis plant leaf each pixel where gray level of pixel exceeds offset threshold value for disease spot. In advance method color level RGB of spot used to prepare background normal spot (BNS) with 0 background color 1 normal leaf and 2 for spot. Histogram analysis of the image is carried out to determine normal leaf and disease leaf with spots. Colored histogram of three types is used for the analysis of specific disease powdery mildew. Leaves with the same disease are analyzed by using blue histogram to identify lobe peak that crosses specific threshold starting and end of lobe used to rank disease based statistical analysis is done in order to develop feature set. Future work suggested in the paper is disease signature on more platforms like Android, iOS, etc.

In the same year, DeChant et al. [4] combined different classifiers in three stages instead of using a single classifier. Real field images of maize plants are acquired for Northern leaf blight (NLB). In the first stage, 224*224 image segments are identified from real field images. Five CNN algorithms were trained for identifying infected and non-infected regions on plants. The first stage generates a heat map for the next stage. The second stage works on locating the region of infection by understanding the heat map. The final stage combines all the results and diagnoses lesions by stacking all the heat maps.

Singh et al. [5] carried out work for identifying leaf disease based on genetic algorithms (GA). Genetic algorithms are based on the principle of clustering chromosomes based on interest. GA works in iteration till it is satisfied. Segmentation of leaves done using GA later features extracted by color contrast, Energy and Homogeneity. SVM classifier is used to classify five classes of bacterial spots, Frog eye, sunburn, fungal and early scorch on leaves of lemon, rose and beans.

Brahimi et al. [6] proposed an architecture to classify nine diseases for tomatoes using pre-trained ImageNet architecture which initialized weights for the next stage. The authors replaced the last classification layer for nine image classes instead of 1000 classed in standard architecture. After classification, the area of the affected part is visualized for an inexperienced user.

Amara et al. [7] proposed a method using LeNet for bananas that uses preprocessed RBG and grayscale images in 60*60 size. The first stage of LeNet carries out feature extraction by pooling layer and ReLU activation function. Max pooling layer extracts reduced features. The fully connected layer uses the Softmax activation function to map extracted features to classes.

In 2018, GoogLeNet and CIFAR-10 model based on deep learning were proposed by Xihai Zhang et al. [12]. The two models were tested on eight kinds of maize leaf diseases. Two models were trained for nine kinds of maize leaf images by adjusting model parameters, changing the pooling combination adding dropout operation and ReLU—rectified linear unit function and reducing the number of classifiers. A dataset was prepared for object recognition algorithm research at all stages from training to the evaluation phase. The maize image dataset contains 3060 out of which 2248 were for 80% training and 612 20% were for testing.

In 2019, work carried out by Michael Gomez et al. [14] helps farmers with pest and disease detection using convolutional neural networks. A dataset of 5 different classes of banana disease was collected from South Africa and India. Diagnosis was carried out not only on leaves but also on other parts of the plant. These images are captured from plants without detaching from plants in various environmental conditions and growth stages. To avoid confusion between dried and old leaves images are also considered, firstly, labeling of the bounding box on captured infield images for diseases part completed. Six different models were developed for three different convolution neural network architectures. The model was trained with ResNet50, MobileNetX1 and Inception V2 architecture using TensorFlow object detection API with GPU. Observation after study shows that ResNet150- and Inception V2-based models performed well compared to MobileNet V1, and an accuracy of more than 90% was observed. To give mobile capability, the results were compared on single shot detector (SSD) Mobile Net V1.

Jakjoud et al. [17] proposed to create 5 models with different optimizers and tested them with the Raspberry Pi model. Plant leaf images from different sources like websites, free datasets and captured images with smartphone cameras were used. Augmentation and rotation methods were used to expand the existing dataset. D-CNN model with multiple layers has first part feature extraction for input image. The first layer is the convolution layer that holds a learnable filter for each application for raw pixel values with RGB color intensities convolution of these with filter create 2D feature map. It performs subsampling from a number of convolution layers that act as pooling layers after applying the convolution kernel. The second part of the model is a fully connected layer of neurons that acts as a classifier by connections of learned maps later put in the output layer in predefined classes; healthy or infected leaf is predicted by a sigmoid activation function. It was found that SGD models are fast and robust, 90% efficient and accurate on laptops. In contrast, AdaDelta is best for automated detection using Raspberry Pi.

3.1 Analysis

After a detailed study of papers, evaluation based on various parameters are presented in the following (Table 3).

3.2 Performance Evaluation of Various Parameters

Here, the lists of the parameters that have been considered in the entire contributions are given in the tabulation for considering it in future works for developing an efficient crop disease classification and identification technique (Table 4).

3.3 Important Observations Toward Developing Optimized Model

The following are some of the important observations made after literature analysis which will be useful for future work.

Table 3 Comparative analysis of literature

Author [citation]	Methodology	Features	Challenges/future scope
Mohanty et al. [1]	PlantVillage images preprocessed by segmentation for removing extra background information. Evaluated deep learning model AlexNet and GoogLeNet 364 classes. Transfer learning and training from scratch used	Standardization of hyper-parameters like stochastic gradient descent, learning rate, momentum, weight decay and batch size. Improved results compared to hand engineering features	Accuracy decreases to 31% if image taken outside the dataset. All images obtained under controlled environment. Future work can be extended to images acquired by smartphone with additional information as location and time
Sladojevic et al. [2]	Fifteen disease classes used for three fruit crops. Augmentation process applied to increase size of dataset. ReLU activation function used to speed up CNN	Fine-tuning of hidden layer and hyper-parameter carried out. Results are significantly improved by augmentation of data	In future server-side systems with mobile clients can be developed. Dataset can be extended with wide-area images and aerial photographs
Petrellis [3]	Handcrafted symptom signatures based on color area and number of spots. Background normal spot (BNS) with 0 background color 1 normal leaf and 2 for spot. Histogram analysis of image is carried out to determine normal leaf and disease leaf with spot	Lightweight mobile application. GPS location used to extract historical weather data	Used only 85 images to identify 4 classes. Future work suggested disease signatures on more platforms like Android, iOS, etc.
DeChant et al. [4]	Image segmentation by windowing method for real field maize images used. Three stages of five CNN models used for classification of disease	ReLU activation function used. Fine-tuning of parameter done by hand after each stage	Dead leaf tissue resembles NLB disease to CNN and normal eyes. Thousands of images required for higher accuracy. In future, systems can be mounted on aerial vehicles

(continued)

Table 3 (continued)

Author [citation]	Methodology	Features	Challenges/future scope
Singh et al. [5]	Images of five classes and three plants used. Image segmentation based on genetic algorithms used for plant leaves	Color contrast, Energy and Homogeneity used for feature extraction and SVM for classification. It helps in early-stage identification of disease	In future, Bayes classifiers, ANN, fuzzy logic and hybrid algorithms can be used
Brahimi et al. [6]	Classification and visualization of nine diseases of tomato crop using pre-trained AlexNet and GoogLeNet	Caffe framework is used to parallelize deep learning models on GPU. Extraction of biological features based on symptoms and visualization for inexperienced users	Future work would be reducing computation and size of deep learning model. Visualization of features would be the next level of research
Amara et al. [7]	LeNet seven-layer model is compact compared to AlexNet and GoogLeNet used for classification. Input image size is small 60*60	Hyper-parameter tuning using stochastic gradient descent, learning rate, momentum, weight decay and batch size	In future, work can be extended for more number of diseases and estimating severity of disease
Bai et al. [8]	Morphological features and watershed algorithm used remove background from real field leaf images. Integrity of leaf checked by solidity parameter which is ratio of extracted area and convex area. For identification infected area gray pixel center used for clustering. Fuzzy C means (FCM) algorithm determines membership fuzziness based on grayness of pixel	Authors modified FCM by adding spatial features of Euclidean distance to gray color scale. Good extracted images give 86.24% accuracy	This method is constrained by Image acquisition condition. Farmer has to keep the infected leaf at the center. Failed extraction gives 2% accuracy

i. Complex real field images could be segmented effectively with clustering techniques based on morphological features and interactive techniques based on contours [18, 35].
ii. Selection of CNN architecture based on time and space required for training. Initial architectures like AlexNet, GoogLeNet and VGG improved accuracy

Table 4 Performance evaluation using Parameters

S. no.	Research papers	Performance measures					
		Type of crop and dataset	Preprocessing technique	Feature extraction	Classifier	Accuracy	Training time
1	Sharada Mohanty et al. [1]	14 crops, 54,306 images	Mask based on color, saturation and lightness	Color, Grayscale, leaf segment	AlexNet and GoogLeNet	98.21%	Multiple hours on GPU
2	Srdjan Sladojevic et al. [2]	3 fruit 4483 original and 30,880 augmented images	Augmentation	Five layers of CNN for extraction	Softmax function for identifying image class	96.3%	40 h on GPU
3	Petrellis [3]	Grapevines and Citrus	5 images used for feature extraction	Histogram analysis. Thresholding	Handcrafted feature Background Normal Spot	90%	Smartphone
4	Chad DeChant et al. [4]	Maize Northern leaf blight (NLB) 1796 images	Image rotation and sliding window approach for segmentation	Windowing for segmentation of image	Five CNN models at three stages	97.8%	NVIDIA Titan GPU takes 3 days per stage. At run stage 1 image classified in 2 min
5	Vijai Singh et al. [5]	Five disease classes for lemon, rose and beans. 106 images used	Genetic algorithm for segmentation	Color co-occurrence	SVM	95.71	Very less computation time compared to NN
6	Brahimi M et al. [6]	Nine disease classes of tomato, 14,828 images	Image resized by 256*256 to 227*227	Pre-trained AlexNet and GoogLeNet model used for feature extraction	CNN model AlexNet and GoogLeNet last layers modified for 10 classes	AlexNet 97.9% GoogLeNet 96.5%	Few hours with Intel Xeon Processor with Quadro GPU

(continued)

Table 4 (continued)

S. no.	Research papers		Performance measures				
		Type of crop and dataset	Preprocessing technique	Feature extraction	Classifier	Accuracy	Training time
7	Amara et al. [7]	Two disease classes for Banana, 3700 images	Image resized by 60*60	LeNet used for feature extraction from gray and color images	Last fully connected layer modified for classification	Gray images 94.44% and color images 98.61%	Deep learning 4j library used to speed up on GPU
8	Bai et al. [8]	129 cucumber disease images	Leaf extraction by background removal by using morphological features	Solidity, Image neighborhood and grayscale used for clustering	Iterative use of marked watershed algorithm based on FCM	Good extraction 86.24% Failed extraction 2%	–
9	Alvaro F. Fuentes [9]	Six disease data for tomato crop 13,262 images	Image resized by 256*256 to 227*227 AlexNet and 224*224 for VGG 16	Algorithm performed well at minibatch size 32. Learning rate 30 gives best results	AlexNet and VGG16 Net	AlexNet 97.49% and VGG16 97.23%	CUDA enabled 4 GB GPU
10	Michael Gomez et al. [14]	18,000 original field images Banana plant	Labeling bounding box	R-CNN used with CNN model to differentiate between background and object	R-CNN ResNet50, MobileNetX1 and R-CNN Inception V2	Performed equally with accuracy 90%	GPU NVIDIA Tesla

compared to the handcrafted approach. Present work analyzed these models for images taken in the lab. Later research work on real field images was carried out with architectures like ResNet and Inception. These models have an increased number of hidden layers and reduced kernel size [1, 6, 9, 10, 14].

iii. Simplification of CNN architecture based on the principal of pruning. Pruning makes the network smaller and faster. In the present research, modification by pruning higher fully connected layers and kernel size is practiced by some of the researchers [2].

iv. Hyper-parameter tuning controls the number of parameters required by CNN architecture. Fine-tuning of these hyper-parameters like filter, input size, hidden layer and batch size reduces computational time [17].

v. Speedup for training achieved by graphical processing unit (GPU) instead of central processing unit (CPU) [4, 7, 9, 14].

vi. Accuracy of CNN depends on activation function, ReLU proves to be better than Tanh. Optimizer is another important hyper-parameter; Adagrad, Adadelt and Adam give better result in present review papers [17].

vii. Compact models like MobileNet, ShuffleNet replace expensive convolution layer by depthwise separable convolution block. These lightweight models have 3*3 and pointwise 1*1 filters [14, 36, 38].

4 Challenges for Future Research Work

After analyzing present research work, the following limitations of existing work could be addressed by researchers in the future:

i. Various capture conditions of in-field images affect the accuracy of training and testing models. Real field images are complex with green and soil in the background. Most of the time manual segmentation is done to separate leaves from the background [1, 19, 21].

ii. Performance of CNN models is high compared to traditional models. But it requires a large dataset, more training time in hours even on GPUs. Resource and accuracy trade-offs need to be evaluated [4, 7, 9, 14].

iii. Developing a model from scratch is a very complex and time-consuming task. There is limited availability of computation power and data on mobile and embedded devices. Transfer learning needs to explore more for plant images by customizing existing models for such devices [1, 17].

iv. Very few annotated image datasets are available. Most research was carried out on images on the PlantVillage dataset. Such models are reducing in accuracy for real field images [18, 19].

v. Automatic identification and visualization of disease features for inexperienced end-users is not addressed together in present research [19, 20].

vi. Present algorithms assume one disease on the image. factors like visual symptoms that vary with stages of crop development, meteorological conditions, and nutritional deficiency affect crop condition [19, 21].

5 Conclusion

Traditional models like Support Vector Machine (SVM) are efficient models for multi-class classification. With the availability of high processing power even with mobile devices, new approaches for automatic disease detection need to be explored. Convolution Neural Network (CNN) is a promising model which has attained more demand in the research field nowadays. Various architectures of CNN need to be evaluated for various applications. The hybrid approach will be useful for real field crop images due to complex real field images. Extracting leaves from real field images needs to be carried out using interactive segmentation methods. More efficient, lightweight models for mobile devices need to be evaluated for computation power and accuracy.

References

1. Mohanty S.P., Hughes D.P., Salathe M.: Using deep learning for image-based plant disease detection. Front. Plant Sci. (2016). https://doi.org/10.3389/fpls.2016.01419
2. Sladojevic S., Arsenovic M., Anderla A., Culibrk D., Stefanovic D.: Hindawi: deep neural networks based recognition of plant diseases by leaf image classification. Comput. Intell. Neurosci. (2016). https://doi.org/10.1155/2016/3289801
3. Petrellis N.: Mobile application for plant disease classification based on symptom signatures. Assoc. Comput. (2017). ACM ISBN 978-1-4503-5355-7/17/09
4. DeChant C., Wiesner-Hanks T., Chen S., Stewart E.L., Yosinski J., Gore M.A.: Automated identification of northern leaf blight infected maize plants from field imagery using deep learning. Phytopathology **107**(11), 1426–1432 (2017)
5. Singh V., Misra A.K.: Detection of plant leaf diseases using image segmentation and soft computing techniques. Elsevier. Inform. Process. Agricult. **4**, 41–49 (2017)
6. Brahimi M., Boukhalfa K., Moussaoui A.: Deep learning for tomato diseases: classification and symptoms visualization. Int. J. Appl. Artif. Intell. **31**(4), 299–315 (2017)
7. Amara J., Bouazizi B., Algergawy A.: A deep learning-based approach for banana leaf diseases classification. In: Lecture notes, in informatics (LNI), pp. 79–88 (2017)
8. Bai X., Li X., Zetian Fu, Lv X., Zhang L.: A fuzzy clustering segmentation method based on neighbourhood grayscale information for defining cucumber leaf spot disease images. Comput. Electron. Agric. **136**, 157–165 (2017)
9. Fuentes A.F., Yoon S., Lee J., Park D.S.: High-performance deep neural network-based tomato plant diseases and pests diagnosis system with refinement filter bank. Front. Plant Sci., 1 August 2018
10. Rangarajan A.K., Purushothaman R., Ramesh A.: Tomato crop disease classification using pre-trained deep learning algorithm. ScienceDirect, Procedia Comput. Sci. **133**, 1040–1047 (2018)
11. Sabrol H., Satish K.: Tomato plant disease classification in digital images using classification tree. IEEE International Conference on Communication and Signal Processing, April 2016

12. Zhang X., Qiao Y., Meng F.: Identification of Maize leaf disease using improved Deep CNN. IEEE Access (2018). https://doi.org/10.1109/Access.2018.2844405
13. Toda Y., Okura F.: How convolutional neural networks diagnose plant disease. Plant Phenomics (2019) Article ID 9237136. https://doi.org/10.34133/2019/9237136
14. Selvaraj M.G., Vergara A., Ruiz H., Safari N.: AI powered banana diseases and pest detection. Open Access J. Plant Methods (2019)
15. Shen H., Kaiya Y.: Distinction of vegetable diseases by image processing. ACM (2019). ISBN 978–1–4503–6843–8/19/09. https://doi.org/10.1145/3338840.3355653
16. Khan M.A. et al.: An optimized method for segmentation and classification of apple diseases based on strong correlation and genetic algorithm based feature selection. IEEE Access (2019), Special Section On New Technologies For Smart Farming, Research Challenges and Opportunities
17. Jakjoud F., Hatim A., Bouaaddi A.: Deep learning application for plant diseases detection. Assoc. Comput. Mach. ACM (2019) ISBN 978–1–4503–7240–4/19. https://doi.org/10.1145/3372938.3372983
18. Ferentios K.P.: Deep learning models for plant disease detection and diagnosis. Comput. Electron. Agric. Elsevier (2018)
19. Arsenovic M., Karanovic M., Sladojevic S., Anderla A., Stefanovic D.: Solving current limitations of deep learning based approaches for plant disease detection. Symmetry **11**, 939 (2019). https://doi.org/10.3390/sym11070939
20. Garcia J., Barbedo A.: Digital image processing techniques for detecting, quantifying and classifying plant diseases. Springer Open J. (2013)
21. Garcia J., Barbedo A.: A review on the main challenges in automatic plant disease identification based on visible range images. Sci. Direct Biosyst. Eng. 144, 52e60 (2016)
22. Zhou G., Zhang W., Chen A.: Rapid detection of rice disease based on FCM-KM and faster R-CNN fusion. Open Access J. IEEE Access (2019)
23. Liakos K.G., Busato P., Moshou D.: Simon Pearson and Dionysis Bochtis . machine learning in agriculture: a review. Sensors, 1–29 (2018). https://doi.org/10.3390/s18082674
24. Hungilo G.G., Emmanuel G., Emanuel A.W.R.: Image processing techniques for detecting and classification of plant disease. Rev., ACM ISBN 978-1-4503-6269-6/19/04
25. Pongnumkul S., Chaovalit P., Surasvadi N.: Applications of smartphone-based sensors in agriculture: a systematic review of research. Hindawi Publishing Corporation, J. Sens., 1–18 (2015). https://doi.org/10.1155/2015/195308
26. Hang J., Zhang D., Chen P., Zhang J., Wang B.: Classification of plant leaf diseases based on improved convolutional neural network. Sensors (2019). https://doi.org/10.3390/s19194161
27. Jha K., Doshi A., Patel P., Shah M.: A comprehensive review on automation in agriculture using artificial intelligence. Sci. Direct Artif. Intell. Agric., 2589–721 (2019). https://doi.org/10.1016/j.aiia.2019.05.004
28. Potgieter J.: Plant Disease Detection and Classification by Deep Learning. Muhammad Saleem, Plants (2019)
29. Asmita, A., Pawar, V.R.: Machine learning regression technique for cotton leaf disease detection and controlling using IoT. IEEE International Conference on Electronics, Communication and Aerospace Technology ICECA 2017 978-1-5090-5686-6/17. IEEE (2017)
30. Nagaraju M., Chawala P.: Systematic review of deep learning technique in plant disease detection. Int. Syst. Assur. Eng. Manag., Springer (2020)
31. Thorat A., Kumari S., Valakunde N.D.: An IoT based smart solution for leaf disease detection. International Conference on Big Data, IoT and Data Science (BID) 2017, 978-1-5090-6593-6/17 IEEE (2017)
32. Patil S.S., Thorat S.A.: Early detection of grapes diseases using machine L system earning and IoT. Second International Conference on Cognitive Computing and Information Processing (CCIP) 978-1-5090-1025-7/16, IEEE (2016)
33. Saravanan M., Perepu S.K.: Focusing social media based analytics for plant diseases in smart agriculture. ACM (2018). ISBN 978-1-4503-6465-2/18/07. https://doi.org/10.1145/3227696.3227720.

34. Foughalia K., Fathallah K., Frihida A.: Using Cloud IOT for disease prevention in precision agriculture. Sci. Direct Procedia Comput. Sci. **130,** 575–582 (2018)
35. Ngugi L., Zahhad M.: Recent advances in image processing techniques for automated leaf pest and disease recognition—a review. Inform. Process. Agric. Sci. Direct, April 2020
36. Howard A.G., Zhu M., Chen B.: MobileNets: efficient convolutional neural network for mobile vision application. Dmitry. ACM 17 Apr 2017. arXiv:1704.04861v1 cs.CV
37. Michele A., Colin V., Santika D.: MobileNet convolutional neural Network and support vector machine for Palmprint recognition. Sci. Direct Procedia Comput. Sci. **157**, 110–117 (2019). https://doi.org/10.1016/j.procs.2019.08.147
38. Zhang X., Sun J.: ShuffleNet an extremely efficient convolution neural network for mobile devices. Open Access Comput. Vis. Found. IEEE XPlore (2017)
39. Angelica E.J., Pascual V., Mhar J., Plaza J., Lorenzo J., Tesorero L., De Goma J.C.: Disease Detection of Asian Rice (Oryza Sativa) in the Philippines Using Image Processing. Association for Computing Machinery, ACM ISBN 978-1-4503-7290-9/19/10. https://doi.org/10.1145/3366650.3366676
40. Saradhambal G., Dhivya R., Latha S., Rajesh R.: Plant disease detection and its solution using image classification. Int. J. Pure Appl. Math. **119**(14), 879-884 (2018). ISSN: 1314–3395

Application of Transfer Learning with CNNs for Pneumonia Detection in Chest X-rays

Piyush Batra and Imran Hussain

1 Introduction

Deep learning techniques have a vast scope in medical diagnostics. One of the medical fields where deep learning methods can shine is radiology and imaging. These techniques can revolutionise disease diagnosis by performing time-consuming radiograph analysis quickly. Deep learning generally requires a large number of labelled samples for a model to perform competently. But, the cost of collecting vast amounts of medical data can be expensive and time-consuming. To overcome this issue, we made use of a concept called transfer learning. Transfer learning is a technique where a pre-trained model is adapted and reused as initialisation or a fixed feature extractor for a new model to solve the task at hand. Generally, this pre-trained model is trained for a similar type of problem on millions of data samples utilising large compute resources. There are many applications to explore in machine learning in the context of medical science; one of those is the Detection of Pneumonia using X-Ray of the Lungs.

Pneumonia is an infectious disease that inflames the alveoli in one or both lungs, which can give rise to many problems such as chest pain, difficulty in breathing, high fever, and nausea. It causes fluid exudation in and around air sacs and increases the effort of breathing. If not timely treated, it can lead to permanent fibrosis and lung damage. A diversity of pathogens, including viruses, bacteria, and fungi, can cause pneumonia. It is severely consequential in infants, immunocompromised patients, the elderly, and people with known medical problems. It mostly influences the lungs, but sometimes complexities can prompt issues in different parts of the body, as well. The infection from the lungs can spread via the blood to the entire body, causing septicemia, which can be fatal. Vulnerability, treatment, and recuperation time depends upon the cause of pneumonia. Traditionally, its diagnosis includes

P. Batra · I. Hussain (✉)
Department of CSE, SEST, Jamia Hamdard, New Delhi, India

B. Agarwal et al. (eds.), *Proceedings of International Conference on Intelligent Cyber-Physical Systems*, Algorithms for Intelligent Systems,
https://doi.org/10.1007/978-981-16-7136-4_21

Blood tests, chest X-ray examination, pulse oximetry, a sputum test, and various other tests depending on the patient's medical history and current condition. According to the National Heart, Lung, and Blood Institute [1], test and analysis of chest X-rays are considered as the most effective method for pneumonia diagnosis. But, the process of reading an X-ray can be very time-consuming and requires domain expertise and experience. This delay means a possible hold-up at the start of treatment for very sick pneumonia patients. Sometimes admission of patients in hospitals may exceed the medical expertise available in the hospital. So, a computer-aided diagnosis may help in the timely treatment of patients and hence control the casualties associated with the disease.

In this study, we implemented a robust and efficient deep learning pneumonia diagnosis framework. This implementation uses transfer learning to get optimal results using the limited data that we have and helps us train our models on small compute resources. All the pre-trained models used in this paper were initially trained on ImageNet, which is a large dataset consisting of millions of labelled images from thousands of categories. This system can rapidly analyse immense amounts of chest X-rays and assist a medical expert in the diagnosis process.

We took the dataset from Kaggle named "Chest X-Ray Images (Pneumonia)" [1]. Although the dataset consisted of 5856 images, the size of images was larger than 1000 pixels per dimension, and the total dataset weighed around 1.15 gigabytes. This paper's work includes the implementation of different neural network models that were trained and fine-tuned for the best performance. Implementation for this paper also provides for the use of different techniques and functions like a learning-rate reduction on the plateau, early stopping, and checkpointing. Since the data was imbalanced, performance measures like precision and recall were taken into account to give more reliable outputs for the classification. The comparison of all the models implemented for this research, including methods and their implementation, is briefly explained in this paper.

2 Related Work

Over recent years, Deep learning techniques have achieved human-level accuracy in pattern recognition and computer vision applications [2]. There are many applications of deep learning and computer vision in healthcare which include drug discovery and management, medical imaging, medical image analysis, predictive analysis and therapy, healthcare monitoring, nuclear medicine, etc. When it comes to disease diagnosis using medical images, deep learning techniques have the potential to perform complex classification tasks smoothly with a rapid review to screen multiple radiographs in a matter of time.

Medical image diagnosis is now being automated using these state-of-the-art deep learning techniques. The timely examination of chest X-rays for signs of pneumonia plays a crucial role in the treatment course of the patient. This is helpful, especially in infants, where the fatality rate of pneumonia is quite high. If the disease is detected earlier, they can receive a targeted treatment sooner, resulting in fast recovery. But, analysis of chest radiographs is a time-consuming task and requires medical experts. As the number of patients increases, the availability of expert radiologists may decrease, which means a possible delay in the treatment process. Recently, several frameworks for pneumonia detection using chest X-rays have been proposed to solve this problem.

For example, Rajpurkar et al. [3] developed CheXNet, which is a 121-layer convolutional neural network to detect pneumonia trained on over 100,000 X-ray images with 14 diseases. For validation, they compared CheXNet with expert radiologists and found out that it surpasses average radiologist performance on the F1 metric. They further developed ChestX-ray14 and achieved substantial results on all 14 diseases. Kermany et al. [4] developed a deep neural network model based on Mask-RCNN, which includes global and local features combined with image augmentation for pixel-wise segmentation, for pneumonia diagnosis. Their generalised AI system got an accuracy of 92.8% with a recall (sensitivity) of 93.2% for pneumonia detection. Jaiswal et al. [5] developed a deep neural network model based on Mask-RCNN, which includes global and local features combined with image augmentation for pixel-wise segmentation, for pneumonia diagnosis. Toğaçar et al. [6] proposed an approach based on 300 deep features combining 100 features each from AlexNet, VGG16, and VGG19 architectures. They then used a decision tree, KNN, linear discriminant analysis, linear regression, and SVM as classifiers. They achieved the most promising accuracy with an LDA classifier. Ge et al. [7] used temporal sequence information of 13,930 patients with acute ischaemic stroke to implement a deep learning model based on MLP neural network and RNNs for post-stroke pneumonia prediction. This method outperformed standard machine learning algorithms like logistic regression, support vector machines, and extreme gradient boosting. With a deep-learning-based model, they achieved an AUC of 0.928 and an AUC of 0.905 for pneumonia prediction within 7 days and 14 days, respectively. Varshni et al. [8] proposed a deep learning framework for a pneumonia detection system. They observed the performance of different architectures like DenseNet-169 [9], VGGNets [10], Xception [11], and ResNet50 [12] as feature extractors and different classifiers including SVM (RBF), Naive Bayes, k-nearest neighbours, and Random Forest for classification. According to their findings, DenseNet-169 [9] with SVM performed the best with an AUC of 0.80. Ayan and Ünver [13] used transfer learning to compare VGG16 and Xception architectures. They achieved an accuracy of 87% on the VGG16 model and an accuracy of 82% on the Xception model.

Stephen et al. [14] constructed a convolutional neural network model from scratch to extract features from a given chest X-ray image and classify it to determine if a person is infected with pneumonia. They also employed several data augmentation methods and achieved a validation accuracy of 93.73%. Chouhan et al. [15] proposed a deep transfer learning ensemble model which combines the output from five different architectures including AlexNet, ResNet18, Inception V3, DenseNet121, and GoogLeNet. They achieved an accuracy of 96.4% with a recall of 99.62%. Liang and Zheng [16] proposed a framework based on deep learning that combines residual thought and dilated convolution to detect childhood pneumonia. They achieved a recall of 96.7%, and an f1-score of 92.7%. This approach tends to solve the low image resolution and blockage of inflammation area in children's chest radiography. Bhandary et al. [17] proposed two different deep learning frameworks to examine lung pneumonia and cancer. The first model that they suggested was a modified version of AlexNet (MAN) with an SVM classifier validated against pre-trained models including AlexNet, VGG16, VGG19, and ResNet50. The second approach implements a fusion of handcrafted and learned features in the MAN to improve classification accuracy during lung cancer assessment. They evaluated this model on lung cancer CT images from the cancer imaging archive and achieved an accuracy of 97.27%.

3 Material and Methods

We have incorporated different techniques like transfer learning, data augmentation, and model-based data pre-processing in this paper. The following subsections contain the methodology, dataset, data preparation, and performance metrics used in this study.

3.1 Transfer Learning and Pre-trained Neural Networks

Deep learning models are generally repurposable. This feature allows practitioners to utilise a model trained on a larger dataset to solve a similar problem at hand. Due to the lack of labelled data and computational resources, we included a design methodology called Transfer learning into this study. Transfer learning is a concept in machine learning that centres around putting away information picked up while taking care of one problem and applying it to an alternate yet related problem. In this technique, we generally freeze initial and middle layers which capture general features like colour blobs, patches, edges, etc. and just retrain the latter layers. It helps leverage the information learned from the task for which the model was initially trained on. Figure 1 shows a simple transfer learning workflow and how it helps us speed up the training process as well as getting excellent performance without the need for massive datasets.

In this study, we employed Keras Applications API [18] along with some standard Python libraries including Scikit-Learn [19], Matplotlib [20], NumPy [21], etc. to implement and test various deep transfer learning models. All the models used in this study were models pre-trained on the ImageNet dataset, which consists of millions of labelled images from thousands of categories. We made use of nine different neural network architectures which offer high results and accuracies. These architectures include VGG16 [10], VGG19 [10], Xception [11], MobileNetV2 [22], ResNet50V2 [12], InceptionV3 [23], InceptionResNetV2 [24], NASNetMobile, and

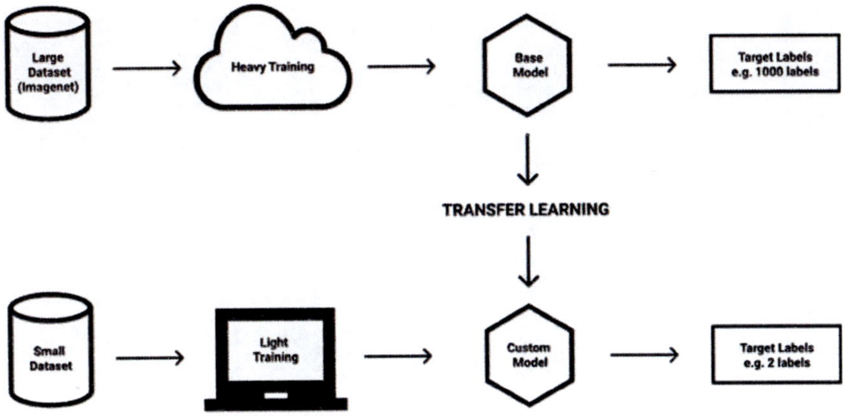

Fig. 1 Simple transfer learning workflow

NASNetLarge [25]. All these models were trained and fine-tuned to our computational resource limit. To get the best results, we trained these models using different optimisers [26], activation functions, and losses along with checkpointing techniques like Early stopping and Reduce learning rate on the plateau.

3.2 Dataset

For this study, we used Guangzhou Women and Children's Medical Center dataset [27], which is available on Kaggle as "Chest X-Ray Images (Pneumonia) [28]". It contains 5,856 labelled images of chest X-rays divided into Pneumonia and normal categories. These chest radiographs were taken from paediatric patients of age one to five. To avoid any misclassification in the dataset, these chest radiographs were checked and labelled by two specialist clinicians and further by a third-party radiologist [29]. Figure 2 represents some samples belonging to each class.

The data was already divided into three folders for training, validation, and testing. Each folder contains radiographs from both pneumonia and normal categories. The training set contains a total of 5216 (89.07%) chest X-ray images which includes 3877 X-rays belonging to children with pneumonia and 1341 healthy lung X-rays with no pneumonia/normal class. The validation set contains 16 chest X-ray images with an equal class split. The test set comprises almost 10.66% of the total data, with 390 X-rays from the pneumonia class and 234 X-rays from the normal category. We used the data as it is without tinkering with the classification split. Although, before feeding it to our model, we pre-processed all the images followed by data augmentation. Table 1 summarises the data split used in our study.

For further understanding of the nature of data and determining the skewness of data, we processed the data and plotted them into each category. From Fig. 3, we

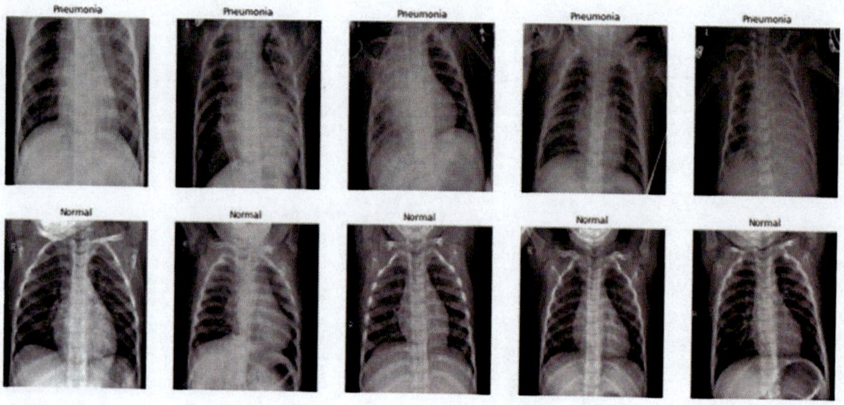

Fig. 2 Samples from the dataset [28]

Table 1 Dataset details

	Train	Validation	Test	Total
Pneumonia	3875	8	390	**4273**
Normal	1341	8	234	**1583**
Total	**5216**	**16**	**624**	**5856**

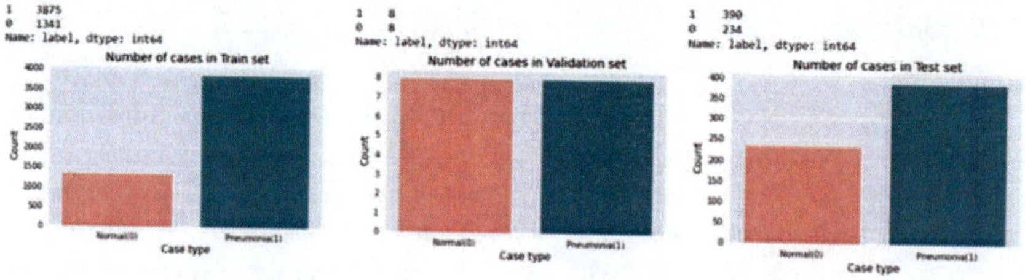

Fig. 3 Nature of data for each category

can see that there is a considerable imbalance in the ratio of classes for the training set. This data imbalance is a ubiquitous challenge that people face when it comes to medical data.

The data available for the pneumonia class is almost thrice when compared to the normal class. Hence, we can't use accuracy as a model performance measure alone; as accuracy is the number of correctly predicted data points out of all the data points in the dataset. An option in contrast to using accuracy is to use precision and recall measurements. Performance metrics used are explained in more detail in the following subsections.

3.3 Methodology

In this study, we trained, fine-tuned, and compared nine different deep learning CNNs by applying transfer learning methods. Figure 4 shows an overview of the methodology used in this study. First, we rescaled the chest X-ray images to the default input size of respective models. Then we applied data augmentation techniques to the given dataset followed by transfer learning using VGG16, VGG19, Xception, InceptionV3, InceptionResNetv2, MobileNet, and NASNetMobile architectures. The following sections give out in detail the steps of our methodology (Fig. 5).

In this approach, we are leveraging the features learned from a larger dataset and applying them to our use. Utilising such learned features allows us to build a robust and high-performing model from the available data. We will train, fine-tune, and

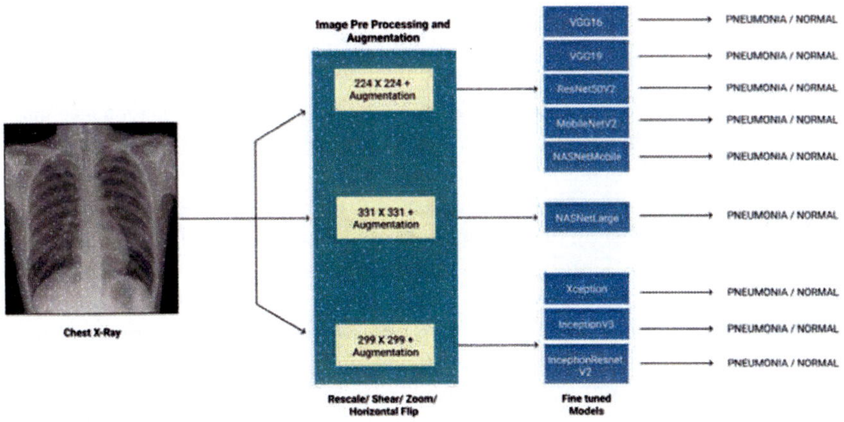

Fig. 4 Overview of the methodology

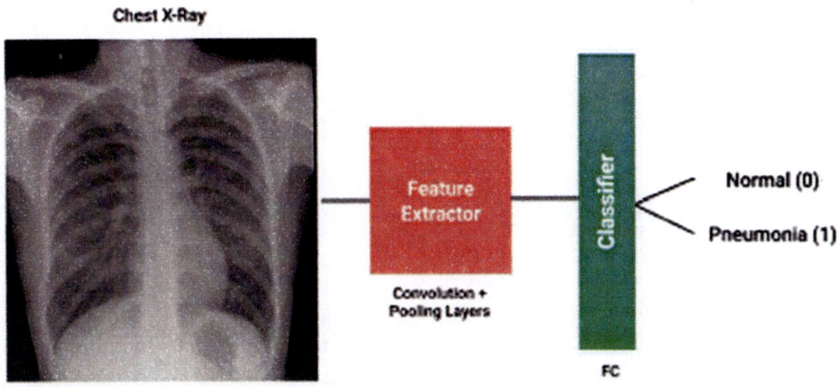

Fig. 5 Demonstration of the proposed approach

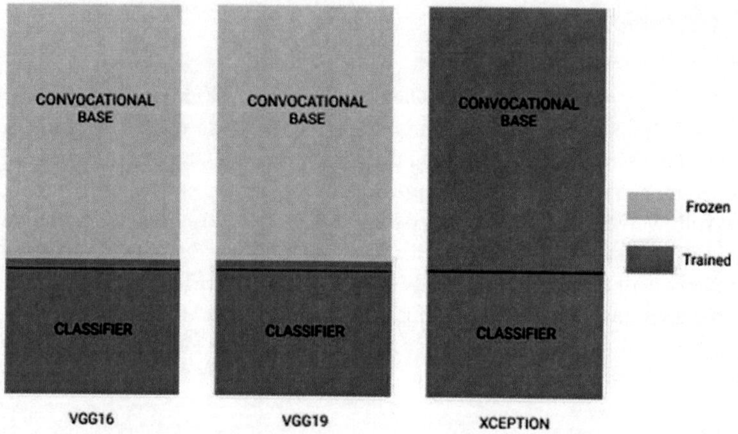

Fig. 6 Demonstration of trainable and frozen layers

compare nine different reliable architectures pre-trained on the ImageNet dataset. As ImageNet contains millions of images, our new model would not need to learn low-level features which are relevant to our classification problem. These feature extractors can then be used with a new classifier to perform classification.

For training and testing of different models, we employed a cloud instance with Intel(R) Xeon(R) CPU @ 2.20 GHz (Intel Corporation, Santa Clara, CA, USA), NVIDIA(R) T4 GPU (NVIDIA Corporation, Santa Clara, CA, USA) with 15 GB available GPU memory and 13 GB of RAM. Using Keras applications API, we imported the pre-trained model without the fully connected output layers and used it as a feature extraction module. Then we flatten the output from this model to a single vector and define a new classifier with a final dense output layer followed by an activation function to generate probabilities for each label. This classifier was then trained from scratch on top of the pre-trained model to get meaningful predictions.

Figure 6 shows our transfer learning approach by freezing and unfreezing different layers of the neural network architecture. For some of the architectures, we used the model by freezing all the pre-trained layers, and for some architectures, we had to fine-tune the model to get the best performance.

3.4 Data Pre-processing

The original images in the dataset that we used were 8 bit JPEG images with a resolution of more than 1200 pixels. To feed these images as input to our model, we resized the images in accordance with the default input shape on which the model was trained on. This also helped in making computations easier and train models faster. For VGG16, VGG19, ResNet50V2, MobileNet, and NASNetMobile models, we resized the X-ray images to 224 × 224 × 3 dimensions, which means 224 wide, 224 high, and three colour channels.

For Xception, InceptionV3, and InceptionResNetV2 models, we resized the X-ray images to 299 × 299 × 3 pixels whereas for the NASNetLarge model, we resize the images to an input size of 331 × 331x 3. In addition to resizing the input data, we also normalised the image matrix, so we target floating-point values between 0 and 1 instead of RGB coefficients 0 to 255.

3.5 *Data Augmentation*

Image augmentation is a very significant technique used for deep learning in the medical domain. This method covers for lack of a sufficient number of medical samples for a model to learn from. This technique creates variations in current images so that the model would never observe the exact same sample twice. This also helps in expanding the dataset artificially to prevent overfitting and hence improving generalisation. Figure 7 shows different random augmentation operations on a single chest X-ray image.

In this study, we used three augmentation methods, Random shearing, Random Zooming, and randomly flipping the images horizontally. These operations helped us create a vibrant and expanded form of the original small dataset and also helped us with the data imbalance problem.

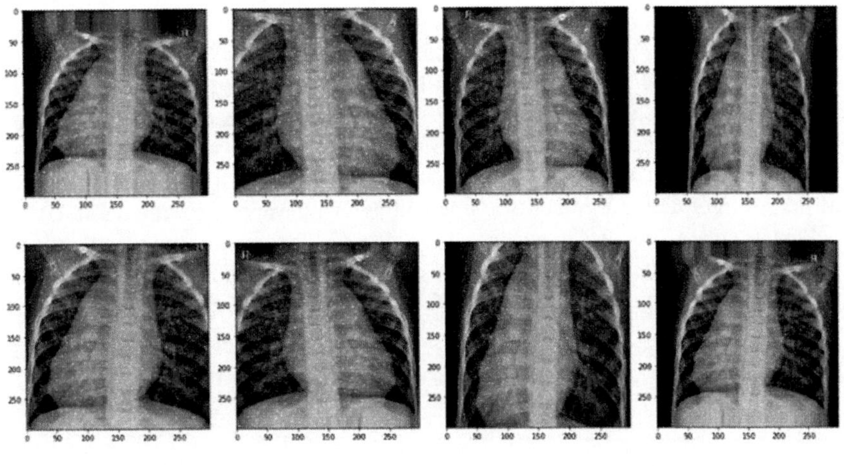

Fig. 7 Shear/Zoom/Horizontal flip Augmentation of a resized X-ray sample

3.6 Performance Metrics

Accuracy is the widely used performance metric for the evaluation of the model. But for imbalance data, it could be severely misleading as it is given by the number of correctly predicted data points out of all the data points in the dataset.

$$\text{Accuracy} = (TP + TN)/(FP + FN + TP + TN) \quad (1)$$

A common alternative to this problem is to use metrics like precision and recall instead of accuracy. Precision evaluates the number of positive class predictions that actually belong to the positive class. In contrast, Recall evaluates the number of positive class predictions made out of all the correctly classified positive labels and all the incorrectly classified negative labels, in other words, all actual positive examples in the dataset.

$$\text{Precision} = TP/(TP + FP)$$

$$\text{Recall} = TP/(TP + FN)$$

$$F1\ Score = 2 * (\text{Recall} * \text{Precision})/(\text{Recall} + \text{Precision}) \quad (2)$$

where

True-positive (TP): Correctly classified pneumonia cases.
True-negative (TN): Correctly classified normal cases.
False-positive (FP): Incorrectly classified normal cases.
False-negative (FN): Incorrectly classified pneumonia cases.

Therefore, to obtain impartial and more insightful results, we used sensitivity/recall and precision as our main performance metrics. We also used the Area under the receiver operating characteristic curve (AUC) with a default threshold of 0.5 for interpreting probabilistic predictions. All these metrics are unbiased and helped us create a robust framework. The further study is moreover based on recall metric as our aim here is to increase the true-positives as well as decrease as many false-negatives as we possibly can. A false-negative classification can lead to the further advancement of the disease and decrease the chance of survival.

Table 2 Results from various VGG16 models

Model	Recall (%)	Precision (%)	F1 Score (%)	AUC (%)	Accuracy (%)
Model 1	98.46	92.31	95.29	97.77	93.91
Model 2	97.69	92.70	95.13	97.83	93.75
Model 3	97.18	92.89	94.99	96.14	93.59
Model 4	**98.97**	90.40	94.49	96.44	92.79

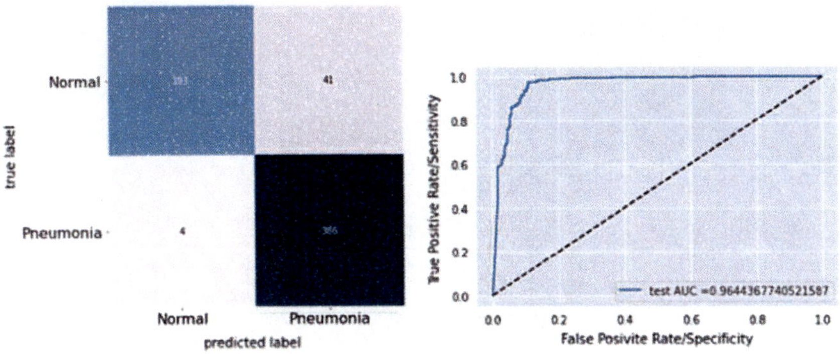

Fig. 8 Confusion matrix and ROC curve for VGG16 model

4 Results and Discussion

4.1 Results of Various Proposed Architectures

During our study, we trained and fine-tuned many different models with different configurations for all the nine architectures. This was done to create a robust model which predicts as few false-negatives as possible. In the following subsections, the performance of the best models from each neural network architecture is mentioned along with the confusion matrix and ROC curve for the best recall model.

4.1.1 VGG16

Table 2 and Fig. 8 show the results of the best models from VGG16 architecture and the confusion matrix and ROC curve for the best recall model among them.

4.1.2 VGG19

Table 3 and Fig. 9 show the results of the best models from VGG19 architecture and the confusion matrix and ROC curve for the best recall model among them.

Table 3 Results from various VGG19 models

Model	Recall (%)	Precision (%)	F1 Score (%)	AUC (%)	Accuracy (%)
Model 1	**98.21**	89.70	93.76	94.46	91.83
Model 2	97.95	89.88	93.74	94.66	91.83
Model 3	97.69	90.93	94.19	94.24	92.47
Model 4	96.92	92.42	94.62	95.96	93.11

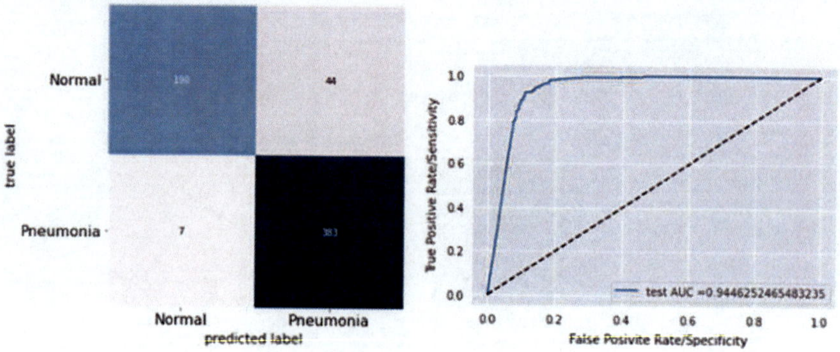

Fig. 9 Confusion matrix and ROC curve for VGG19 model

Table 4 Results from various Xception models

Model	Recall (%)	Precision (%)	F1 Score (%)	AUC (%)	Accuracy (%)
Model 1	**99.74**	87.81	93.40	96.64	91.19
Model 2	99.49	91.08	95.10	97.47	93.59
Model 3	99.23	91.27	95.09	96.87	93.59
Model 4	98.46	94.12	96.24	98.70	95.19
Model 5	97.44	95.24	96.32	98.96	95.35

4.1.3 Xception

Table 4 and Fig. 10 show the results of the best models from Xception architecture and the confusion matrix and ROC curve for the best recall model among them.

4.1.4 MobileNetV2

Table 5 and Fig. 11 show the results of the best models from MobileNetV2 architecture and the confusion matrix and ROC curve for the best recall model among them.

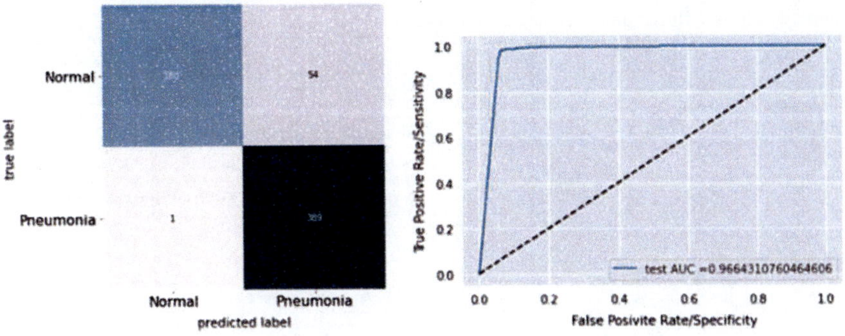

Fig. 10 Confusion matrix and ROC curve for Xception model

Table 5 Results from various MobileNetV2 models

Model	Recall (%)	Precision (%)	F1 Score (%)	AUC (%)	Accuracy (%)
Model 1	100	76.92	86.96	91.42	81.25
Model 2	**99.74**	83.48	90.89	93.46	87.50
Model 3	99.49	84.35	91.29	96.45	88.14
Model 4	99.23	85.43	91.81	96.54	88.94
Model 5	98.97	86.94	92.57	95.62	90.06

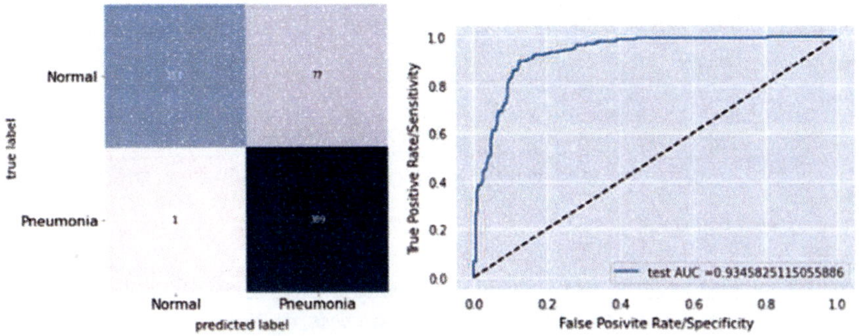

Fig. 11 Confusion matrix and ROC curve for MobileNetV2 model

4.1.5 ResNet50V2

Table 6 and Fig. 12 show the results of the best models from ResNet50V2 architecture and the confusion matrix and ROC curve for the best recall model among them.

Table 6 Results from various ResNet50V2 models

Model	Recall (%)	Precision (%)	F1 Score (%)	AUC (%)	Accuracy (%)
Model 1	**98.97**	91.69	95.19	97.93	93.75
Model 2	98.46	92.09	95.17	98.17	93.75
Model 3	97.95	92.49	95.14	98.24	93.75
Model 4	97.44	93.83	95.60	98.53	94.39

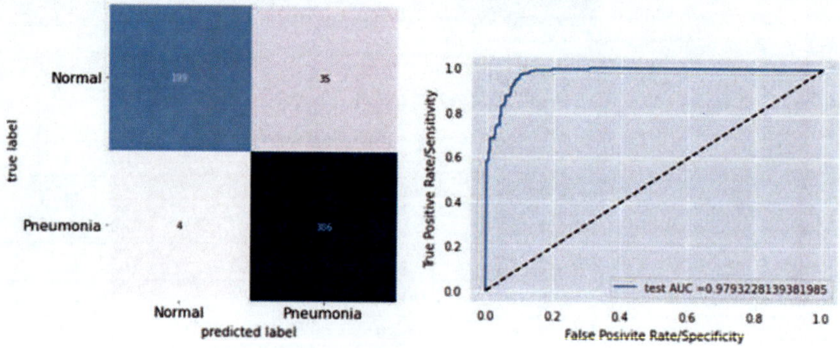

Fig. 12 Confusion matrix and ROC curve for ResNet50V2 model

Table 7 Results from various InceptionV3 models

Model	Recall (%)	Precision (%)	F1 Score (%)	AUC (%)	Accuracy (%)
Model 1	**99.49**	87.98	93.38	97.98	91.19
Model 2	97.95	92.42	95.38	98.35	94.07
Model 3	96.67	95.69	96.17	98.59	95.19

4.1.6 InceptionV3

Table 7 and Fig. 13 show the results of the best models from InceptionV3 architecture and the confusion matrix and ROC curve for the best recall model among them.

4.1.7 InceptionResNetV2

Table 8 and Fig. 14 show the results of the best models from InceptionResNetV2 architecture and the confusion matrix and ROC curve for the best recall model among them.

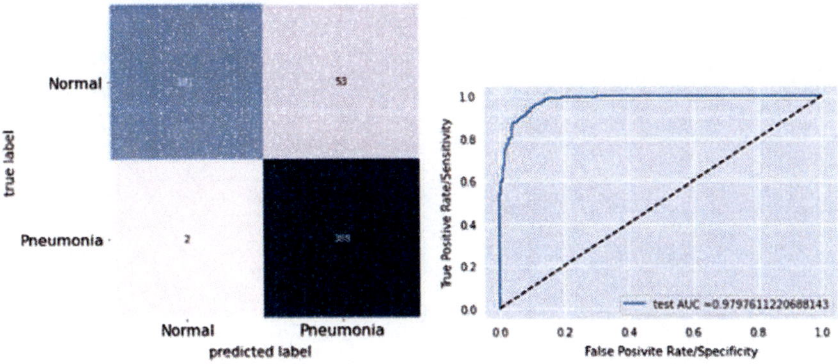

Fig. 13 Confusion matrix and ROC curve for InceptionV3 model

Table 8 Results from various InceptionResNetV2 models

Model	Recall (%)	Precision (%)	F1 Score (%)	AUC (%)	Accuracy (%)
Model 1	**98.46**	88.07	92.98	96.81	90.71
Model 2	98.21	85.30	91.30	95.12	88.30

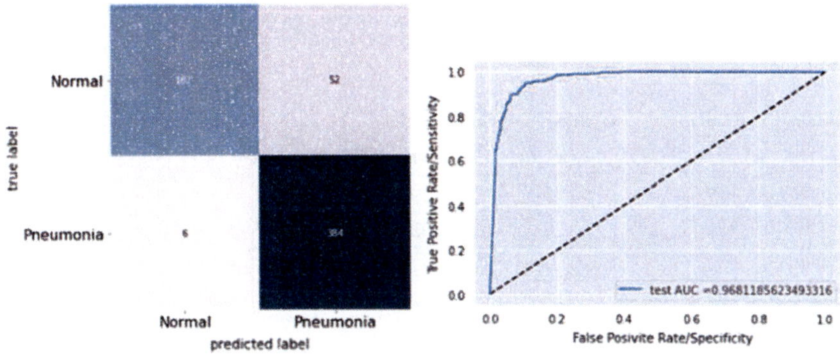

Fig. 14 Confusion matrix and ROC curve for InceptionResnetV2 model

4.1.8 NasNetMobile

Table 9 and Fig. 15 show the results of the best models from NASNetMobile architecture and the confusion matrix and ROC curve for the best recall model among them.

Table 9 Results from various NASNetMobile models

Model	Recall (%)	Precision (%)	F1 Score (%)	AUC (%)	Accuracy (%)
Model 1	**99.49**	83.62	90.87	92.23	87.50
Model 2	99.23	83.05	90.42	97.33	86.86
Model 3	98.97	90.40	94.49	98.57	92.79
Model 4	98.21	91.85	94.92	95.41	93.43
Model 5	97.69	95.01	96.33	97.67	95.35

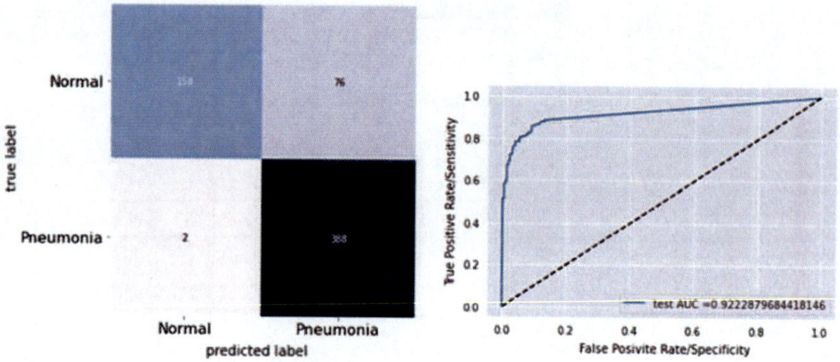

Fig. 15 Confusion matrix and ROC curve for NASNetMobile model

Table 10 Results from various NASNetLarge models

Model	Recall (%)	Precision (%)	F1 Score (%)	AUC (%)	Accuracy (%)
Model 1	**99.23**	80.96	89.17	95.59	84.94
Model 2	98.72	86.13	92.00	96.23	89.26
Model 3	97.95	88.43	92.94	96.99	90.71
Model 4	94.62	93.18	93.89	98.06	92.31

4.1.9 NasNetLarge

Table 10 and Fig. 16 show the results of the best models from NASNetLarge architecture and the confusion matrix and ROC curve for the best recall model among them.

4.2 Results

The results suggest that the Xception model performs the best for the given dataset with a recall of 99.74%, an F1 score of 93.40%, and an AUC of 96.64%. We set to train

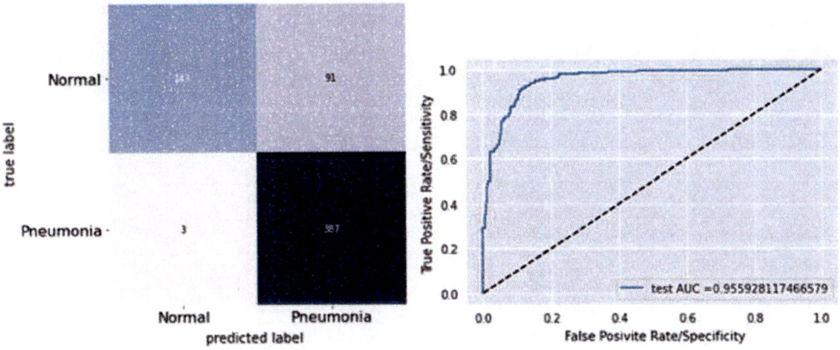

Fig. 16 Confusion matrix and ROC curve for NASNetLarge model

Table 11 Comparative results for each model

Model	Recall (%)	Precision (%)	F1 Score (%)	AUC (%)	Test accuracy (%)
VGG16	98.97	90.40	94.49	96.44	92.79
VGG19	98.21	89.70	93.76	94.46	91.83
Xception	**99.74**	**87.81**	**93.40**	**96.64**	**91.19**
MobileNetV2	99.74	83.48	90.89	93.46	87.50
ResNet50V2	98.97	91.69	95.19	97.93	93.75
InceptionV3	99.49	87.98	93.38	97.98	91.19
InceptionResNetV2	98.46	88.07	92.98	96.81	90.71
NASNetMobile	99.49	83.62	90.87	92.23	87.50
NASNetLarge	99.23	80.96	89.17	95.59	84.94

this model for 100 epochs with early stopping patience of 5. For training, we used the binary cross-entropy loss function and SGD optimiser with an initial learning rate of 0.001 and momentum of 0.9. We also added a callback named ReduceLROnPlateau with a factor of 0.2 using Keras API to reduce the learning rate when validation loss stops improving. Table 11 shows the results of the best models from each neural network architecture.

4.3 Comparison

In this section, we are comparing our results with the other work that has been done using similar pre-trained models and datasets. Here also, we are comparing the outcome based on the recall metric (if available).

Table 12 shows that our models outperform past works done using similar architectures and datasets. Apart from these significant works, Chouhan et al. [15] proposed

Table 12 Comparing results with similar models and dataset

Model	Method	Recall (%)	Precision (%)	AUC (%) AUC (%)	Accuracy (%)
Ayan et al. [13]	VGG16	85.90	91.34	–	87.98
Asnaoui et al. [30]	VGG16	85.22	87.73	–	86.26
Bharadwaj et al. [31]	VGG16	–	–	49.00	–
Thakur et al. [32]	VGG16	98.71	87.69	–	90.54
This paper	**VGG16**	**98.97**	90.40	96.44	92.79
Asnaoui et al. [30]	VGG19	90.43	80.39	–	85.94
Bharadwaj et al. [31]	VGG19	–	–	48.00	–
This Paper	**VGG19**	**98.21**	89.70	94.46	91.83
Ayan et al. [13]	Xception	93.59	81.66	–	82.85
Asnaoui et al. [30]	Xception	76.45	95.77	–	83.14
Bharadwaj et al. [31]	Xception	–	–	69.8	–
Luján-García et al. [33]	Xception	99.23	84.31	96.80	87.98
This paper	**Xception**	**99.74**	87.81	96.64	91.19
Kermany et al. [4]	InceptionV3	93.20	90.10	–	92.80
Asnaoui et al. [30]	InceptionV3	95.59	93.75	–	94.59
This paper	**InceptionV3**	**99.49**	87.98	97.98	91.19
Asnaoui et al. [30]	MobileNetV2	94.61	98.06	–	96.27
This Paper	**MobileNetV2**	**99.74**	83.48	91.93	87.50
Asnaoui et al. [30]	ResNet50	94.92	98.49	–	96.61
Bharadwaj et al. [31]	ResNet	–	–	80.00	–
This Paper	**ResNet50V2**	**98.97**	91.69	97.93	93.75
Asnaoui et al. [30]	InceptionResNetv2	93.88	98.61	–	96.61
This Paper	**InceptionResNetv2**	**98.46**	88.07	96.81	90.71

an ensemble model of five different architectures that reached a recall of 99.62% with an AUC of 99.34%. Liang and Zheng [27] used a deep residual networks-oriented approach to achieve a recall of 96.70% with 89% precision.

For Pneumonia detection, no other author used NASNetMobile or NASNetLarge neural networks on which we achieved a recall of 99.46% and 99.23%, respectively. NASNetLarge is a vast architecture with over 1040 layers and 85 million trainable parameters. Due to this huge size, fine-tuning this architecture on a single GPU system is a cumbersome task. On our configuration, it took almost 1085 s per epoch on average to train this model. Accordingly, in future work, we would fine-tune this architecture so that it generalises better and other future plans are shared in the next section.

5 Conclusion and Future Work

This study aimed to develop a framework to aid medical professionals in the detection of pneumonia using chest x-ray. We applied concepts of transfer learning and retrained nine different pre-trained models and compared them based on recall. We chose recall as our main performance metrics because we want to reduce as many false-negatives as we can. This means the less incorrect classification of pneumonia class and hence timely treatment of the patients. Our results show that the Xception neural network performs best for this problem with a recall of 99.74% and an AUC of 96.64%.

Our current standalone models have achieved state-of-the-art recall, but they still lack precision. The precision results could be improved by using an ensemble model and also by looking into some more pre-processing and image augmentation techniques for the training dataset. In future work, we plan to develop a framework that takes the patient's symptoms into account along with an ensemble model based on the best models from our current study. We also plan to use advanced feature reduction algorithms to remove unimportant features from learned feature sets to increase the performance of existing convolutional neural network models.

However, our work can efficiently detect the pneumonia class with the least false-negatives and add to the development of an automated, reliable, and fast system for Pneumonia diagnosis using chest radiographs.

Acknowledgements We thank Dr. Nitya Batra for her valuable insights during the planning and implementation of this research work. She made us better understand the disease, and all her help is very much appreciated.

References

1. Pneumonia | NHLBI, NIH. (2020). Retrieved 20 July 2020, from https://www.nhlbi.nih.gov/health-topics/pneumonia
2. Aradhya, V.N.M., Mahmud, M., Guru, D.S., et al.: One-shot cluster-based approach for the detection of COVID–19 from chest X–ray images. Cogn Comput **13**, 873–881 (2021). https://doi.org/10.1007/s12559-020-09774-w
3. Rajpurkar, P., Irvin, J., Zhu, K., Yang, B., Mehta, H., Duan, T., Ding, D., Bagul, A., Langlotz, C., Shpanskaya, K., Lungren, M.P., Ng, A.Y.: CheXNet: Radiologist-Level Pneumonia Detection on Chest X-Rays with Deep Learning (2017). Retrieved 20 July 2020, from https://arxiv.org/abs/1711.05225
4. Kermany, D.S., Goldbaum, M., Cai, W., Valentim, C.C., Liang, H., Baxter, S.L., McKeown, A., Yang, G., Wu, X., Yan, F., Dong, J., Prasadha, M.K., Pei, J., Ting, M.Y., Zhu, J., Li, C., Hewett, S., Dong, J., Ziyar, I., Shi, A., Zhang, R., Zheng, L., Hou, R., Shi, W., Fu, X., Duan, Y., Huu, V.A., Wen, C., Zhang, E.D., Zhang, C.L., Li, O., Wang, X., Singer, M.A., Sun, X., Xu, J., Tafreshi, A., Lewis, M.A., Xia, H., Zhang, K.: Identifying medical diagnoses and treatable diseases by image-based deep learning. Cell **172**(5), 1122-1131.e9 (2018). https://doi.org/10.1016/j.cell.2018.02.010

5. Jaiswal, A.K., Tiwari, P., Kumar, S., Gupta, D., Khanna, A., Rodrigues, J.J.: Identifying pneumonia in chest X-rays: a deep learning approach. Measurement **145**, 511–518 (2019). https://doi.org/10.1016/j.measurement.2019.05.076
6. Toğaçar, M., Ergen, B., Cömert, Z., Özyurt, F.: A deep feature learning model for pneumonia detection applying a combination of mRMR Feature selection and machine learning models. IRBM (2019). https://doi.org/10.1016/j.irbm.2019.10.006
7. Ge, Y., Wang, Q., Wang, L., Wu, H., Peng, C., Wang, J., Xu, Y., Xiong, G., Zhang, Y. and Yi, Y.: Predicting post-stroke pneumonia using deep neural network approaches. Int. J. Med. Inform. **132**, 103986 (2019). https://doi.org/10.1016/j.ijmedinf.2019.103986
8. Varshni, D., Thakral, K., Agarwal L., Nijhawan R., Mittal, A.: Pneumonia detection using CNN based feature extraction. 2019 IEEE International Conference on Electrical, Computer and Communication Technologies (ICECCT), Coimbatore, India, pp. 1–7 (2019). https://doi.org/10.1109/ICECCT.2019.8869364.
9. Huang, G., Liu, Z., van der Maaten, L., & Weinberger, K. (2016). Densely Connected Convolutional Networks. Retrieved 12 June 2021, from https://arxiv.org/abs/1608.06993
10. Simonyan, K., Zisserman, A.: Very Deep Convolutional Networks for Large-Scale Image Recognition (2014). Retrieved 25 July 2020, from https://arxiv.org/abs/1409.1556
11. Chollet, F.: Xception: Deep Learning with Depthwise Separable Convolutions (2016). Retrieved 25 July 2020, from https://arxiv.org/abs/1610.02357
12. He, K., Zhang, X., Ren, S., Sun, J.: Deep Residual Learning for Image Recognition (2015). Retrieved 25 July 2020, from https://arxiv.org/abs/1512.03385
13. Ayan, E., Ünver, H.M.: Diagnosis of pneumonia from chest X-ray images using deep learning. 2019 Scientific Meeting on Electrical-Electronics & Biomedical Engineering and Computer Science (EBBT), Istanbul, Turkey, pp. 1–5 (2019). https://doi.org/10.1109/EBBT.2019.8741582
14. Stephen, O., Sain, M., Maduh, U., Jeong, D.: An efficient deep learning approach to pneumonia classification in healthcare. J. Healthcare Eng. **2019**, 1–7 (2019). https://doi.org/10.1155/2019/4180949
15. Chouhan, V., Singh, S.K., Khamparia, A., Gupta, D., Tiwari, P., Moreira, C., Damaševičius, R., de Albuquerque, V.H.C.: A novel transfer learning based approach for pneumonia detection in chest X-ray images. Appl. Sci. **10**(2), 559 (2020). https://doi.org/10.3390/app10020559
16. Liang, G., Zheng, L.: A transfer learning method with deep residual network for pediatric pneumonia diagnosis. Comput. Methods Programs Biomed. **187**, 104964 (2020). https://doi.org/10.1016/j.cmpb.2019.06.023
17. Bhandary, A., Prabhu, G.A., Rajinikanth, V., Thanaraj, K.P., Satapathy, S.C., Robbins, D.E., Shasky, C., Zhang, Y., Tavares, J.M.R., Raja, N.S.M.: Deep-learning framework to detect lung abnormality—a study with chest X-Ray and lung CT scan images. Pattern Recogn. Lett. **129**, 271–278 (2020). https://doi.org/10.1016/j.patrec.2019.11.013
18. Chollet, F.: Keras: the Python deep learning API (2015). Retrieved 20 July 2020, from https://github.com/fchollet/keras
19. Pedregosa, F., Varoquaux, G., Gramfort, A., Michel, V., Thirion, B., Grisel, O.: Scikit-learn: machine learning in python. J. Mach. Learn. Res., 2825–2830 (2011)
20. Hunter, J.D.: Matplotlib: A 2D graphics environment. Comput. Sci. Eng. **9**(3), 90–95 (2007)
21. Oliphant, T.E.: A guide to NumPy (Vol. 1). Trelgol Publishing, USA (2006)
22. Sandler, M., Howard, A., Zhu, M., Zhmoginov, A., Chen, L.: MobileNetV2: Inverted Residuals and Linear Bottlenecks (2018). Retrieved 25 July 2020, from https://arxiv.org/abs/1801.04381
23. Szegedy, C., Vanhoucke, V., Ioffe, S., Shlens, J., Wojna, Z.: Rethinking the Inception Architecture for Computer Vision (2015). Retrieved 25 July 2020, from https://arxiv.org/abs/1512.00567
24. Szegedy, C., Ioffe, S., Vanhoucke, V., Alemi, A.: Inception-v4, Inception-ResNet and the Impact of Residual Connections on Learning (2016). Retrieved 25 July 2020, from https://arxiv.org/abs/1602.07261
25. Pneumonia in Children—UNICEF Data. (2019). Retrieved 20 July 2020, from https://data.unicef.org/topic/child-health/pneumonia

26. Kingma, D., Ba, J.: Adam: A Method for Stochastic Optimization (2014). Retrieved 25 July 2020, from https://arxiv.org/abs/1412.6980
27. Kermany, D., Zhang, K., Goldbaum, M.: Labeled Optical Coherence Tomography (OCT) and Chest X-Ray Images for Classification. Mendeley Data, v2 (2018). https://doi.org/10.17632/rscbjbr9sj.2
28. Chest X-Ray Images (Pneumonia). (2018). Retrieved 20 July 2020, from https://www.kaggle.com/paultimothymooney/chest-xray-pneumonia
29. Ansari, N., Faizabadi, A., Motakabber, S., Ibrahimy, M.: Effective pneumonia detection using ResNet based transfer learning. Test Eng. Manag. **82**, 15146–15153 (2020)
30. Asnaoui, K., Chawki, Y., Idri, A.: Automated Methods for Detection and Classification Pneumonia based on X-Ray Images Using Deep Learning. Retrieved 20 July 2020, from https://arxiv.org/abs/2003.14363
31. Bharadwaj, G., S.S.J.: Pneumonia detection using transfer learning. Int. J. Adv. Sci. Technol. **29**(3), 986–994 (2020). Retrieved from http://sersc.org/journals/index.php/IJAST/article/view/4178
32. Thakur, S., Goplani, Y., Arora, S., Upadhyay, R., Sharma, G.: Chest X-ray images based automated detection of pneumonia using transfer learning and CNN. Proceedings Of International Conference On Artificial Intelligence And Applications, 329–335 (2020). https://doi.org/10.1007/978-981-15-4992-2_31
33. Luján-García, J., Yáñez-Márquez, C., Villuendas-Rey, Y., Camacho-Nieto, O.: A transfer learning method for pneumonia classification and visualization. Appl. Sci. **10**(8), 2908 (2020). https://doi.org/10.3390/app10082908
34. Dadonaite, B., Roser, M.: Pneumonia. Our World In Data (2018). Retrieved from https://ourworldindata.org/pneumonia
35. Deng, J., Dong, W., Socher, R., Li, L.-J., Li, K., Fei-Fei, L.: Imagenet: A large-scale hierarchical image database. In 2009 IEEE Conference on Computer Vision and Pattern Recognition, pp. 248–255 (2009)
36. Zoph, B., Vasudevan, V., Shlens, J., Le, Q.: Learning Transferable Architectures for Scalable Image Recognition (2017). Retrieved 25 July 2020, from https://arxiv.org/abs/1707.07012v4

An Analysis of Various Text Segmentation Approaches

Sumit Kumar Daroch and Pardeep Singh

1 Introduction

Text segmentation is method of separating written text into sections or parts, and these parts are known as segments. The text can be broken into sentences, topics, and words. Each section has a relevant meaning. The concept refers both to conceptual mechanisms used by humans when interpreting text, and to artificial processes that are the subject of natural language processing (NLP) applied in computers. Text segmentation is the procedure of splitting down a document into constituent portions based on its semantic structure. The difficulty in text segmentation varies depending on what is the type of that text and how it is written: informative, talkative, descriptive, and so on. The capacity of section archives dependent on theme would empower access and investigation of the subtopic in a report instead of access and examination of entire record. Section/Segment give us a superior comprehension of the archives.

We would like to draw your attention to the fact that it is not possible to modify a paper in any way, once it has been published. This applies to both the printed book and the online version of the publication. Every detail, including the order of the names of the authors, should be checked before the paper is sent to the Volume Editors.

S. K. Daroch (✉) · P. Singh
National Institute of Technology Hamirpur, Hamirpur, Himachal Pradesh 177005, India
e-mail: cs16mi518@nith.ac.in

P. Singh
e-mail: pardeep@nith.ac.in

B. Agarwal et al. (eds.), *Proceedings of International Conference on Intelligent Cyber-Physical Systems*, Algorithms for Intelligent Systems,
https://doi.org/10.1007/978-981-16-7136-4_22

1.1 Why Text Segmentation

There are many reasons why we would want to do this, but maybe the most apparent is that surfing or looking for the results makes things much simpler for a person. Consider a long, continuous recording of a news program or a business conference. It can be difficult to find a specific news article or discuss a specific subject, especially if someone do not want to view or listen to the complete program. The first option is to check for similar words and keywords that are related to your interest, now a person can find the keyword, but it won't tell where is the starting of that section or topic and also there is no guarantee that someone will always be able to find the keywords or word you have selected, particularly if the error rates of the words are high. If the whole document is segmented according to the topic, then it is easier to find out the topic of your interest.

We will go far deeper than this: someone may be like to evaluate and identify the material of every section so that he can connect subjects from one session to another or record the evolution of news reports through multiple newscasts. He may wish to create a concise overview with the main points of every issue. In these cases, text segmentation helps a lot.

Text segmentation is a fundamental NLP challenge that is used in a wide range of activities including summarization, passage extraction, context comprehension, and emotion extraction, among others. Fine-grained text segmentation into many sections makes for a more accurate understanding of the construction of particular document which can be used to produce improved document representations.

Let us take a real-world example of a newspaper. In the newspaper, all the news are divided according to the topics like there is a separate page for the news related to sports. And this page is further divide into section which contain different news related to different sport like news related to Cricket and Football. This is all done to provide the better understanding of the news to the readers.

1.2 Type of Text Segmentation

Text segmentation is of mainly three types:

- **Word Segmentation**: The method of splitting a string or a text which is written in a specific language into its constituent words is known as word segmentation. It is the method by which computer algorithms decide the word borders in a sentence or text. For most higher level NLP functions, parsing and machine translation, POS tagging, word segmentation is the initial phase. It can be seen as the issue of correctly defining word types from a string of characters.

When we listen to speech, we hear a sequence of sentences, but when we talk, we cannot discern words through pauses. Then, eliminating vocabulary from continuous speech is the first step in learning a language's words. Our flow of speaking is also continuous, so for a machine to understand what the person is going to say, machine have to broke that text into meaning full words. Let us take an example that someone is asking that:

Example: Are you a "male or female"?

To give the correct answer to this question, person must have the understanding of each unit word. Here maleorfemale is a single string, machine can predict it as "maleor female" or "male or female". So, word segment help to correctly identify the words, so that one can understand proper meaning of given text. It is widely used in speech to text conversion and also used in to understand the language which doesn't have any delimiter to separate the words.

- **Sentence Segmentation**: The method of deciding the longer processing units composed of one or more words is sentence segmentation. This role includes the detection of sentence limits in various sentences between words. Since most written languages have punctuation marks that appear at sentence borders, phrase segmentation is sometimes referred to as identification of sentence boundaries, disambiguation of sentence boundaries, or recognition of sentence boundaries. All these words refer to the same task: to decide how a text can be separated for further processing into sentences.

In this form of segmentation, we break the written string or text into its Sentences part. We divide text into sentences for better understanding. In English language, we use full stop (.), to determine the ending of sentences, but in English language, we also use full stop (.) for the abbreviation. So with the help of full stop we cannot correctly identify the sentence ending. So sentence segmentation is used for this purpose. Let us take an example a text is written as follow:

Mr. Sumit is a student of NIT Hamirpur.

Here Mr. Sumit is a single entity. If we divide the sentence according to the full stop then Mr. is in different sentence and Sumit is in different sentence, which is wrong. So, to identify correct ending of sentence, firstly we have to understand the text and then divide accordingly. So here text segmentation plays an important role.

- **Topic Segmentation**: Topic segmentation is an important activity for semantic text analysis, which seeks to find the borders between topic blocks in a text. In this type of segmentation, we broke the written string or text into its component topics. This segmentation consists of two prime functions:

  - Topic Recognition.
  - Text Segmentation.

A document may contain multiple topics, we have to identify the different type of topic present in the document and then create segment accordingly. Numbers of segments that we create is equal to the numbers of topics in that document. Each segment contains a different topic. Sentences belong to the same topic are in same segment. Segmenting the text into topics or discourse turns might be useful in information retrieval. A whole book, for example, can be interpreted as one topically coherent portion. In a nutshell, chapters are subsegments of the book, and paragraphs are sub-segments of the chapters. A topically coherent portion is often formed by a single sentence or n-gram. As a consequence of the use of segmentation, the exact definition of a subject varies.

2 Various Approaches

Utiyama and Isahara [27] proposed an analytical procedure to determine the highest probability segmentation of a given text. Since it calculates probabilities from the provided document, this approach does not involve training data. Therefore, any text in any domain may be added to it. As a result, it can be used on any text in any domain. The experiment demonstrated that the procedure outperforms, or is at least as effective as, a cutting-edge text segmentation scheme. This approach determines the most likely segmentation of a given document. The probability of words in segments are naturally calculated using this approach. These probabilities, known as word densities, have been used to detect critical word meanings in documents. This approach is established on the assumptions that a word's density is large in a section where the word is addressed in detail. They experiment this approach on Choi's dataset and find Pk [1], in % is 10. Figure 1 shows the variation of results, where data is divided into various sets according to the number of sentences in a document. But in this method, error rate is high. LDA [22] can improve the result.

Brants et al. [4] introduce a new approach for topic-related text segmentation that incorporates the use of the Probabilistic Latent Semantic Analysis (PLSA) structure

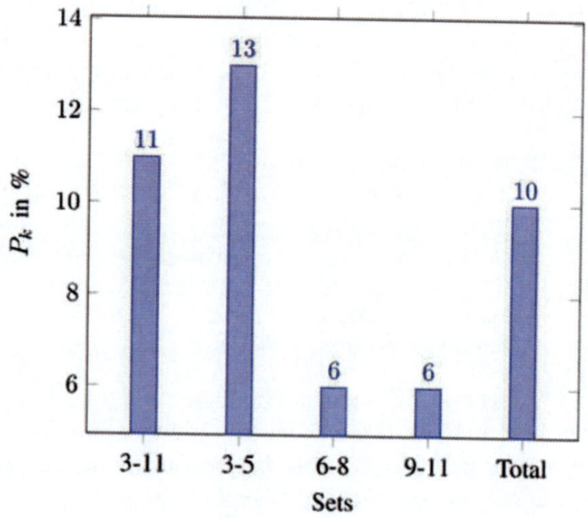

Fig. 1 Variation of Pk with respect to sets [27]

or representation with a procedure for choosing segmentation marks that depend on similarity values among neighboring lines. Latent means hidden, so we have to find the hidden features. Here, topics are hidden. PLSA authorize for a greater comprehension of fragmented knowledge in a chunk of text, like a single sentence or a list of sentences. Connecting or linking distinct occurrence of the identical model, by using various random occurrences or a various number of latent groups, improves segmentation efficiency even more. They evaluated this segmentation method on two corpus. The first corpus is Brown Corpus is which consists of 500 text samples with an average length of 2000 words each. They collect data by training and research, as defined in [3]. The second corpus is Router-21578, which was developed using the process described in [18]. The error reduction in Brown corpus is 31% and in Router-21578 is 71%. This model, however, cannot be used in real-world applications due to its high memory requirements and long execution time.

Athanasios Kehagias [16] presented a new text segmentation algorithm that is based on dynamic programming. This approach is based on unsupervised learning. It performs linear text segmentation by globally minimizing a segmentation cost function that integrates two factors: (a) similarity of the term within the segment and (b) prior segment length information. The algorithm's segmentation accuracy is measured by accuracy, recall, and Beeferman's segmentation metric. They experiment this approach on Choi's dataset and find Pk, in % is 6.40 (Fig. 2 where sets shows number of sentences in a document). Then they compared this methods with C99 [6], U00 [2], and CWM [7] methods and find out this method gives better results. The performance of this algorithm is very satisfactory. But this algorithm performs poor when number of sentences in a set is not between 9 and 11 as compared to topic modeling [22].

Chiru [5] purposed an unsupervised learning approach for text segmentation known as unsupervised cohesion-based segmentation. In this technique, they also

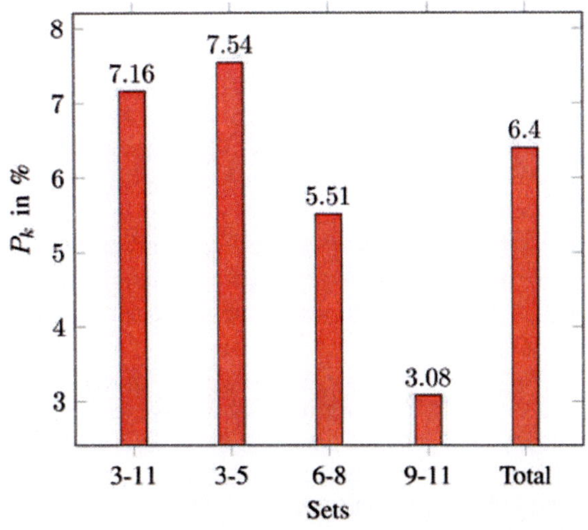

Fig. 2 Variation of Pk with respect to sets [16]

propose a voting system that improve the segmentation results. In this work they presented three modules, each with different methods and a voting mechanism is implemented in order to boost the outcomes achieved from the three processes. All the three modules are implemented using an lexical cohesion approach based on unsupervised learning and used to compare the results of various lexical cohesion-based approaches to find out that how they can be strengthened by integrating their outputs applying the voting system. The text is thought to be divided into subsections, with each column devoted to a single subject. Firstly, part-of-speech tagging is done. The first approach examines applicant limits to determine which are true limits. Any time a new candidate subject cap is encountered, it will be assessed to see if it is a genuine topic change. If it isn't, the present paragraph will be considered part of the previous subject, so this candidate topic cap will be skipped, and the review will go on to the next paragraph. In second module they use the concept of lexical chains and clustering algorithm. After part-of-speech tagging, the locations of the most.

Significant indexes are consumed into account for clustering. Then cohesion chains are made using repetition of words, synonyms, and words relationships. The clustering algorithm produces the first lexical chain with the first token, and so on. Topic changes have already been established, with a high density of chain begins and ends suggesting that the topic has shifted. Third module approach is same as second module approach, but in this module number of chains is equals to number of cluster/segment. An effort is made in voting to merge the results received from all the three modules to find out if some improvement in the last conclusion can be accomplished. The techniques will be tested to see whether they are similar, and whether scaling variables may be used to improve the probability of accurate outcomes. The scaling factors would be calculated empirically.

Eisenstein and Barzilay [9] presented an unsupervised learning-based approach that are using a novel Bayesian Lexical for text segmentation. In this method, unsupervised systems are guided by lexical cohesion: the propensity to cause a compact and coherent lexical distribution by well-formed segments. They demonstrate that we can put the lexical consistency in a Bayesian sense with help of modeling the terms in each subject section as generate from a multinomial language structure linked with the segment, and that maximizing the conclusional probability in such a model that generates lexically cohesive segmentation. This is in contrast to previous techniques, which focused on handcrafted harmony metrics. But this paradigm allows for the inclusion of additional functionality such as cue words, an important predictor of discourse architecture that has not traditionally been used in unsupervised learning-based segmentation frameworks. For both text and speech datasets, this model consistently outperforms a variety of state-of-the-art schemes. They further demonstrate that an entropy related testing and a previously existing method can be performed as particular instances of the Bayesian system. On all metrics, Bayesian schemes produce a raw output benefit of 2–3% over all baselines on the medical textbook corpus. On the ICSI meeting corpus, Bayesian structures outperform the best benchmark by 4–5% on the Pk metric and obtain a smaller gain on the WindowDiff metric. Table 1 shows the results for two datasets. This model performs better as compared to UI [27], LCSEG [12] and MCS [20].

Table 1 Values of Pk and WindDiff for different datasets [9]

Dataset	WinDiff (WD)	Pk Value
Medical Textbook	0.339	0.353
ICSI meeting	0.258	0.312

Misra [22] approached the task of text segmentation from a topic/subject modeling standpoint. They look at how the Latent Dirichlet Allocation (LDA) subject system can be used to splitting the text into the segments from a document. One important advantage of the suggested solution is that it outputs the subject distribution associated with each segment in addition to the segment boundaries. This data may be helpful in applications, for example, section retrieval and discourse inspection. The LDA-based approach introduced in this paper is depend upon the following assumption: if a section is build from just one story, there will be small amount of active topics, while if a segment is build from numerous story, there will be a considerably large amount of active subjects or topics. Expanding this logic, if a segment is rational (the subject circulation for that segment contains just a small amount of active topics), the log likelihood for that segment is normally high, as opposed to segment that is not coherent [21]. This discovery is vital to the effectiveness of the suggested LDA-based method for task of text segmentation, and it has remained undetermined excluding for its primary work in recognizing document consistency [21]. They experiment this approach on Choi's dataset and find Pk, in % is 15.5 (Unadapted LDA) for 3–11 sentences (Fig. 3, where sets shows number of sentences in a document). The main disadvantage of this approach is vocabulary Mismatch. To fix the issue of vocabulary overlap, they split Choi's dataset into two parts, and with the help of Part A (called it Adapted LDA) they find Pk, in % = 2.3 for 3–11 sentences (Fig. 4, where sets shows number of sentences in a document). This method performs better than Choi

Fig. 3 Variation of Pk with respect to sets (Unadapted LDA)

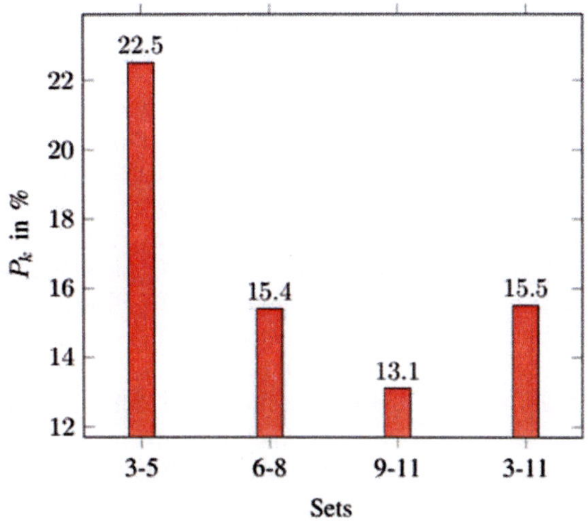

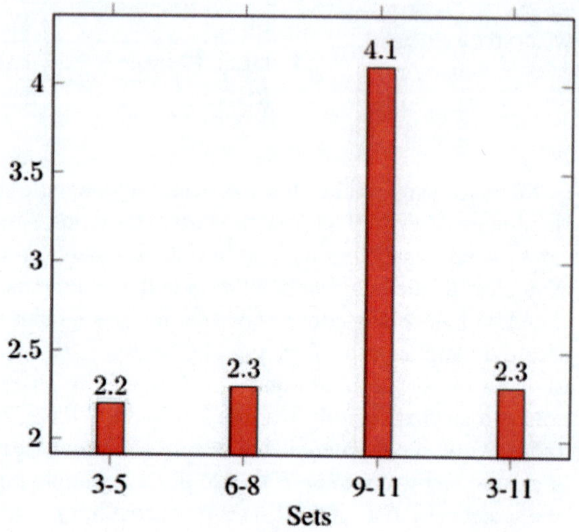

Fig. 4 Variation of Pk with respect to sets (Adapted LDA)

[6], Utiyama and Isahara [27], Choi [7], and Brants [4] methods. But Fragkou's [10] method perform better for a set where sentences are between 9 and 11.

Kazantseva and Szpakowicz [15] introduced a new linear text segmentation algorithm. It is a type of Affinity Propagation, a cutting-edge clustering algorithm in the context of factor graphs. Affinity Propagation for Segmentation (APS) takes a series of pairwise connection between data points and generates segment borderlines and segment centers data points that better represent all other data points inside the segment. APS transfers messages iteratively through a cyclic factor graph before convergence. Each iteration uses data from all available similarities to generate high-quality performance. For practical segmentation functions, APS scales linearly. This algorithm is derived from the original Affinity Propagation formulation and equate it to two state-of-the-art segmenters when it comes to topical text segmentation. In this technique they made four matrices: First matrix is similarity matrix, second is responsibility matrix, third is availability matrix, and last is criterion matrix. The higher values of each row of criterion matrix is set up as exemplar. Text whose exemplar are same are in same cluster. Sentences are data points in this context and exemplars are segment centers. They allocate each sentence in a text to a section center in order to optimize net similarity. The developers also made freely accessible implementations of Java. On three datasets, they tested the APS algorithm's accuracy. The first dataset contains AI lectures, the second dataset is collection of chapters from medicinal textbooks, and the third dataset contains 85 works related to fiction. For AI WindowDiff = 0.404, for Fiction dataset, it gives WindowDiff = 0.350, and for Clinical Dataset, WindowDiff = 0.371 (Table 2). They compared this method with BayesSeg [9] and MinCutSeg [20]. For clinical dataset, BayesSeg [9] perform better, where WindowDiff = 0.353.

Table 2 Values of WindowDiff for different datasets [15]

Dataset	WindowDiff
AI	0.404
Clinical	0.371
Fiction	0.350

Glavas et al. [13] presented a novel algorithm for linear text segmentation (TS) based on unsupervised learning that constructs a semantic relatedness graph. In this approach, all the sentences become nodes of the relatedness graph G. The semantic similarity of all pairs of sentences in a given text is then determined. In this method, they used a graph data structure. A sentence is represented as node and if there is relation between two sentences or sentences are in same segment then there is edge between the nodes. To find the edges or semantic relation between sentences, they use greedy lemma alignment. Greedy Lemma Alignment use resemblance of their propagation vectors and greedily match background terms between sentences. If the words of two sentences have similar vector distribution, then it create the pair of the words. But this approach doesn't give optimal result, so they use new method that is building a weighted complete bipartite graph [11] between each pair of words of two sentences. A linked edge links each word in one sentence to different word in another sentence. And then running this algorithm [11], similarity between the words is calculated and then make pair of the words. Then create the set of the words pairs of two sentences and find out that there is a relation between two sentences or not. If there is a relation then edge is passes through that sentences or nodes. GRAPHSEG [13] was tested on four subsets of the Choi dataset, each with a different number of sentences. It gives Pk = 7.2 and WindowDiff = 9.0 for 3–11 sentences set as shown in Fig. 5 (where sets shows number of sentences in a document). On a synthetic dataset, this approach performs competitively with the best-performing LDA-based approaches [22]. But Riedl and Biemann [25] perform better for all the sets.

Sehikh et al. [26] proposed a new method for topic segmentation in speech recognition transcripts that uses bidirectional Recurrent Neural Networks (RNNs) to calculate lexical cohesion. The bidirectional RNNs catch meaning from the previous set of words as well as the next set of words. To identify subject transitions, the prior and subsequent contexts are contrasted. This method does not use a segmented corpus subject design for preparation, unlike previous research focused on arrangement and discriminative systems for topic segmentation. This model is learned by reading news stories on the internet. Based on DNN-HMM [19] acoustic models, ASR transcriptions are obtained from French ASR method. The feasibility of this solution is shown by this ASR transcripts. The bidirectional RNNs gathered meaning in the previous and subsequent sets of words, and compared the two sets of contexts to identify subject shifts. Concatenating news stories from the internet was used to train these models discriminatively. This RNN models outperformed the C99-LSA [6] and TopicTiling [25] models on ASR transcripts of French television news programs. They use the traditional subject segmentation appraisal steps Pk and WindowDiff [23] for comparison of the proposed and base line approaches. WindowDiff score is

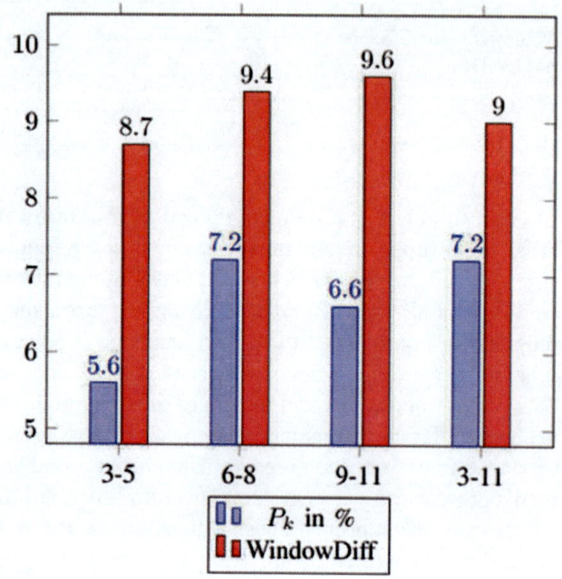

Fig. 5 Variation of Pk and WindowDiff with respect to sets for semantic relatedness graph model

evaluated for TV5 dataset and found 0.34 and Pk is 0.26, which means it has less loss of information. Figure 6 shows the result for two techniques in this paper.

Koshorek et al. [17] presented a massive new dataset for text segmentation that is automatically take out and tagged from Wikipedia, and formulated text segmentation as a supervised learning problem. They also build a model (Fig. 7) that established on this dataset and prove that it generalizes effectively to common text that hasn't been used before. In this dataset, input x is a document which consist n sentences

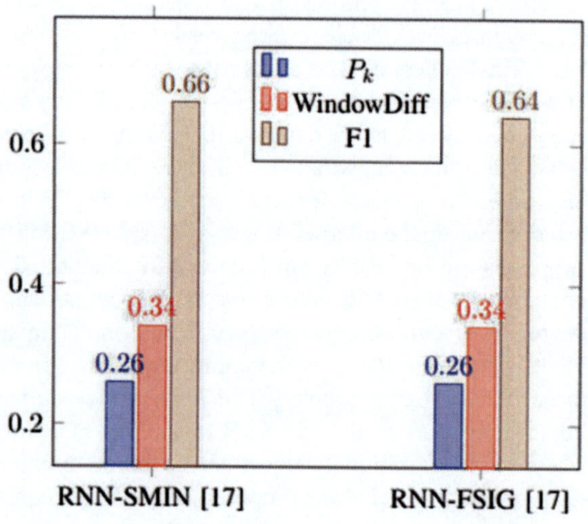

Fig. 6 Results for two different techniques in [26]

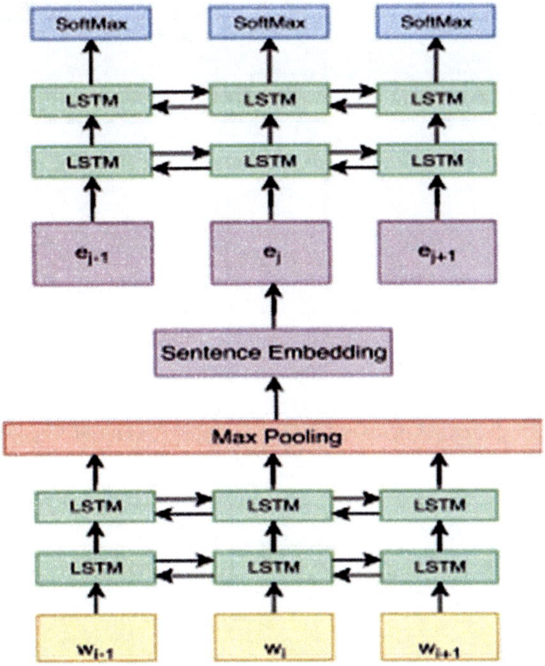

Fig. 7 Architectural Diagram of LSTM to based Supervised Model [17]

and n-1 binary value which represent whether that sentence is end of segment or not. Model (Fig. 7) has two sub-networks. First sub-network is used to generate sentence representation and second sub-network is used to predict the text segmentation. First sub-network generates sentence representation using BiLSTM cells, it take words from sentence as input and max-pooling over the LSTM outputs yields the final sentence representation. Then these embedding fed up to the second sub-network as an input and this sub-network feeds a two-layer bidirectional LSTM with a series of sentence embeddings as data. After that, they added a fully connected layer to each of the LSTM outputs to generate a series of n vectors. To obtain n-1 segmentation probabilities, they disregard the last vector and use a softmax function. This model can be run on modern GPU hardware and has a linear runtime in terms of text length. They tested this approach on WIKI-50 and CHOI datasets and find out the Pk value = 18.24 for WIKI-50 dataset. Figure 8 shows the variation of Pk for two models for CHOI's dataset. But GRAPHSEG [13] provides better results on the synthetic Choi dataset on comparing with this approach, but this performance does not carry over to the natural Wikipedia data, where they underperform the random baseline.

Pinkesh Badjatiya [1] proposed an attention-dependent bidirectional LSTM [14] model for in which CNNs are used to learn sentence embeddings and the segments are concluded based on contextual data (Fig. 9). Variable-sized background information can be managed dynamically by this model. This model takes sentence s and its K-sized left context (K numbers of left sentences) and K-sized right context (K

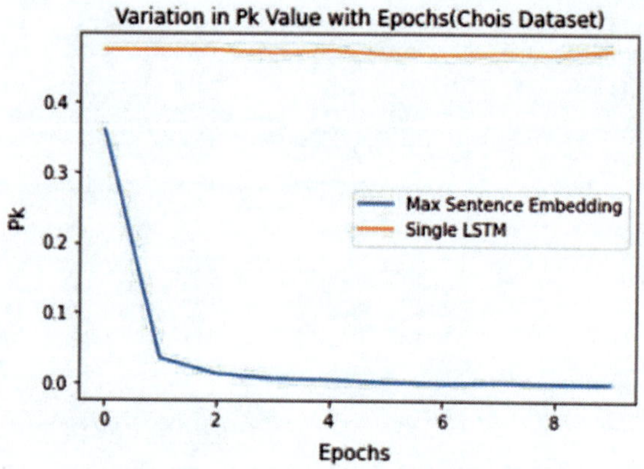

Fig. 8 Variation of Pk with respect epochs for models in paper [17]

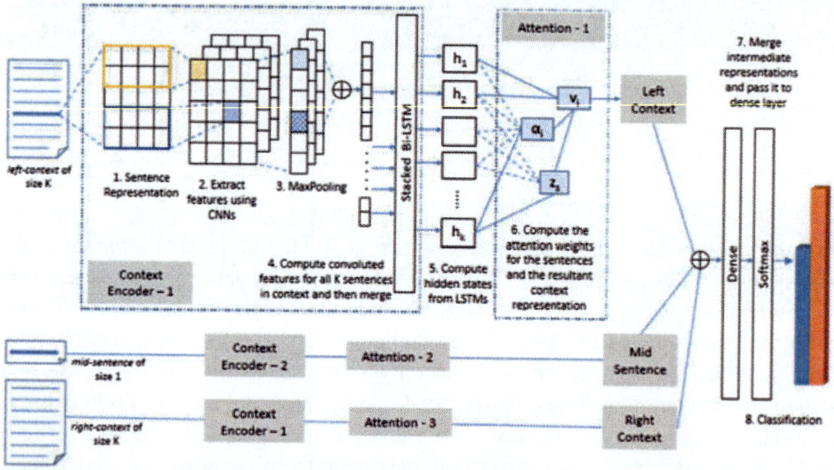

Fig. 9 Architectural diagram of attention-based model [1]

number of right sentences) as input and with help of this information it predicts that whether this sentence s is a beginning of new segment or not. In this model, sentences embedding is done with the help of word2vec model and that feature are extracted by using CNNs. Max-pooling is used to maximize the result. All the features from all the K sentences are merged and then stacked-BiLSTM is used to compute hidden states. After this, results are fed into attention layer.

Attention layer help to get more information from a text. And finally, to produce results, they used a softmax layer as an activation function and to get the highest value from the target attribute, they use the arg max function. They trained and tested this model on three different datasets Clinical [20], Fiction [15] and Wikipedia [1].

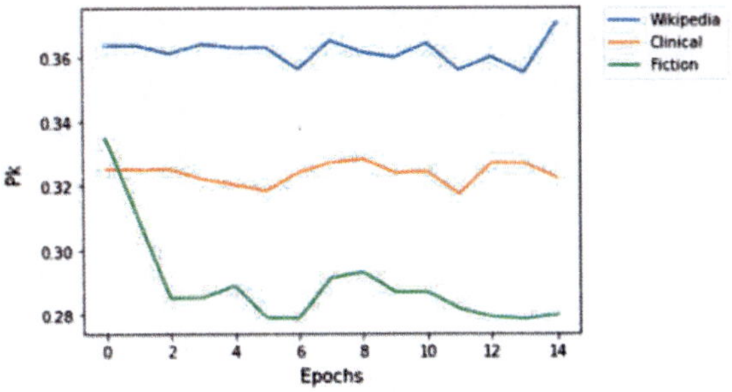

Fig. 10 Pk value's variation with respect to epochs for different datasets for model in [1]

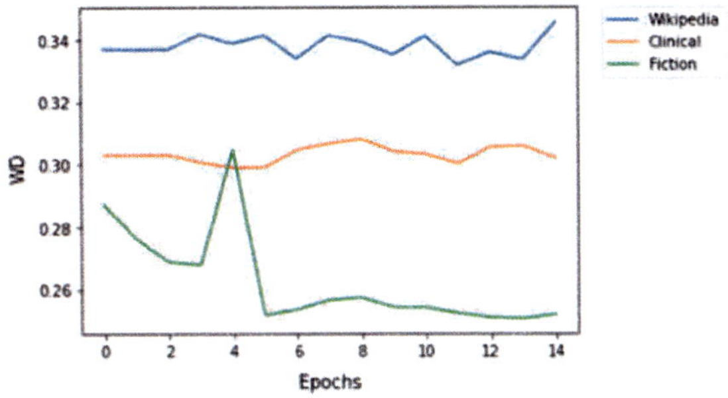

Fig. 11 WD value's variation with respect to epochs for datasets for model in [1]

Table 3 Results of different datasets [1]

Dataset	WindowDiff	Pk Value
Clinical	0.294	0.318
Fiction	0.308	0.378
Wikipedia	0.315	0.344

Figure 10 and Fig. 11 shows the variation of Pk and WindowDiff for different datasets respectively. Table 3 shows the results for various datasets. They also compare this model with some baseline models like PLDA [24], TSM [8] and find out this model gives better WindowDiff for three datasets, but some models perform better for Pk values.

3 Conclusion

We compare the performance of the various models that are trained using unsupervised learning and use same Choi's Dataset. Figure 12 shows the variation in Pk values for this comparison, where x-axis shows the references of various paper accordingly and x-axis shows the Pk value. From this we conclude that for Choi's dataset, best model was proposed in [22]. But in this method they divided the dataset

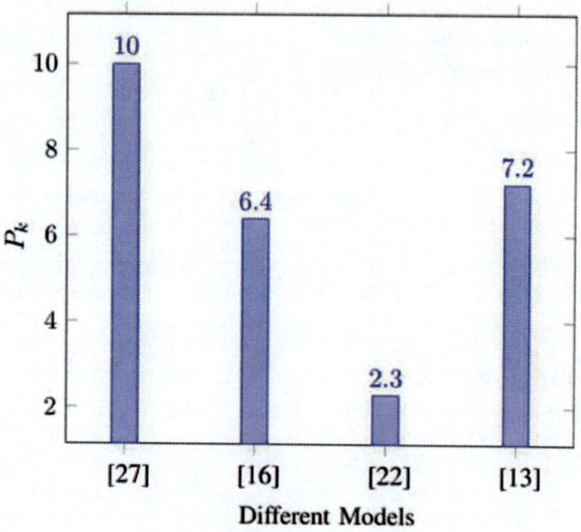

Fig. 12 Variation of Pk with respect to different models (Choi's Dataset)

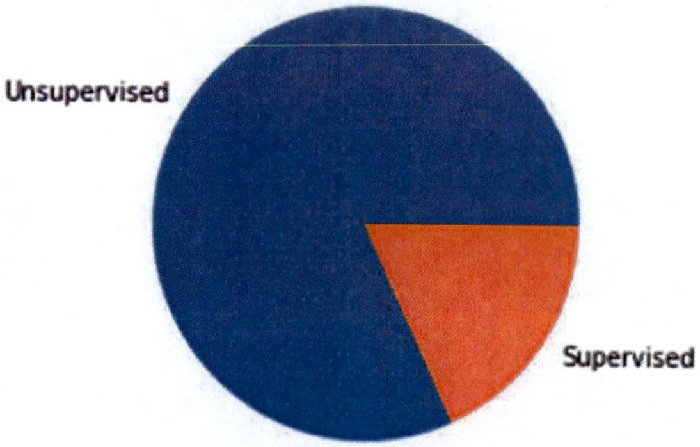

Fig. 13 Pie chart showing Unsupervised and Supervised techniques compared in this paper

to remove vocabulary mismatch and use only half dataset. So we can say that best result is given by model in [16].

Text segmentation is critical for activities like context Fig. 11. WindowDiff (WD) value's variation with respect to epochs for different datasets for model in [1] comprehension, summarization record indexing, and noise elimination in NLP. There are many techniques available for text segmentation but most of the text segmentation approaches are unsupervised learning-based (as shown in Fig. 13), this may be due to the unavailability of supervised data. After comparing these approaches (from Table 4) for text segmentation, we find out that Attention-based text segmentation is best approach for text segmentation. This experiment gives best values.

Table 4 Comparison table for various approaches for text segmentation

References	Year	Learning approaches	Methodology	Performance	Shortcomings
[27]	2001	Unsupervised Learning	Maximum-probability segmentation. In terms of the likelihood specified by a statistical model, it selects the optimum segmentation	Choi's dataset: Pk in %: 10	LDA [22] can improve the result. Error rate is high
[4]	2002	Unsupervised Learning	Use of Probability Latent Semantic Analysis system with the procedure of choosing segment point	Brown corpus: Error reduction: 31% Router-21578 Error Reduction :71%	Big memory constraints and large execution time make this approach impractical to implement in the real world
[10]	2003	Unsupervised Learning	Dynamic programming algorithm universal minimization of the cost function of segmentation	Choi's dataset: Pk in %: 6.40	Perform poor when number of sentences in a set is not between 9 and 11
[5]	2007	Unsupervised Learning	Three different module based on cohesion. Proposed a Voting System	Proposed a Voting System	Due to the use of statistical technique, this system give inaccuracy. Error rate high
[9]	2008	Unsupervised Learning	By modeling the terms in each topic segment, lexical cohesion is put in a Bayesian space	Medical textbook corpus Pk: 0.339 ICSI meeting corpus: Pk: 0.258	Due to the use of statistical technique, this system give inaccuracy. Error rate high
[22]	2009	Unsupervised Learning	Usage of the subject model of latent Dirichlet allocation (LDA) to produce the segment from the text	Choi's dataset: Pk in %: 2.3	Dynamic programming-based [10] approach for text segmentation performs well than LDA. Vocabulary Mismatch

(continued)

Table 4 (continued)

References	Year	Learning approaches	Methodology	Performance	Shortcomings
[15]	2011	Unsupervised Learning	Adaptation of Affinity Propagation, takes series of pairwise connections between data points and generates segment borderline	Fiction WindowDiff: 0.350	The choice for objects to use as examples needs to be defined
[13]	2016	Unsupervised Learning	Builds a semantic relatedness graph, all the sentences become nodes and if there is similarity between sentences then a edge is created between them	Choi's Dataset: Pk: 7.2 WinDiff: 9.0	It is used for short text. Topic Modeling give better results
[26]	2017	Unsupervised Learning	Bidirectional RNN with LSTM cells to calculate cohesion. Word Embedding	Choi's Dataset: Pk: 7.2 WinDiff: 9.0	The entire utterance may not be usable for such implementations, such as real-time speech recognition, and Bidirectional RNN may not be satisfactory
[17]	2018	Unsupervised Learning	First sub-network generates sentence representation using BiLSTM cell. Second sub-network feeds a two-layer BiLSTM with a series of sentence embeddings as data	Wiki-50 Dataset Pk: 18.24	Not perform good in Choi's Dataset as compared from GraphSeg [26]
[1]	2018	Unsupervised Learning	Extract features using CNNs MaxPooling, BiLSTM, Attention	Fiction: WinDiff: 0.308 Clinical WinDiff: 0.294 Wikipedia: WinDiff: 0.315	We have to decide number of sentences (Right and Left Context) for training. Due to this documents dropout may occur

References

1. Badjatiya, P., Kurisinkel, L.J., Gupta, M., Varma, V.: Attention-based neural text segmentation. In: European Conference on Information Retrieval, pp. 180–193. Springer (2018)
2. Blei, D.M., Moreno, P.J.: Topic segmentation with an aspect hidden markov model. In: Proceedings of the 24th Annual International ACM SIGIR Conference on Research and Development in Information Retrieval, pp. 343–348 (2001)
3. Brants, T.: Test data likelihood for plsa models. Inf. Retrieval **8**(2), 181–196 (2005)
4. Brants, T., Chen, F., Tsochantaridis, I.: Topic based document segmentation with probabilistic latent semantic analysis. In: Proceedings of the Eleventh International Conference on Information and Knowledge Management, pp. 211–218 (2002)
5. Chiru, C.-G.: Unsupervised cohesion-based text segmentation. EUROLAN 2007 Summer School Alexandru Ioan Cuza University of Ias¸i, p.93
6. Choi, F.Y.Y.: Advances in domain independent linear text segmentation. arXiv preprint cs/0003083 (2000)
7. Choi, F.Y.Y., Wiemer-Hastings, P., Moore, J.D.: Latent semantic analysis for text segmentation. In: Proceedings of the 2001 Conference on Empirical Methods In Natural Language Processing (2001)
8. Du, L., Buntine, W., Johnson, M.: Topic segmentation with a structured topic model. In: Proceedings of the 2013 Conference of the North American Chapter of the Association for Computational Linguistics: Human Language Technologies, pp. 190–200 (2013)
9. Eisenstein, J., Barzilay, R.: Bayesian unsupervised topic segmentation. In: Proceedings of the 2008 Conference on Empirical Methods in Natural Language Processing, pp. 334–343 (2008)
10. Fragkou, P., Petridis, V., Kehagias, A.: A dynamic programming algorithm for linear text segmentation. J. Intell. Inform. Syst. **23**(2), 179 197 (2004)
11. Frank, A.: On kuhn's hungarian method a tribute from hungary. Naval Res. Logist. (NRL) **52**(1), 2–5 (2005)
12. Galley, M., McKeown, K., Fosler-Lussier, E., Jing, H.: Discourse segmentation of multi-party conversation. In: Proceedings of the 41st Annual Meeting of the Association for Computational Linguistics, pp. 562–569 (2003)
13. Glavas, G., Nanni, F., Ponzetto, S.P.: Unsuper-vised text segmentation using semantic relatedness graphs. Association for Computational Linguistics (2016)
14. Huang, Z., Xu, W., Yu, K.: Bidirectional lstm-crf models for sequence tagging. arXiv preprint arXiv:1508.01991 (2015)
15. Kazantseva, A., Szpakowicz, S.: Linear text segmentation using affinity propagation. In: Proceedings of the 2011 Conference on Empirical Methods in Natural Language Processing, pp. 284–293 (2011)
16. Kehagias, A., Fragkou, P., Petridis, V.: Linear text segmentation using a dynamic programming algorithm. In: 10th Conference of the European Chapter of the Association for Computational Linguistics (2003)
17. Koshorek, O., Cohen, A., Mor, N., Rotman, M., Berant, J.: Text segmentation as a supervised learning task. arXiv preprint arXiv:1803.09337 (2018)
18. Lee, L.: Measures of distributional similarity. arXiv preprint cs/0001012 (2000)
19. Li, L., Zhao, Y., Jiang, D., Zhang, Y., Wang, F., Gonzalez, I., Valentin, E., Sahli, H.: Hybrid deep neural network–hidden markov model (dnn-hmm) based speech emotion recognition. In: 2013 Humaine Association Conference on Affective Computing and Intelligent Interaction, pp. 312–317. IEEE (2013)
20. Malioutov, I.I.M.: Minimum cut model for spoken lecture segmentation. Ph.D. thesis, Massachusetts Institute of Technology (2006)
21. Misra, H., Cappe, O., Yvon, F.O.: Using lda to detect' semantically incoherent documents. In: CoNLL 2008: Proceedings of the Twelfth Conference on Computational Natural Language Learning, pp. 41–48 (2008)

22. Misra, H., Yvon, F., Jose, J.M., Cappe, O.: Text' segmentation via topic modeling: an analytical study. In: Proceedings of the 18th ACM Conference on Information and Knowledge Management, pp. 1553–1556 (2009)
23. Purver, M.: Topic segmentation. Spoken language understanding: systems for extracting semantic information from speech, pp. 291– 317 (2011)
24. Purver, M., Kording, K.P., Griffiths, T.L., Tenenbaum, J.B.: Unsupervised topic modelling for multi-party spoken discourse. In: Proceedings of the 21st International Conference on Computational Linguistics and 44th Annual Meeting of the Association for Computational Linguistics, pp. 17–24
25. Riedl, M., Biemann, C.: Topictiling: a text segmentation algorithm based on lda. In: Proceedings of ACL 2012 Student Research Workshop, pp. 37–42 (2012)
26. Sehikh, I., Fohr, D., Illina, I.: Topic segmentation in asr transcripts using bidirectional rnns for change detection. In: 2017 IEEE Automatic Speech Recognition and Understanding Workshop (ASRU), pp. 512–518. IEEE (2017)
27. Utiyama, M., Isahara, H.: A statistical model for domain independent text segmentation. In: Proceedings of the 39th Annual Meeting of the Association for Computational Linguistics, pp 499– 506 (2001)

A Study of Moving Vehicle Detection and Tracking Through Smart Surveillance System

Manoj Kumar, Susmita Ray, Dileep Kumar Yadav, and Rohit Tanwar

1 Introduction

Nowadays, the transportation network have become an integral part of day-to-day human's lives. Around 40% of the world's population spends at least an hour on the roads/streets per day [1]. Not only this, the National Highways play a major role in connecting cities and towns, hence they are extensively used for transportation systems as well as for traveling between cities and towns. Nowadays, in parallel with the increasing population, the increase in the number of vehicles increases the time people spend in traffic [2]. Travel and transportation issues have become a difficult task when the system and the behavior of users are very difficult to frame and to predict travel patterns/behaviors. Therefore, machine learning and deep learning algorithms may be helpful to conquer the demurral of simultaneously increasing travel demand, increasing congestion, road safety, traffic prediction, etc. These challenges emerge due to continued growth of personal and transportation vehicles and due to the exponential growth in population and mainly in developing countries like India, China, etc. [3].

Deep learning and machine learning (DL & ML) is an extensive vicinity of computer technology and engineering area that makes machines work just like the human being. It is applied to look for problems which might be difficult to make clear utilizing traditional experiential techniques. Artificial intelligence primarily based totally techniques may be applied as Knowledge-Based Systems (KBS) and as an Artificial Neural Network (ANN). The KBS structures are the ones in which

M. Kumar · S. Ray
Manav Rachna University, Faridabad, India

D. K. Yadav
Galgotias University, Greater Noida, India

R. Tanwar (✉)
School of Computer Science, University of Petroleum and Energy Studies, Dehradun, India

B. Agarwal et al. (eds.), *Proceedings of International Conference on Intelligent Cyber-Physical Systems*, Algorithms for Intelligent Systems,
https://doi.org/10.1007/978-981-16-7136-4_23

AI works, primarily based totally at the predetermined regulations described with inside the set of rules via way of means of humans. The ANNs, on the other hand, are systems of neurons connected and designed onto various layers, their working is very similar to the human brain, they take some input list and, based on the input list, the ANNs produce required outputs [3].

Amazing progress in the field of vehicle detection has been achieved by the use of deep convolutional networks (CNNs). CNNs have a high ability to learn image features and can perform various related tasks, such as bounding box classification and regression [8]. It is possible to split the method of detection into two classes in general. The two-stage method generates a candidate box of the object via way of means of various algorithms after which classifies the object via way of means of a CNN. The one-stage method does now no longer create a candidate box, however, turns the positioning hassle of the object bounding box at once right into a processing regression trouble. For the two-degree method, Region-CNN (R-CNN) [29] makes use of selective location seek with inside the picture. The convolutionary network must have a fixed-size image supply, and the deeper network structure requires a long training period and uses a large amount of memory for storage. Deep learning is based on ANN, but requires more hidden neurons and hidden layers than conventional ANNs. Deep learning has proven to be a huge success in the aspects of speech recognition, computer vision, NLP, and item recommendation. They also achieve domain efficiency in multiple classification and prediction task in transport scenarios. As long as enough training data and GPU resources are available, it is possible that conventional machine learning methods can be overridden by deep learning models. There are several deep learning models used in highways and transportation system and those will be given below in the diagram. According to the WHO Global Road Safety Report 2018, there were over 1.5 lakhs of casualties in India due to road accidents alone. With 1.94% of the overall traffic, national highways accounted for 30.2% of road injuries and 35.7% of fatalities in 2018. For 2.97% of the length of the routes, state highways account for 25.2% and 26.8% of injuries and fatalities, respectively. About 5.8% of deaths attributable to collisions are due to driving on the wrong side of the road. The use of cellphone accounts for 2.4% of deaths, and another 2.8% of casualties were due to drunken driving [4]. The machine learning- and artificial intelligent-based applications are used in transportation for intelligent buses, connected busses, smart roads, and computer vision-enabled vehicles [18–22]. Apart from these, IoT-enabled devices to communicate in wireless network along with messaging system, alert, voice-based system is inbuilt in vehicles now a days to upgrade the transportation system (Fig. 1).

Another application which is highly recommended in smart vehicles to check ECG, EEG, EDA, etc., for driver or other passengers. Such ITS-based applications monitor driver's (i) health and feelings monitoring, (ii) sleepy cautioning, and (iii) alert control (Fig. 2).

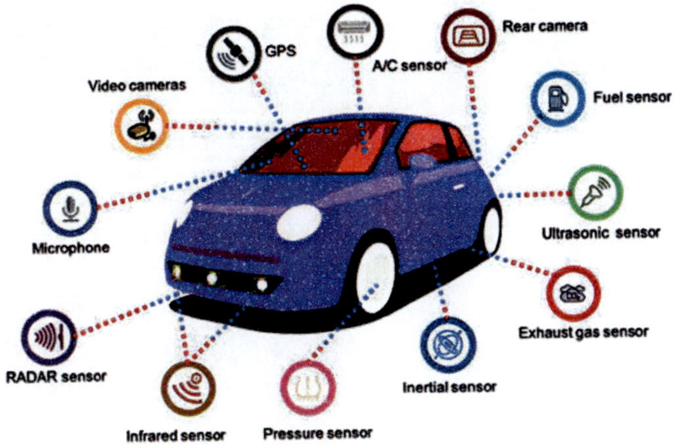

Fig. 1 Different types of vehicle sensors in smart car [17]

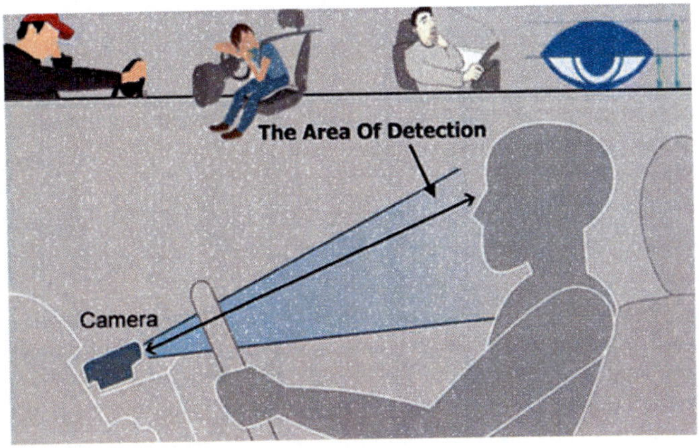

Fig. 2 ITS User monitoring applications [18]

2 Literature Survey

The scientists and researchers have done their research on the problems on highways and transportation systems, based on the data that is being collected from different sources. They have developed numerous machine learning as well as deep learning models for accident prediction, highway safety, designing and controlling transport network structures, intelligent transport systems, traffic flow prediction, travel demand prediction, automated driving (self-driving cars), traffic signal control, crack condition of the roads on highways, etc.

For real-time assessment of highway traffic monitoring, a system has been proposed. In this, the onboard vehicle equipment and the roadside units (RSUs)

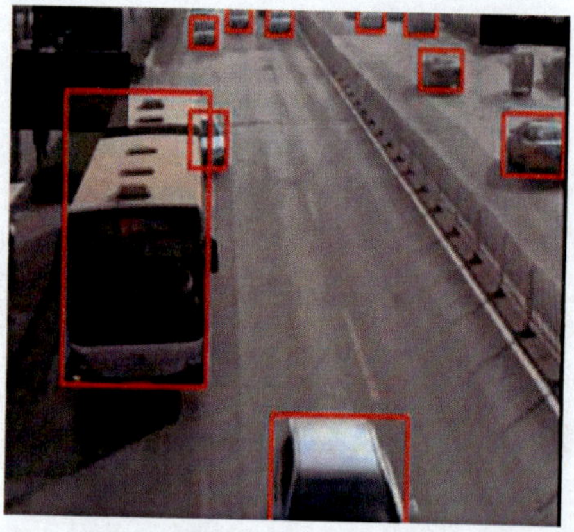

Fig. 3 Moving object detection [31]

work together to assess the contingency of an occurrence under the artificial intelligence criterion. They specifically focus on two paradigms of AI, i.e., (1) vector support machines (SVMs) and (2) ANNs [5] (Fig. 3).

Data on the number of vehicles and vehicle categories play a crucial role in the management of highways. Because of the several types of vehicles like Bus, truck, Bike, tractor, etc., types of vehicles and their identification remains challenge that directly affects the precision of the vehicle count. A vehicle recognition system based on computer vision has been proposed to address this issue. YOLOv3 network is used to detect the type of vehicle and their trajectories are found with the help of ORB algorithm.

The number of vehicles and the form of vehicles play a vital role in the management of highways. This issue must be resolved using some M/L and D/L models. A vehicle identification system based on computer vision has been proposed to address this issue. The basic concept of the early models of YOLO algorithms is continued by the YOLOv3 algorithm. The convolutional neural network is used to extract the characteristics of the input image. According to the size of the feature diagram, such as M*N, the input image is split into M*N grids. The core of the object label box is located in a grid unit, and that M*N grid is responsible for predicting objects or vehicles. YOLOv3 network is used to detect the type of vehicle and their trajectories are found with the help of ORB algorithm. In the ORB algorithm, the characteristic factors obtained with inside the item detection field with the aid of using the car detection set of rules are extracted. The extraction of the object aspect from the entire road surface area is not carried out, which reduces the amount of the measurement considerably. In the tracking of objects, when the vehicle object changes to a continuous frame, the object's prediction box is drawn in the next frame [6] (Fig. 4).

Traffic flow prediction is a very crucial step in designing a successful intelligent transport system. The deep learning models have been able to predict the traffic density with the help of big data on the highways. "The techniques used are Long Short Time Memory (LSTM), Recurrent Neural Network (RNN), Stacked Long Short Term Memory (S-LSTM), Gated Recurrent Unit (GRU) and Bidirectional

Fig. 4 Moving vehicle counting

Long Short Term Memory (B-LSTM) neural networks" [2]. A deep learning model is developed that mix up the linear model computed using L1 regularization and layer-based sequences. Prediction of the traffic flow is a challenge because of the irregularities occur due to congestion, breakdown and transition between free flows. They demonstrate that deep learning structures can capture these non-linear spatio-temporal effects [7]. Better outcomes had been received by Advanced CNN in item detection, however, CNN is prone to modifications in item detection size [17, 18].

As a different approach to regression models used for crash modeling, a common deep learning approach known as the Deep Belief Network (DBN) has been implemented [8]. These techniques are effective, but mistakes, ambiguity and overfitting are highly probable. A research has been conducted to predict pavement crack ratings using recursive partitioning and ANNs and deep learning frameworks and has proven to be a good technique for detecting cracks [9]. Peng et al. [19] has investigated a method for haze removal in the colored images. This method uses air light white correction along with the local light filtering technique. Song et al. [20] has focused on the literatures and examined possible work for moving motor vehicle identification and incorporate system on highways/roads scene using the deep learning approach. Such work enhances the real-time-based techniques for intelligent transportation system. Kukkala et al. [21] proposed a method for evaluating the path for vehicles running on the highways in transportation system. It also assist the drivers through the driver-assistance systems along with autonomous system. Phillip et al. [22] explored a real-time prediction-based method that automatically analyzed the collision risk from the monocular video data to automate the transportation system. Sudha et al. [27] proposed a five-layer method that consists of preprocessing layer, feature extraction layer, training layer, testing layer, and vehicle detection layer. The work predicts the moving objects behavior more precisely than the EKF and UKF (Table 1).

In short, the technique of vehicle detection may be taken into consideration to were transferred from studies on traditional strategies to that on deep convolutionary community strategies. In addition, for particular traffic scenes, there are fewer public datasets. Small item detection is unreliable due to the sensitivity of convolutional neural networks to scale shifts. It is tough to carry out multi-item tracking

Table 1 Challenges focused by various methods

Method/dataset	Challenges	Dataset
Peng et al. [19]	Haze removal, noise handling in image	RGB Image
Song et al. [20]	Dynamic situation on highway *or* roads	Colored Video
Kukkala et al. [21]	Identification of path of moving vehicle	Colored Video
Phillip et al. [22]	Cluttered, varying illumination on highway	Colored Video
Sudha et al. [25]	Behaviour of moving object in video frames	Colored Video
Suri et al. [26]	Detect moving object through bounding box	Colored video
Sharma et al. [27]	Motion and illumination variation in background	Thermal, Colored frame

and next traffic evaluation while motorway cameras are used. Similarly, various methods for moving object detection and tracking in video have been discussed in [22, 24, 25, 27, 28].

3 Available Vehicle Dataset

A large number of cameras have been widely used for roads and transportation surveillance, but because of copyright, privacy, and security concerns, traffic photos are only published publicly. From the factor of view of picture acquisition, the traffic photograph dataset may be divided into three groups: snap shots taken through a vehicle digital digicam, snap shots taken through a surveillance digital digicam, and snap shots taken through non-tracking cameras (Table 2).

4 Usage of Deep Learning and Machine Learning in Highways and Transportation Systems

This section describes the different ways of applying machine Learning, deep learning, and ANN models in highway management and transportation systems and how these models are making lives easier and efficient.

4.1 Highway Crash Detection and Crash Prediction

To reduce the adverse effect of crashes on highways, it is becoming very essential for the traffic management centers to timely get the information about the crash, preferably before a crash. To avoid unusual highway traffic and secondary crashes, it is very essential to get the crash prediction on time. The researchers investigated many

Table 2 Description of publicly available datasets

Dataset	Sequence	Source: URL
Tsinghua-Tencent Traffic-Sign [11]	This dataset have 100,000 images from car cameras covering various lighting as well as weather condition	https://cg.cs.tsinghua.edu.cn/traffic-sign/
Stanford Car [14]	It includes 19,618 images of various categories of vehicles covering the brands, models, and production years of the vehicles	https://ai.stanford.edu/~jkrause/cars/car_dataset.html
The Comprehensive Cars [4]	It have 27,618 images include the vehicle's maximum speed, number of doors, number of seats	http://mmlab.ie.cuhk.edu.hk/datasets/comp_cars/
BIT-Vehicle Dataset—Vehicle Model Identification Dataset [16]	This dataset contains 9,850 images. The database is divided into six categories. SUV, Car, Truck, etc.	http://iitlab.bit.edu.cn/mcislab/vehicledb/
TRANCOS dataset [17]	TRANCOS dataset contains 1244 images	http://agamenon.tsc.uah.es/Personales/rlopez/data/trancos/
The KITTI benchmark [23]	KITTI contains images of 3D object detection and tracking	http://www.cvlibs.net/datasets/kitti/eval_object.php?obj_benchmark=3d

systems for timely and reliable identification of crashes to assist in the management of road accidents. Nowadays we have different sources of real-time data available and therefore we can use them and build very precise models. Huang et al. conducted a study on using D/L models to detect collision risks and collision detection. For crash detection, CNNs were used and found to be performing better than the regular models, when provided with stable training data without overfitting. Data from different time slots were checked for prediction to further explore the model prediction power model. The result of their analysis shows that better collision detection and very close collision prediction results are available in the deep learning model [10].

4.2 *Crash Hotspot Identification on Highways*

The hotspot are the roadway sites or places where the frequency of occurring of a crash is high [11]. Expected Equivalent Property Damage Only (EEPDO) has been used for identification of crash hotspot. There are various techniques for crash detection and crash prediction but it is very difficult to correctly predict for a crash to happen. The crashes are random and rare, i.e., fluctuating with time and space. Wu et al. compared the performance of machine learning algorithms, KNN algorithm, and

Negative Binomial (NB) to find an estimate of EEPDO. Negative binomial assumes that the primary data follows a certain gamma distribution that is not commonly retained for crash data. KNN is supposed to provide a better approximation of the crash data compared to being a non-parametric predictor and therefore does not require data consideration. Both the KNN and NB algorithms were given the same primary data to find the EEPDO and the performance assessment of two algorithms was calculated on the basis of Mean Squared Error (MSE). The one with least MSE performs better than the other with larger MSE value. The result of their experiment showed that the K-Nearest Neighbor algorithm outperformed the NB in finding accurate value for EEPDO that helps in identifying crash hotspots [12].

4.3 Monitoring Driver Behavior

On highways, the drivers drove their vehicle at very high speed which might lead to dangerous accidents, these accidents can cause property damage, life damage, and other damages. These accidents largely depend on the driver's behavior. The Vice President of General Administration of PTT Global Chemical Public Company Limited (GC), explored how to reduce the risks of road/highway travel. He proposed a framework that utilizes artificial intelligence for facial recognition along with data and video analytics to track the actions of a driver in real time. This system's mainly focused to detect if the driver driving the vehicle is feeling distracted or sleepy. The working of this application can be outlined as: the company's vehicles are equipped with driver-focused cameras and a GPS (Global Positioning System) for detecting speed. The information on facial recognition is gathered and moved to the cloud where machine learning is used to analyze it. If the driver showed the sign of risks then they will intimate by an immediate alarm, and then a new driver may be sent off by the fleet manager, if necessary. Machine learning from the data collected can continuously enhance the identification of sleepiness as the device is used more and may also be able to identify specific behavioral signals. Over a time, the approach gets intelligent and helps to predict and avoid accidents even more accurately [13].

4.4 Intelligent Transport System (ITS)

Intelligent transport systems (ITS) are more likely in the future to be a major component of smart cities. It is today's demand. ITS makes use of the information technology and sensing technology to improve the transportation and transit systems. Few of the applications and services of the ITS systems are in public transit system management, traffic management, self-driving vehicles and traveler information systems, etc. Nowadays we have abundant data available collected from various sources like in vehicle sensors, cameras, etc. These collected data can be very useful in making Machine Learning and Artificial Intelligence models for ITS [15]. Mirialys Machin et al. discussed about the use of various AI techniques that can improve the ITS systems. They conducted their study into three main areas of ITS: (1) Vehicle Control, (2) Traffic forecast and Control, and (3) Road Safety and Accident prognostic. The selected AI techniques that were used are (1) ANNs, (2) GAs, (3) FLs, and (4) ESs. The result of their study can be summarized as: For vehicle control systems the most widely used AI technique was Genetic Algorithm and also GAs are suitable for multi-objective improvements. ANNs were used for traffic prediction and traffic control services. For road safety and accident prediction, it was found that for estimating the accident frequency FL seems suitable and for injury severity in traffic accident ANNs was performing well.

4.5 *Crack Condition of Highway Pavements*

Pavements are the surfacing of a road that helps in absorbing or transmitting the load to the sub-base and underlying soil. Heat and cold weather causes the pavement to expand and contract that eventually causes cracks in pavement. So it has become a necessity to make sure that the pavements are in correct shape. Because it is very expensive to make a new pavement rather than to maintain it [14]. Sylvester Inkooma et al. investigated the use of M/L algorithms to predict the crack rating of pavements. ANNs and recursive partitioning were the algorithms used. They found in their results that both the ANNs & recursive partitioning can be used to predict crack conditions. Based on their quality of-fit statistics, mean absolute deviation (MAD $<$ 0.4) and root-mean square errors, crack ratings were observed (RMSE between 0.30 and 0.65) [9].

4.6 *Traffic Surveillance and Traffic Flow*

Traffic surveillance means monitoring the traffic flow on highways and roads. Traffic surveillance can help in monitoring of the roads for accidents, closures and also in highway management not only this, it is useful in making decisions regarding future road development and constructions [6, 16].

Now a days, researches are trying to focus on IoT and sensor enabled vehicles which uses computer vision technology and work in cloud environment through wireless network. For example, a sensing technology-based application that can be combined with information and communication technology (ICT) to improve the intelligent transport system, such as when a vehicle is involved in a road accident due to a sudden chuckhole opening and the vehicle is stuck inside. So, such applications are very helpful for drivers to get out of such kind of sudden danger zones. Various technological aspects and methods have been depicted in [28] (Fig. 5).

Song et al. [6] "For example, an application based on sensing technology that can be combined with information investigated a vehicle recognition system based on

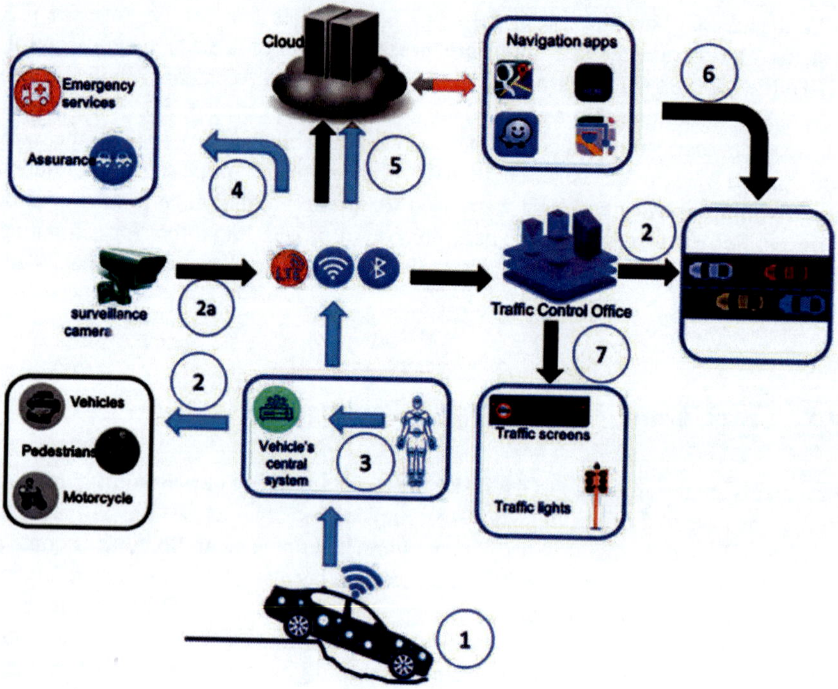

Fig. 5 ICT-based sensing technology enabled intelligent transportation [17]

computer vision and an intensive counting system based on deep learning models. In the proposed traffic sensing the main highway in the diagram is 1st drawn out and further categorized into the remote and the proximal region in the proposed traffic sensing and counting method by using the newly proposed image segmentation system, an important way to improve vehicle detection results. Then, to decide the type and position of the vehicle, the above two positions are put on the YOLOv3 network. For example, a sensing technology-based application that can be combined with information and communication technology (ICT) to improve the intelligent transport system, finally, the vehicle's trajectories are obtained by the ORB algorithm, which can be used to measure the vehicle's driving direction and to estimate the number of unique vehicle trajectories. Then, to decide the type and position of the vehicle, the above two positions are put on the YOLOv3 network. To authenticate and confirm the proposed method of segmentation, highway surveillance image sequences focused on different scenes are used. The test results confirm that high detection accuracy can be given by using the proposed segmentation process, especially for smaller vehicles.

5 Benefits of Highways and Transportation Surveillance Cameras

Video surveillance is the key factor for highways and transportation surveillance system. Now a days every highway or road is equipped with surveillance cameras. These cameras includes the static as well as PTZ cameras for the purpose of surveillance. Highways and transportation system has many benefits and some of these are given below.

- It can help us for the planning and traffic measurements
- It can help us for the enforcement of laws i.e. they enforce red light jump or over speeding issues and catch violations and issue tickets to the concerned.
- It can detect unwanted or unauthorized parking of the vehicle or tailgating of the toll plazas.
- Traffic surveillance cameras encourages us for the safe driving and to follow the traffic rules.
- Traffic surveillance cameras help us in getting traffic news feed on radio as well as on TV.
- Traffic surveillance cameras help us in making future highways/roads management systems.

6 Conclusion

This paper discusses various machine learning- and artificial intelligence and deep learning-based methods for motion-based vehicle identification, detection, and tracking, especially for autonomous as well as smart surveillance system. This paper work considered various methods for moving vehicle detection and tracking and focusses on various parameters as well as metrics for analysis. This paper work has depicted the real-time-based methodologies to automate the transportation system. This paper has discussed several highways and transportation publicly available datasets as mentioned in the table for result analysis. The literature reveals various methods for vehicle counting, lane detection and tracking, pre-crash alert system, etc. Such work are very helpful for reducing the number of crashes, pile-up issues occurred on highways. In smart cities, such work proved to be very convenient in making intelligent transportation systems.

References

1. Machin, M., Sanguesa, J.A., Garrido, P., Martinez, F.J.: On the use of artificial intelligence techniques in intelligent transportation systems. IEEE Wireless Communications and Networking Conference Workshops (WCNCW) (2018)

2. Özdağ, M.E., Atasoy, N.A.: Analysis of Highway Traffic Using Deep Learning Techniques. ISAS WINTER-2019, Samsun, Turkey, vol. 4
3. Abduljabba, R., Dia, H., Liyanage, S., Bagloee, S.A.: Applications of artificial intelligence in transport: an overview. Sustainability **11**(1), 189, 1–24 (2019) https://doi.org/10.3390/su11010189
4. Yang, L., Luo, P., Loy, C.C., Tang, X.: A large-scale car dataset for fine-grained categorization and verification. Computer Vision and Pattern Recognition (CVPR) (2015) http://mmlab.ie.cuhk.edu.hk/datasets/comp_cars/
5. Ma, Y., Chowdhury, M., Sadek, A., Jeihani, M.: Real time highway traffic condition assessment framework using vehicle infrastructure integration (VII) with artificial intelligence (AI). IEEE Transactions on Intelligent Transportation Systems, Vol. 10, Issue: 4 (2009)
6. Song, H., Liang, H., Li, H., Dai, Z., Yun, X. : Vision-based vehicle detection and counting system using deep learning in highway scenes. Europ. Transport Res. Rev. **11**(1). https://doi.org/10.1186/s12544-019-0390-4
7. Polson, N.G., Sokolov, V.O.: Deep learning for short-term traffic flow prediction. Transp. Res. Part C: Emerg. Technol. **79**, 1–17 (June 2017)
8. Pan, G., Fu, L., Thakali, L.: Development of a global road safety performance function using deep neural networks. Int. J. Transp. Sci. Technol. **6**(3), 159–173 (2017)
9. Inkoom, S., Sobanjo, J., Barbu, A., Niu, X.: Prediction of the crack condition of highway pavements using machine learning models. Struct. Infrast. Eng. Maintenan. Manag. Life-Cycle Design Perform. Taylor & Francis **15**(7), 940–953 (March 2019). https://doi.org/10.1080/15732479.2019.1581230
10. Huang, T., Wang, S., Sharma, A.: Highway crash detection and risk estimation using deep learning. Accident Anal. Prevent. **135** (2019). https://doi.org/10.1016/j.aap.2019.105392
11. Zhu et al.: Traffic-sign detection and classification in the wild. The IEEE Conference on Computer Vision and Pattern Recognition (CVPR) (2016) https://cg.cs.tsinghua.edu.cn/traffic-sign/
12. Wu, D., Wang, N., Wang, F., Hong, S.: Applying machine learning algorithms to highway safety EEPDO. CCSCI 2017, vol. 1, pp. 1421–1426, 2017. https://doi.org/10.1109/CSCI.2017.248
13. Microsoft Asia News Center: Artificial Intelligence and road safety: A new eye ont the highway, 4-March-2019, Online: https://news.microsoft.com/apac/features/artificial-intelligence-and-road-safety-a-new-eye-on-the-highway/
14. Krause, J., Stark, M., Deng, J., Fei-Fei, L.: D object representations for fine-grained categorization. 4th IEEE Workshop on 3D Representation and Recognition, at ICCV 2013 (3dRR-13). Sydney, Australia. Dec. 8, 2013. https://ai.stanford.edu/~jkrause/cars/car_dataset.html
15. Yuan, T. et al.: Harnessing machine learning for next-generation intelligent transportation systems: a survey. HAL Id: hal-02284820. https://hal.inria.fr/hal-02284820v2
16. Dong, Z., Wu, Y., Pei, M., Jia, Y.: Vehicle type classification using a semi-supervised convolutional neural network. In: IEEE Transactions on Intelligent Transportation Systems, vol. 16, no. 4, pp. 2247–2256, Aug. 2015. http://iitlab.bit.edu.cn/mcislab/vehicledb/
17. Ricardo, G. et al.: Extremely overlapping vehicle counting. Iberian Conference on Pattern Recognition and Image Analysis (IbPRIA), 2015. https://gram.web.uah.es/data/datasets/trancos/index.html
18. Doudou, M., Bouabdallah, A., Berge-Cherfaoui, V.: Driver drowsiness measurement technologies: current research, market solutions, and challenges. Int. J. Intell. Transp. Syst. Res. **18**, 297–319 (2020)
19. Peng, Y., Lu, Z., Cheng, F., Zheng, Y., Huang, S.: Image haze removal using airlight white correction, local light filter, and aerial perspective prior. IEEE Trans. Circuits Syst. Video Technol. **30**(5), 1385–1395 (May 2020). https://doi.org/10.1109/TCSVT.2019.2902795
20. Song, H., Liang, H., Li, H., Dai, Z., Yun, X.: Vision-based vehicle detection and counting system using deep learning in highway scenes. Europ. Transp. Res. Rev. **11**, article 51, 1–16, 2019. https://doi.org/10.1186/s12544-019-0390-4

21. Kukkala, V.K., Tunnell, J., Pasricha, S., Bradley, T.: Advanced driver-assistance systems: a path toward autonomous vehicles. IEEE Consumer Electron. Mag. **7**(5), 18–25 (Sept. 2018). https://doi.org/10.1109/MCE.2018.2828440
22. Girshick, R., Donahue, J., Darrell, T., Malik, J.: Rich feature hierarchies for accurate object detection and semantic segmentation. In: 2014 IEEE Conference on Computer Vision and Pattern Recognition (2014). https://doi.org/10.1109/cvpr.2014.81
23. Geiger, A., Lenz, P., Urtasun, R.: The KITTI vision benchmark suite. Conference on Computer Vision and Pattern Recognition (CVPR) (2012)
24. Uijlings, J.R.R., van de Sande, K.E.A., Gevers, T., Smeulders, A.W.M.: Selective search for object recognition. Int. J. Comput. Vision **104**(2), 154–171 (2013)
25. Sudha, D., Priyadarshini, J.: An intelligent multiple vehicle detection and tracking using modified vibe algorithm and deep learning algorithm. Soft Comput. **24**, 17417–17429 (2020)
26. Suri, A., Sharma, S.K., Yadav, D.K.: Moving object detection using optical flow and fuzzy algorithm. J. Adv. Res. Dyn. Control Syst. 11(11), 840–847 (2019)
27. Sharma, L., Yadav, D.K. : Histogram based adaptive learning rate for background modelling and moving object detection in video surveillance. Int. J. Telemed. Clin. Pract. Inderscience **2**(1), 74–92 (2017)
28. Suri, A., Sharma, S.K., Yadav, D.K.: Comprehensive Study of Various Techniques for Object Detection and Tracking in Video. Elsevier, International Conference on Innovative Advancement in Engineering and Technology, JNU, Jaipur, pp. 1–6, Feb. 2020.

Privacy and Security Issues in Vehicular Ad Hoc Networks with Preventive Mechanisms

Shally Nagpal, Alankrita Aggarwal, and Shivani Gaba

1 Introduction

The advanced wireless networks have expanded drastically in delivering mobile service in the last decade from a global perspective. Mobile ad hoc networks do not need an administrator and are one of the key areas of implementation. MANET can be used in locations where equipment cannot be installed, such as military installations. Vehicular wireless ad hoc is a newer class of vehicular ad hoc. When the VANET is being used on the road, cars are both mobile devices and nodes. Vehicle-to-vehicle connectivity is critical for operational protection and active transportation. More difficult routing protocols need to be developed, such as BINA, because of VANET, the fluidity of the nodes, and network heterogeneity. These mobility models demonstrate the way that nodes on the road are moving. During the simulation and application of protocols, the templates are used [1]. They should be designed in such a way that their patterns are comparable to vehicle movements in the real world.

Traffic protection is an essential objective of VANET and because improved conditions minimize collisions, and this means lives as well. However, such additional facilities as a weather outlook and position can also make the vacation more comfortable and infotainment [2].

The WAVE architecture and its associated protocol collection of services and interfaces are described by the IEEE 1609 group. The WAVE model specifies the protection requirements for message sharing as well. The WAVE standard offers a foundation for VANET applications in many different verticals, such as surveillance, automated tolls, traffic alerts, and more [3]. Let us see in Fig. 1 how the 1609 architecture with the OSI model compares with various norms.

For the topic of privacy and security, having both is a good idea. Every piece of software has the ability to impact your digital security. Security and protection

S. Nagpal · A. Aggarwal (✉) · S. Gaba
Panipat Institute of Engineering and Technology, Samalkha, Haryana 132101, India

B. Agarwal et al. (eds.), *Proceedings of International Conference on Intelligent Cyber-Physical Systems*, Algorithms for Intelligent Systems,
https://doi.org/10.1007/978-981-16-7136-4_24

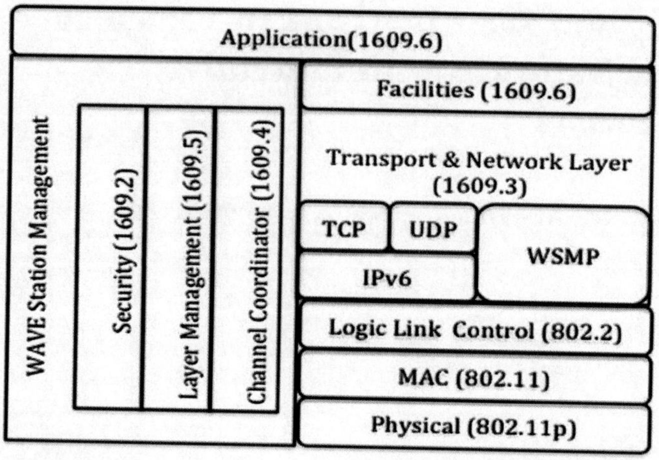

Fig. 1 Protocol analytics

are closely linked. Privacy involves controlling your personal information, as well as how it is used [4]. Pay attention to the security mechanisms you find yourself lured to when you add new personal digital assistant (PDA) apps to your phone. In contrast, security refers to how your personal information is safeguarded. As you probably already know, your personal data may be found in many different places. It presents a challenge to both your security and your protection. "Protection" and "security" are two persons who believe that defence is, in effect, the same as protection [5]. Since they are sometimes present in the same place that is in light of the fact that the two have once in a while joined in the same planet. The findings, however, do not compare the cards exactly, and understanding how they work together might assist you to deal with situations that are highly correlated.

1.1 Privacy Versus Security

Here's a model. You may impart individual data to your bank when you open a financial record. What occurs after that? Here are three potential results, all identified with your own data (not to the cash you may have stored in the financial records).

(a) Your privacy and security are kept up. The bank uses your information to open your record and give you things and organizations. They continue to guarantee that data.
(b) Privacy is undermined and security is kept up. The bank offers a bit of your information to a promoter. Note: You may have assented to this in the bank's assurance disclosure. The result? Your own information is in a more noteworthy number of hands than you may have required.

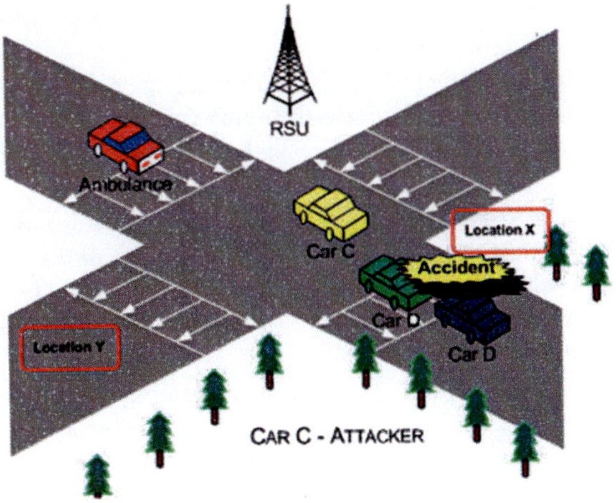

Fig. 2 Typical VANET environment

(c) Both your privacy and security are undermined. The bank gets hit by a data break. Cybercriminals enter a bank database, a security break. Your information is revealed and could be sold on the diminished web. Your insurance is no more. You could transform into the overcomer of computerized blackmail and information extortion [6].

It would be incredible if your dangers started and finished with that hypothetical bank. Be that as it may, your own data is likely everywhere throughout the associated world—in government workplaces, at human services suppliers, at stores and cafés, and in a considerable lot of your online records [7]. You may state it's all over—not actually; however, it's surely in enough places that it's out of your control (Fig. 2).

1.2 VANET Characteristics

VANETs are improvised frameworks, particularly incredible, and strong and offer various organizations, anyway with compelled access to the framework establishment. VANETs have novel characteristics when stood out from MANET, and these characteristics are incredibly fundamental [8] for security and assurance perspectives in VANET, which are discussed underneath:

Dynamic Network Topology: The topology of VANET isn't steady and can change quickly because of the high versatility of vehicles. Accordingly, it makes VANET progressively powerless against the assaults and hard to perceive the speculated vehicles.

High mobility: Compared to MANET, VANET has higher portability. Vehicles are travelling so quickly that it might delay V2V communication. Likewise, when hubs increase in mobility, the smaller number of work hubs in the system becomes increasingly apparent.

In the distant access of vehicular condition (WAVE), transmission power is extraordinarily constrained, ranging from 0 to 28.8 dBm and limited to separation of up to 1 km. This might be considered a consequence of the constricted force transfer, which limits the area that VANET may include.

The VANET data must have arrived at the hubs in a certain timeframe in order to enable fast decisions and swift subsequent progress.

Even with VANET, vehicle relationships might be eliminated or remain dynamic at a few distant locations. Therefore, individual security in VANET cannot be guaranteed in this way.

The complexity of computing and storage becomes more problematic in VANET when data trading across cars and foundations is routine.

1.3 VANET Security and Challenges

The internet has additional problems that are vital for researchers to examine, such as fewer fundamental problems, flexibility, and lack of distant accessibility. The security of VANETs checks to make sure that the messages that have been transported have not been tampered with by the attackers. In addition, the driver is accountable for ensuring that people are aware of any traffic conditions required by law within the allotted time limit [9]. The unquestioned attributes of VANET cause it to be dynamically unstable to attackers. Thus, security issues should be handled suitably; otherwise, it might negatively impact numerous VANET aims.

1.4 VANET Security Services

(a) **Availability**: Accessibility is a significant piece of security administrations that require consideration since it is straightforwardly connected with all the well-being applications. The fundamental obligation of accessibility is to oversee usefulness, and its security must guarantee that the system and different applications must stay useful if there should arise an occurrence of defective or malignant conditions.

(b) **Confidentiality**: In view of testaments and shared open keys, classification guarantees that the assigned collector approaches the information while outside hubs will be unable to gain admittance to that information until the private information were gotten by the assigned client.

(c) **Authentication**: Verification assumes an indispensable job in VANETs. It forestalls the VANETs against suspected substances in the system. It is critical to have the related data of transmission modes, for example, client recognizable proof and sender address.

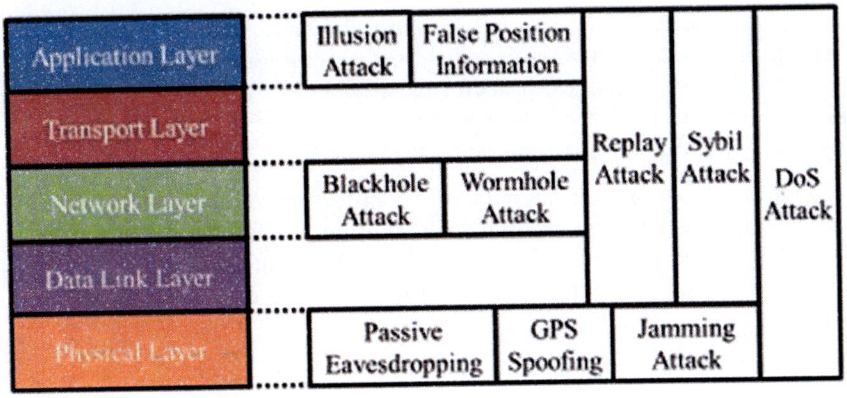

Fig. 3 Layered-based assaults in VANET

(d) **Non-repudiation**: It guarantees that, if there should be an occurrence of the question, the sender and the collector of the message don't decline to take part in transmission and gathering.

(e) **Data integrity**: It guarantees the content of the message isn't changed during the correspondence procedure. In particular, in VANET, it tends to guarantee while utilizing open key framework and cryptography renouncement processes (Fig. 3).

2 Security Attacks and Threats in VANETs

2.1 *Attacks on Availability*

The accessibility of data is a significant piece of the VANET framework if there should arise an occurrence of an absence of accessibility that may prompt a decrease in the productivity of VANETs [10]. Right now, it will clarify the dangers and assaults in VANETs.

Jamming Attack: Right now, the aggressor upsets the correspondence divert in VANETs by utilizing an intensely fuelled sign with equal recurrence. This is the most hazardous assault for security applications since it did not follow the substantial wellbeing alert.

Greed approach with misuse of message confirmation code: Normally, this occurs when malicious devices use the MAC convention to extend transmission capacity and drive up costs for customers.

Malicious software attack: Attacks may be performed using the product portions that interact with the OBU and RSU to operate the OBUs and RSUs. In the case of a malware attack taking place in VANETs, the system will encounter anomalies in various areas of the VANET framework.

Vehicle broadcast tampering attack: Vans may send comparable messages by making changes to the message or adding another message to the VANET.

VANET's denial-of-service (DOS) attacks are one of the fundamental assaults in the field. The attacker places the correspondence between cars, which stops everything you can think of, from happening.

Blackhole Attack: This is the primary assault that objectives accessibility in the specially appointed system and furthermore exists in VANETs. This assault is generally brought about by an enlisted VANET client. The speculated hub gets the parcels from the system, yet it decreases to add to the systems administration activity.

Spamming Attack: Right now, the measure of spam messages was infused by the aggressor, for example, and in the VANET framework, which causes a crash by using more data transmission.

Grayhole Attack: It is the variation of blackhole assault, and it happens when conniving vehicles select a portion of the information bundles to advance and drop the other parcel without being followed.

2.2 *Attacks on Confidentiality*

Classification assurances can be encoded while utilizing the declarations and sharing open keys to all trade messages, and the main assigned vehicles can get to. Along these lines, the vehicle which is outside the hubs cannot get private and classified data among the vehicles. Classification is ensured through cryptographic arrangements. Right now, will examine the regular dangers on classification, which are talked about underneath.

Traffic Analysis Attack: An attack that affects personal privacy might be referred to as traffic analysis. When receiving a message transmission, the adversary is on the lookout for a recurrence and compiles as much of the most advantageous data as possible.

Social Assaults on Networks: The first step in a social attack is to seize the driver's attention. Corrupt and unethical messages are sent by the aggressor. An aggressor's intention is to prompt a reaction from the drivers.

Eavesdropping Attack: Spying is exceptionally normal in remote correspondence innovation, for example, MANET and VANET. The point of this assault is to get private data from the secured information.

Man-in-the-Middle Attack: This assault happens in V2V correspondence to check intently and change the messages. The aggressor can gain access and power to the whole V2V correspondence; however, the correspondence substances believe that they can speak with one another legitimately in private.

2.3 *Attacks on Authentication*

Verification is a significant piece of the VANET framework, which is utilized to secure against the assaults in view of the malevolent hubs entering the framework. The validation is answerable for shielding VANETs from inward and outside assaults. This segment features the dangers and assaults on validation in VANET.

Tunnelling Attack: This assault is like a wormhole assault. The assailant utilizes a similar system to start the private discussion, and the aggressor joined two far-away pieces of the VANET by using an additional correspondence channel named burrow.

Free-Riding Attack: This assault is exceptionally normal and starts with a functioning malevolent client by putting forth bogus validation attempts while related to the agreeable message confirmation.

Sybil Attack: The Sybil assault was first talked about. This is the riskiest assault wherein the hub contains many phoney characters to disturb the ordinary method of tasks of the VANETs by communicating various messages.

Node Impersonation Attack: This assault happens by effectively getting the substantial ID of the client and sending it to another approved client in the VANET.

GPS Spoofing: Due to the fact that the hub's location and the area of it are vital in VANETs, it needs to be precise and real. A log file including an area table is maintained on each GPS satellite.

Replay Attack: This assault is an extremely basic assault which is otherwise called a playback assault; this assault happens when legitimate information is deceitfully transmitted or makes postpone produce an unapproved and vindictive impact.

Message Tampering Attack: Message tampering is a typical assault, wherein the assailant can modify the traded messages in V2I or V2V correspondence that is deliberately used to profit fake reactions.

Masquerading Attack: The aggressor utilizes bogus IDs to go about as another vehicle. This assault happens when one client did not demonstrate his own personality and professes to be an alternate client to get unapproved legitimately.

Key or/and Certificate Replication Attack: This assault is brought about by the use of copy keys or potentially testaments of different clients as confirmation of verification to make vulnerability which makes the circumstance most exceedingly terrible for traffic specialists to distinguish the vehicle.

2.4 *Attacks on Data Integrity*

Now, we will examine the regular dangers to uprightness, which are talked about underneath:

Masquerading Attack: The aggressor utilizes bogus IDs to go about as another vehicle. This assault happens when one client did not demonstrate his own personality and professes to be an alternate client to get unapproved legitimately.

Message Tampering Attack: Message tampering assault demonstrates that this assault regularly happens when the aggressor adjusts or changes late message information to be transmitted.

Replay Attack: This assault is an extremely basic assault which is otherwise called a playback assault; this assault happens when legitimate information is deceitfully transmitted or makes postpone produce an unapproved and vindictive impact.

Illusion Attack: This assault got information from radio wires and gathered pernicious information from sensors that produce traffic cautioning messages by utilizing the current street condition which may make a hallucination to the vehicles close by.

2.5 Attacks on Non-repudiation

Repudiation Attack: Repudiation assault happens when an aggressor denies participating in the movement for sending and getting a message if there should be an occurrence of any debate.

3 Authentication Perspectives with Assorted Dimensions

In VANET, verification should be possible in two different ways: right off the bat, at the degree of the hub, called hub validation, and besides, at the degree of message level, called message confirmation. Checking the message trustworthiness assumes a significant job to improve the VANET security framework. Along these lines, message confirmation is viewed as a key parameter in the VANET so as to give secure correspondence in VANET, and a few necessities of verification that must be fulfilled are recorded beneath.

Utilization of Bandwidth

Data transfer capacity use is significant during validation and must be used appropriately via a huge level of security.

Computational and Communication Overhead

The computational expense caused because of a lot of cryptographic activity to be finished by a vehicle or believed expert for checking a verification demand must be abbreviated. Besides, the time required to process a computerized signature in confirmation must be controlled.

Powerful Authentication

The validation plans must have a great capacity to forestall VANETs from assaults.

Scalability

The procedure of validation should be versatile which can deal with various system tasks and correspondences.

4 Research Work on Security Services

Security administrations assume a significant job to guarantee secure correspondence in VANETs. Right now, the work has portrayed the ongoing examination business related to VANET security administrations.

4.1 Confidentiality

Some new security strategies have recently emerged and remain covert in order to safeguard confidential material in the VANET, which includes useful data from non-registered customers. Also, it ensured confidential communication with cryptographic protections in place. We shall detail the present methods for personal privacy in the VANETs for the time being.

The authors of [11] devised another security mechanism by guaranteeing that sensitive data was kept secret by encrypting the data using a shared key that was distributed in a public setting. The primary aim of the technique described is to ensure the client secret data and vehicle authentications are secure, which is feasible using the new security requirements and working out a comprehensive VANET security architecture to prevent unauthorized clients from infiltrating the system. Even still, the cryptic vehicle message is very important for helping automobiles access the internet and RSUs.

Rabieh et al. [12] presented powerful security safeguarding key administration techniques alluded to as DIKE, which is utilized to accomplish and improve the secrecy of information in area-based administrations (LBSs) in the VANET framework. So as to control the listening stealthily assault, privacy must be all around kept up and the administration substance from these sorts of assaults is ensured. Right now, a client does not take part in the VANET framework, at that point the client may not join the current VANET framework and in this way cannot approach the present LBS content.

4.2 Non-repudiation

Gazdar et al. [13] presented a novel system with restrictive protection safeguarding and renouncement (ACPN) for VANETs. This technique used open key cryptography (PKC) to acquire non-repudiation of vehicles by guaranteeing outsiders to get genuine personalities of vehicles. The personality-based mark (IBS) and ID-based on the web/disconnected mark (IBOOS) plans are used for the verification among V2V and vehicle to the side of the road units (V2R). This strategy altogether diminished computational expense. Be that as it may, the treatment of overseeing testaments is mind-boggling due to IBS and IBOOS confirmation plans.

4.3 Availability

A significant amount of recent research has gone into accessibility and provided new conventions to make the accessibility advantages easier to see. It will enhance the current accessibility demonstration tactics in VANETs.

Messaging shipment method was established by Rostamzadeh et al. [14], which helps to boost message flow in rural places. This technique employed automobiles as a conduit to get the most extreme traffic data, such as a location, fuel, and a vehicle's speed. This data was then brought together and communicated to other cars in the neighbourhood. After the test on the NETSTREAM traffic system, the new methodology was evaluated to see how it stacked up against the data spread proficiency and other competing tactics. Although this method did not reference the clear display limitations which were related to and only restricted to the low-thickness zone, the inventor did not shy away from demonstrating his work. Two professors Okamoto and Ishihara presented a strategy for data sharing systems for area subordinate information, which is produced by vehicles utilizing the draw and push technique to adjust the message conveyance and traffic for information spread. This system, called a doling out of populated territories, is used by means of which territory territories are assigned by utilizing the draw and push technique to adjust the message conveyance and traffic for information spread. With this methodology, trustworthy data may be sent regardless of the kind of push and force; however, this will cause an enormous amount of computing costs.

Hussain et al. [15] presented a viable information replication system to oversee information that gets to applications in VANETs, for example, area, fuel, and quickening. Because of high versatility vehicles, the VANETs topology changes progressively, which regularly causes visit detachment. In the event that separation happens much of the time, at that point, the vehicles would not have the option to convey and impart information to one another. In particular, information replication is utilized to improve the exhibition of information access in a dispersed framework. Be that as it may, the vast majority of the hubs in VANETs contain less capacity. In this way, they cannot duplicate overwhelming mp4 records or brief length video cuts.

This issue is essentially improved by creating the solicitation to the vehicles in a detachment to give some piece of their cradles to imitate information while having a similar company and information among different vehicles. On the off chance that, when a vehicle needs to leave a detachment, it moved the supporting information to different vehicles before leaving a company. Accordingly, different vehicles approach the information after it leaves. This strategy has constraints; vehicles habitually leaving and entering a company may require a lot of computational time and bring about calculation charges. Park and Lee acquainted a successful technique with improving the information openness in VANET by using the information copy of the RSU. Right now, the choice of information things is made by utilizing the information get to example and driving example which must be imitated in the RSUs. At that point, the duplicated information is sent legitimately to the encompassing vehicle without including correspondence with RSUs. The primary disadvantage of this methodology is on the off chance that the information size is bigger, at that point the information replication procedure may require a lot of time to deal with the replication procedure.

4.4 Data Integrity

Advanced markings are used to create and integrate with the message so that the transmitting message is guaranteed to be straight. Lately, some work that presents solid data for making sure about information respectability in the VANET framework is examined beneath. When a closed-off getting region allows several parcels to accumulate and progresses conventions wasteful use in VANETs, this becomes the central concern. To fix this problem, Kerrache et al. [16] used STAP to provide area protection and conservation of beneficiaries in VANETs. Vehicles continually went to areas that were previously populated in order to, for example, retail malls and inhabited streets. In order to secure information quality, they provide RSUs to the school's most prominent social location in order to construct a social level with them. The sender immediately registers the MAC address before transmitting the code to the collector. To verify MAC, the collector uses the verification key to get the message. This computation can only be used at the busiest part of the location due to traffic, and because of this, the memory required is likely to be enormous. In an article entitled, "a Productive and Helpful Validation Procedure for the VANET Framework", He et al. reported in [17]. The technique is used to simplify the validation overhead on individual automobiles, as well as to shorten the deferral period. The organization deploys the token methodology to regulate and deal with the verification load that occurs when resisting diverse attacks. When the vehicle has travelled the length of RSUs, it is able to collect the proof token from TA. This token illustrates that the vehicle which was previously acceptable was added to the other confirmation steps. In order to manage the verification problem, this technique has a tremendous computational calculation.

In 2020, Sharma, A et al. proposed a technique built on GSIS that relies on the collection and character-based mark strategies to safeguard security. In the event of debate, the method described above may be used to track all vehicles except for the identity of the sender, which is concealed by TA. Furthermore, Aggarwal et al. [19] have implemented effective software analytics and risk management when working with software analytics and risk management software. The article underscores the data analytics and machine learning used in managing software risk. Kaur et al. [20] proposed remote sensing of images using fuzzy clusters. Mittal et al. [21] explained the use of clustering methods with a threshold. Singh et al. [22] discuss the IoT robot for monitoring disasters and manipulation of data. Singh et al. [24] discussed the use of cloud and sensor networks to control the environment.

5 Conclusion

The automobiles, in the vehicular network domain, are thought to be both unique and personal in assorted ways. Additionally, they should include the basic collection of sensors, which provides information to the driver that the driver is unable to perceive, like front and rear radar. Another consideration is the use of navigation systems such as the global positioning system (GPS) for guiding purposes. Additionally, a smart vehicle should be fitted with a central computing system and a flight data recorder (FDR). It is being proposed that the vehicle be given an electronic license plate, which would allow easier identification, or an electronic chassis number to represent the vehicle identity. Protection is especially important in VANETs because human lives are often at risk; in the conventional network, security only applies to confidentiality, honesty, and availability. Malicious input cannot be altered by an intruder. It should be noted, however, that driver protection is closely linked to the concept of determining driver liability. The loss of timely exchange of information about the vehicles and their drivers can result in a catastrophic catastrophe. The work presents the assorted dimensions of security and privacy aspects in VANET and thereby the need arises to work on the advanced algorithms and approaches.

References

1. Mejri, M.N., Ben-Othman, J., Hamdi, M.: Survey on VANET security challenges and possible cryptographic solutions. Veh. Commun. **1**(2), 53–66 (2014)
2. Pathan, A.S.K. (ed.): Security of self-organizing networks: MANET, WSN, WMN. CRC Press, VANET (2016)
3. Li, F., Song, X., Chen, H., Li, X., Wang, Y.: Hierarchical routing for vehicular ad hoc networks via reinforcement learning. IEEE Trans. Veh. Technol. **68**(2), 1852–1865 (2018)
4. Al-Sultan, S., Al-, M.M., Al-, A.H., Zedan, H.: A comprehensive survey on vehicular ad hoc network. J. Netw. Comput. Appl. **37**, 380–392 (2014)

5. Hossain, E., Chow, G., Leung, V.C., McLeod, R.D., Mišić, J., Wong, V.W., Yang, O.: Vehicular telematics over heterogeneous wireless networks: a survey. Comput. Commun. **33**(7), 775–793 (2010)
6. Raya, M., Hubaux, J.P.: The security of vehicular ad hoc networks. In: Proceedings of the 3rd ACM workshop on Security of ad Hoc and Sensor Networks, pp. 11–21 (2005)
7. Laberteaux, K.P., Haas, J.J., Hu, Y.C.: Security certificate revocation list distribution for VANET. In: Proceedings of the Fifth ACM International Workshop on VehiculAr Inter-NETworking, pp. 88–89 (2008)
8. Wasef, A., Lu, R., Lin, X., Shen, X.: Complementing public key infrastructure to secure vehicular ad hoc networks [security and privacy in emerging wireless networks]. IEEE Wirel. Commun. **17**(5), 22–28 (2010)
9. Hu, Y.C., Laberteaux, K.P.: Strong VANET security on a budget. In: Proceedings of Workshop on Embedded Security in Cars (ESCAR) Vol. 6, pp. 1–9 (2006)
10. Olariu, S., Khalil, I., Abuelela, M.: Taking VANET to the clouds. Int. J. Pervas. Comput. Commun. (2011)
11. Mitola, J., Maguire, G.Q.: Cognitive radio: making software radios more personal. IEEE Pers. Commun. **6**(4), 13–18 (1999)
12. Rabieh, K., Mahmoud, M.M., Younis, M.: Privacy-preserving route reporting schemes for traffic management systems. IEEE Trans. Veh. Technol. **66**(3), 2703–2713 (2016)
13. Gazdar, T., Benslimane, A., Belghith, A., Rachedi, A.: A secure cluster-based architecture for certificates management in vehicular networks. Secur. Commun. Netw. **7**(3), 665–683 (2014)
14. Rostamzadeh, K., Nicanfar, H., Torabi, N., Gopalakrishnan, S., Leung, V.C.: A context-aware trust-based information dissemination framework for vehicular networks. IEEE Int. Things J. **2**(2), 121–132 (2015)
15. Hussain, R., Nawaz, W., Lee, J., Son, J., Seo, J.T.: A hybrid trust management framework for vehicular social networks. In: International Conference on Computational Social Networks, pp. 214–225. Springer, Cham (2016)
16. Kerrache, C.A., Lagraa, N., Calafate, C.T., Lakas, A.: TROUVE: A trusted routing protocol for urban vehicular environments. In: 2015 IEEE 11th International Conference on Wireless and Mobile Computing, Networking and Communications (WiMob), pp. 260–267. IEEE (2015)
17. He, Y., Yu, F.R., Wei, Z., Leung, V.: Trust management for secure cognitive radio vehicular ad hoc networks. Ad Hoc Netw. **86**, 154–165 (2019)
18. Sharma, A., Gaba, S., Singla, S., Kumar, S., Saxena, C., Srivastava, R.: A genetic improved quantum cryptography model to optimize network communication. In: Recent Trends in Communication and Intelligent Systems, pp. 47–54. Springer, Singapore (2020)
19. Aggarwal, A., Dhindsa, K.S., Suri, P.K.: Usage patterns and implementation of random forest methods for software risk and bugs predictions. Int. J. Innov. Technol. Explor. Eng. (IJITEE), **8**(9S), 927–932, Publisher, Blue Eyes Intelligence Engineering & Sciences Publication (2019)
20. Kaur, S., Bansal, R.K., Mittal, M., Goyal, L.M., Kaur, I., Verma, A.: Mixed pixel decomposition based on extended fuzzy clustering for single spectral value remote sensing images. J. Indian Soc. Remote Sens. **47**(3), 427–437 (2019)
21. Mittal, M., Sharma, R.K., Singh, V.P.: Validation of k-means and threshold based clustering method. Int. J. Adv. Technol. **5**(2), 153–160 (2014)
22. Mittal, M., Sharma, R.K., Singh, V.P.: Modified single pass clustering with variable threshold approach. Int. J. Innov. Comput. Inform. Control **11**(1), 375–386 (2015)
23. Singh, R., Gahlot, A., Mittal, M.: IoT based intelligent robot for various disasters monitoring and prevention with visual data manipulating. Int. J. Tomogr. Simul **32**(1), 90–99 (2019)
24. Singh, R., Gehlot, A., Mittal, M., Samkaria, R., Choudhury, S.: Application of icloud and wireless sensor network in environmental parameter analysis. Int. J. Sens. Wirel. Commun. Control **7**(3), 170–177 (2017)

Trends and Sentiment Analysis of Movies Dataset Using Supervised Learning

Shweta Taneja, Siddharth Bhasin, and Sambhav Kapoor

1 Introduction

The motion pictures such as movies and tv shows can be classified into categories which are called genres. However, a movie may have more than one genre. Grouping the movies into broad categories of the genre is yet another challenging classification task to be performed. This activity of grading movies in classes helps both the viewers as well as the critics to draw various conclusions. Labeling the movies into different genres gives more clarity on the type of viewers it shall attract. These trends and results on the genre preferred more by the people will also help the film directors and creators in making the films or shows. In this work, a method of movie classification and sentiment analysis has been proposed.

In our research, the concept of data classification algorithm is used [1]. Classification, being a supervised learning method in machine learning, helps to map the observation to a set of categories which, in this task, helps to identify the genre of a particular movie with the help of its description. Some famous applications of classification are the spam-ham classification of a given email, diagnosing a patient on the basis of the observed characteristics (certain symptoms, blood pressure value, sex etc.) in hospitals etc.

In our work, classification helps in grouping the descriptions of movies into six broad categories or genres which are drama, horror, comedy, western, thriller

S. Taneja (✉) · S. Bhasin
Department of Computer Science, Bhagwan Parshuram Institute of Technology, Guru Gobind Singh Indraprastha University, Dwarka, Delhi, India
e-mail: shwetataneja@bpitindia.com

S. Kapoor
Department of Instrumentation and Control Engineering, Netaji Subhas University of Technology, Dwarka, Delhi, India

B. Agarwal et al. (eds.), *Proceedings of International Conference on Intelligent Cyber-Physical Systems*, Algorithms for Intelligent Systems,
https://doi.org/10.1007/978-981-16-7136-4_25

and documentary. The endeavor is to create a system that can perform rigorous classification.

Multilabel classification techniques involve assigning an instance of an attribute to more than one class label. News articles are a common example of this.

Following are the classifiers that are mostly used in multilabel classification.

One-versus-rest (OvR)

One-versus-rest (OvR) is also called as One-versus-all (OvA) method [2]. In this, a real-valued confidence score is created by the base classifier for its decision, instead of a class label. OvA learner constructed from binary classifier performs a training algorithm where inputs are a learner L, samples X, labels y where yi ∈ {1, ..., K} (for some sample Xi) and the output is a list of classifiers f(k) for k ∈ {1, 2, ..., K}. In order to predict the label k for a classifier, we apply the classifiers to an untold data sample x that gives the maximum confidence score:

$$\hat{y} = \mathrm{argmax} f_k(x), k \in \{1 \ldots K\} \quad (1)$$

OvR with Support Vector Machines

SVM alone supports only binary classification. Therefore, in order to handle the separation of multiple classes, essential parameters and constraints are also added in these extensions [3].

OvR with Logistic Regression

The logistic function is an S-shaped curve (like the sigmoid function) which helps in mapping the values between 0 and 1 by taking real numerical values. It uses Euler's number (e) which is the base of the natural logarithms [4]. Logistic regression is a linear method that predicts the probability and transforming it using a logistic function. The equation can be represented as

$$1/(1 + e^{-\mathrm{value}}) \quad (2)$$

Binary Relevance with Gaussian NB

Binary relevance (BR) is considered as the main baseline for classification in machine learning. The BR method is based on the assumption of independent labels. Hence, the classifier studies each label independently and declares it as irrelevant or relevant. According to several matrices, BR is not only effective in producing ML classifiers but is also computationally efficient. BR together applied with Gaussian Naive Bayes stimulates the model for the multilabel classification. Naive Bayes is based on the principle of MAP (maximum a posteriori) [5]. It is an efficient and popular classifier.

Label Powerset with Logistic Regression

It is used in multilabel classification [6].

Sentiment analysis is an area under natural language processing (NLP) that helps in identifying the sentiment within a text. It is, therefore, also known as opinion mining [7]. Sentiment analysis works on the unstructured data of raw texts and converts them into structured data that can be useful for any brand, organization, politics etc. In our work, we have applied sentiment analysis on the tweets of the Twitter users to find out which genre of movies/tv shows the users like the most and classified the tweets as positive, negative or neutral with the help of TextBlob. TextBlob is a powerful python library that can be used for sentiment analysis and offers a simple API to access its method and accomplish standard NLP operations. TextBlob analyzes an English phrase in the form of a score. Each lexicon has the scores for polarity, subjectivity and intensity with their different specified ranges. The polarity defines if the sentiment for the text is positive, negative or neutral which helps us to understand what people actually think related to movies of a particular genre. In this way, while implementing sentiment analysis we get a general public view over the Twitter platform of the most favorable genre or the trending genre and movies of those genres that can have praising outcomes on the screen which can help in a good critic rating or even can do a good business.

TextBlob is a powerful NLP library for python that can be used for sentiment analysis. It helps in determining the polarity and subjectivity of the text. We have applied TextBlob to the text of tweets to find out about their polarities [8]. The polarity using the TextBlob ranges from -1 to 1. We classified all the tweets whose polarity < 0 as -1 and the tweets whose polarity > 0 as $+ 1$. Then we calculated the number of positive, negative and neutral tweets in the dataset. After that, we computed the percentage of positive and negative tweets using the formula:

$$\text{Percentage of positive tweets} = \left(\frac{\text{Number of positive tweets}}{\text{Total number of tweets}}\right) * 100 \tag{3}$$

Similarly, for negative tweets,

$$\text{Percentage of negative tweets} = \left(\frac{\text{Number of negative tweets}}{\text{Total number of tweets}}\right) * 100 \tag{4}$$

This research work focuses on analyzing trends and sentiments of different movie genres. It covers on following points:

- Using hybrid algorithms to differentiate between multiple labels of the movies. This classification is achieved by using multiple hybrid algorithms such as pipeline for applying SVM with one-versus-rest classifier, binary relevance and Gaussian NB and classifier chains with logistic regression.

- These algorithms are compared by measuring the accuracy.
- For each subsequent genre tweets are extracted using the Twitter API.
- Twitter data is mined for making interpretations.

Sentiment analysis of the genres is done from the extracted tweets. This helps to get insights like evaluating the viewpoint, evaluations, and feelings of a speaker/writer on the social media platform.

The paper is organized as follows: Sect. 2 gives the state of the art in this field. Section 3 highlights the proposed work, which is divided into two subsections. The proposed work includes the flowchart along with pseudo-code. Section 4 shows the dataset used in the work. Section 5 shows the results followed by the conclusion.

2 Related Works

Table 1 shows the summary of work done by different authors.

3 Proposed Work

The proposed methodology is divided into two subsections, namely classification of movie genres and sentiment analysis of the genres.

3.1 Classification of Movie Genres

The flowchart in Fig. 1 shows the stepwise procedure performed in the classification of movie genres.

In the first step, the data extracted from Kaggle is saved into a csv file for further processing. We have considered three attributes of the data stored, that is, the movie name, description and movie genre, out of which description and genre are used in the data classification system. The second step is the most crucial one which is the data pre-processing. In order to influence the results and analysis, pre-processing and mining the data is one of the most crucial parts. The first part in data pre-processing is creating dummies. Here we build a dummy variable or column for each categorical value in the genre. In this way, we store the numerical value against each description representing the genre. Cleaning the data is the process in which initially, the text is tokenized that is segmented into clauses or words, we clean text by removing the unnecessary data in it which may include tags, punctuations, links, emails, phone numbers and other multiple pointless words which are not essential in categorizing the data and does not help the model. Stop words are the words present in the NLTK corpus which are the commonly used words and are unlikely to be useful for

Table 1 Summary of contributions of different authors

S. no.	Authors	Contribution
1	Kadam et al. [9]	The authors have used sentiment analysis as a powerful tool to find the most common features in a program that users generally like in order to increase the success rates of newly proposed TV programs
2	Mhaigaswali and Giri [10]	The authors have stated the importance of social networking in predictive and descriptive analytics
3	Rahim et al. [11]	The authors have stated the importance of YouTube for videos and have used the movie trailers data
4	Zubiaga et al. [12]	The paper has focused on identifying the trends on Twitter by looking at the earliest tweets that produce the trends and categorizing the trends early on. It uses language-independent features relying on the social spreads of the trends to segregate among the trending topics
5	Satyavani et al. [13]	K-Means algorithm has been used by the authors to compare different TV shows on the basis of their popularity
6	Schmit and Wubben [14]	Different classification and regression techniques are used by the authors on tweets from Twitter to make a prediction about the rating of newly released movies. Also, this paper focuses on the importance of textual features in the predictive machine learning tasks
7	Wang and Zhang [15]	The authors have used the Gaussian kernel support vector machine (SVM)model to predict the movie genre preference using customers' behavioral, demographic and social information. Different VC (VapnikChervonenkis) dimensions are used in this paper to compare the error-out-samples
8	Battu et al. [16]	The authors have created a Multi-language Movie Review Dataset and used different techniques such as SVM, random forest, FCNN, LSTM, GRU, hybrid models to predict the movie genre
9	Maloof [17]	The author has shown the application of machine learning in digital investigation
10	Khan and Urolagin [18]	The paper depicts the importance of machine learning in social media data
11	Bhardwaj et al. [19]	For sentiment analysis, a novel approach has been developed for the travel industry
12	Jindal and Taneja [20]	In this paper, the authors have developed an algorithm for multilabel categorization of text documents. It is implemented on five text datasets and has shown promising results

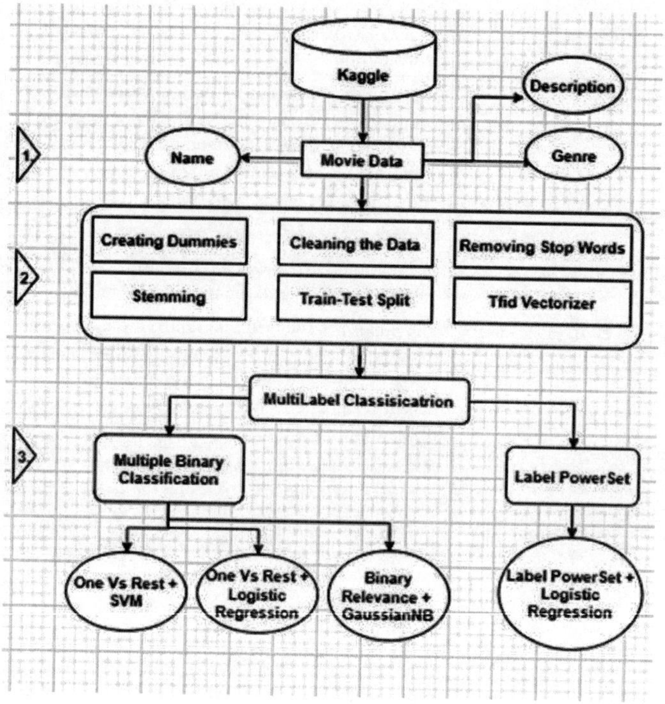

Fig. 1 Flowchart of classification of movie genres

learning. Stemming is the process of generating morphological variants of a base word. A stemming algorithm is the one that reduces or replaces the words to their root word or a common stem. For instance, likes, liking, likely or liking are reduced to like which is their root word.

Stemming helps to reduce redundancy. The data is then segregated into sets: training and testing. Next is the TfidVectorizer, Tfid is the term used for term frequency–inverse document. In TF-IDF, term frequency replicates how often a word appears within a document and marks its frequency and inverse document frequency downscales or removes words that appear a lot across the text. After pre-processing the data, the third step comes in which the system is built for multilabel classification. Since against each movie description we have multiple genres attached to it, we have implemented hybrid algorithms in order to achieve the task of classification. Hybrid algorithms are basically a way of combining models and bringing together the strengths of both knowledge representations.

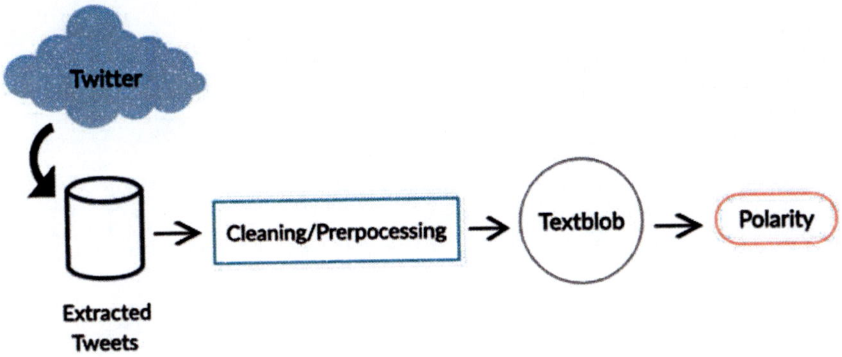

Fig. 2 Flowchart for sentiment analysis starting from extracting Twitter data to the polarity of the genre

3.2 *Sentiment Analysis of Movie Genres*

The second subsection is sentiment analysis of movie genres. The algorithm used for the sentiment analysis of the tweets using TextBlob is given as follows:

1. Use Tweepy to extract tweets from Twitter.
2. Using hashtags of different genres, tweets are extracted for those genres.
3. Merge all the tweets extracted to prepare a dataset consisting of the username along with the text of the tweet.
4. Pre-processing is done to clean all the non-letters in tweets.
5. Analyze the tweets for sentiment analysis using TextBlob.
6. The polarity of each tweet is then found out as $+1$, -1 or 0 and is added to the new dataset.
7. The percentage of positive and negative reviews for each genre is then calculated (Fig. 2).

The pseudo-code for the sentiment analysis is given below. The code demonstrates the method to find the polarity of the sentiment for a given text of lines. Once the polarity of the text is determined by TextBlob, the outcome of positive and negative reviews in tweets is shown and printed for the results.

```
l = list()
for t in texts:
    analysis = TextBlob(t)
    if analysis.sentiment.polarity> 0:
l.append(1)
elifanalysis.sentiment.polarity == 0:
l.append(0)
    else:
l.append(-1)
act = pd.DataFrame({"tcexts":texts,"polarity":l})
sum1=0
for polar in act["polarity"]:
    if polar==1:
        sum1=sum1+polar

sum2=0
for polar in act["polarity"]:
    if polar==-1:

        sum2=sum2-polar
print("The    percentage    of    positive    reviews    is
",(sum1/len(act))*100)
print("The    percentage    of    negative    reviews    is
",(sum2/len(act))*100)
```

4 Dataset Used

For the implementation part, we have used the Movie Dataset from Kaggle [21]. It contains metadata of 45,000 movies present in the Full Movie Lens Dataset. This dataset contains 26 million ratings from users. These are rated are on a scale of 1–5 (Table 2).

Table 2 Dataset used

S. no.	Description	DislikeCount	LikeCount	Title	VideoId	Genre
1	When sibblingsjudy and	84	36,526	Alexa & Katie	AN-Wmg8ByVI	Comedy
2	A family wedding…	94	1356	Tidelands	vI03_g-hlGM	Crime
3	Never lose sight of	7300	156,986	Bird Box	o2AsIXSh2xo	Thriller
4	What happens when …	1609	93,367	The Princess Switch	sP_-iNVjiSE	Romcom

5 Results

Python language has been used for the implementation of the work. Figure 3 shows the percentage of positive reviews tweeted by Twitter users for each genre. According to the graph, the movies/shows of comedy genre received the highest percentage of positive reviews followed by western, thriller, action, documentary, horror and drama.

Figure 4 shows the percentage of negative reviews tweeted by Twitter users for each genre. According to the graph, the movies/shows of the horror genre received the highest percentage of negative reviews followed by thriller, drama, western, action, comedy and documentary. As per the experiment results, the comedy genre is seen to have the highest percentage of positive tweets (66.04%), then western (56.4%), thriller (48.3%), action (47.92%), documentary (40.15%), horror (37%) and drama (23%). The horror genre was observed to have the highest percentage of negative reviews (21.2%) followed by thriller (18.5%), drama (12%), western (10.4%), action (10.05%), comedy (8.63%) and documentary (7.57%).

Table 3 shows the accuracy, precision and recall corresponding to each of the classification algorithms applied to the dataset.

$$\text{Accuracy} = \frac{\text{tp} + \text{tn}}{\text{tp} + \text{fp} + \text{tn} + \text{fn}} \tag{5}$$

$$\text{Precision} = \frac{\text{tp}}{\text{tp} + \text{fp}} \tag{6}$$

$$\text{Recall} = \frac{\text{tp}}{\text{tp} + \text{fn}} \tag{7}$$

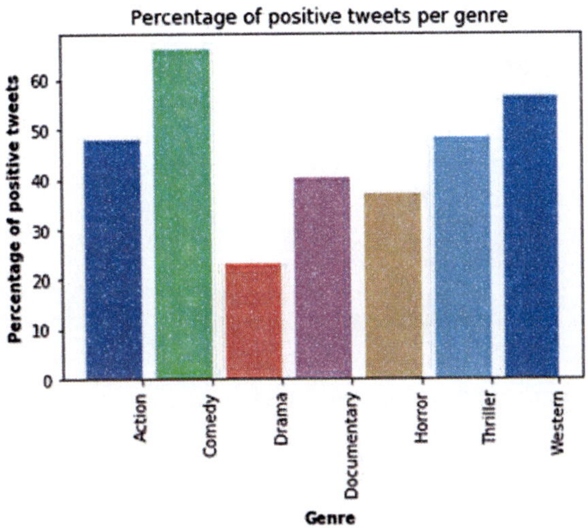

Fig. 3 Percentage of positive tweets per genre

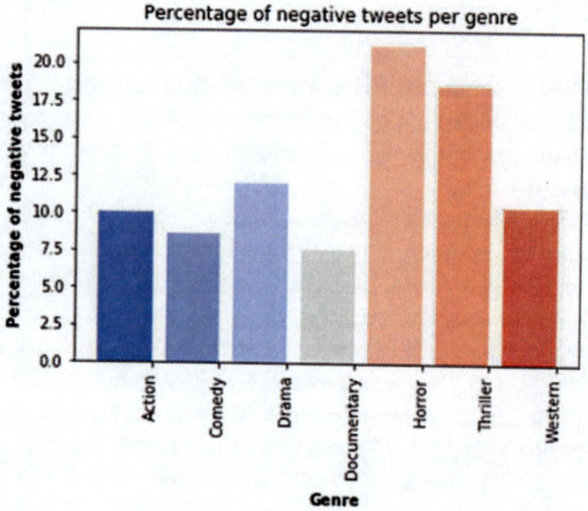

Fig. 4 Percentage of negative tweets per genre

Table 3 Percentage accuracy, precision and recall of the classification algorithms

	Algorithm	Accuracy (%)	Precision (%)	Recall (%)
1	One-versus-rest classifier + logistic regression	74	68.07	41.53
2	One-versus-rest classifier + SVC	83.67	51.16	3834
3	Binary relevance + Gaussian NB	85.33	94.9	79.72
4	Label powerset + logistic regression	80.83	77.14	40.8

where tp stands for true positive, tn is true negative, fn is false negative and fp is false positive. The terms positive and negative suggest the classifier's prediction and the terms true and false allude to whether that prediction belongs to external judgment or observation.

Figure 5 displays the accuracy for each algorithm applied. We can see that the highest accuracy is obtained by using binary relevance plus Gaussian NB algorithm followed by one-versus-rest + SVC, label powerset + logistic regression and One-versus-rest + logistic regression.

6 Conclusion

In this work, we have done classification of movies into various genres. Further, sentiment analysis of the tweets is done using TextBlob. The Movie dataset is used for the experimentation work. We have used different supervised learning algorithms on the dataset and got the best accuracy using binary relevance + Gaussian NB (85.33%). The comedy genre was observed to have the highest percentage of positive tweets, whereas the horror genre was observed to have the highest percentage of negative

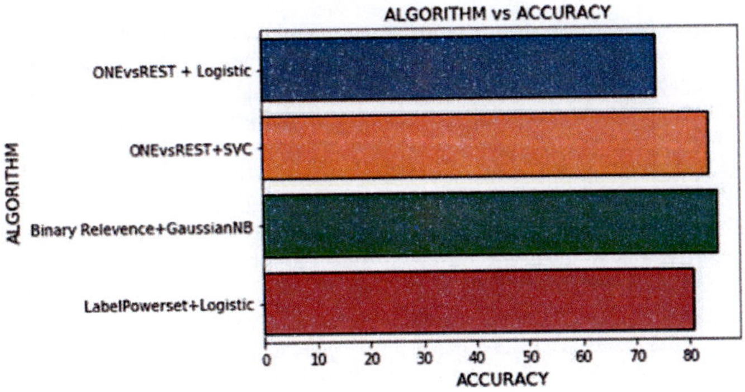

Fig. 5 Comparison of algorithms

reviews. The trends analysis shows that people like comedy shows/movies more than any other genre. It also shows that people are generally more critical of horror movies/tv shows.

References

1. Dangare, C. S., andApte S. S..: Improved study of heart disease prediction system using data mining classification techniques. Int. J. Comput. Appl. **47**(10), 44–48 (2012)
2. Xu, J.: An extended one-versus-rest support vector machine for multi-label classification. Neurocomputing **74**(17), 3114–3124 (2011)
3. Milgram, J., Cheriet, M., Sabourin, R.: "One against one" or "one against all": which one is better for handwriting recognition with SVMs? (2006)
4. Peng, C.Y.J., Lee, K.L., Ingersoll, G.M.: An introduction to logistic regression analysis and reporting. J. Educ. Res. **96**(1), 3–14 (2002)
5. Szymański, P., &Kajdanowicz, T.: Is a data-driven approach still better than random choice with Naive Bayes classifiers?, In: Asian Conference on Intelligent Information and Database Systems, pp. 792–801, Springer (2017)
6. Cheng, W., Hüllermeier, E.: Combining instance-based learning and logistic regression for multilabel classification. Mach. Learn. **76**, 211–225 (2009)
7. Piryani, R., Madhavi, D., Singh, V.K.: Analytical mapping of opinion mining and sentiment analysis research during 2000–2015. Inf. Process. Manage. **53**(1), 122–150 (2017)
8. Munjal, P., Narula, M., Kumar, S., Banati, H.: Twitter sentiments based suggestive framework to predict trends. J. Stat. Manag. Syst. **21**(4), 685–693 (2018)
9. Kadam, T., Saraf, G., Dewadkar, V., &Chate, P. J.: TV show popularity prediction using sentiment analysis in social network. Int. Res. J. Eng. Technol **4**(11) (2017)
10. Mhaigaswali A., Giri N.: Detailed descriptive and predictive analytics with twitter-based TV ratings (IJCAT), vol. 1, pp. 125–130 (2014)
11. Rahim, M. S., Chowdhury, A. E., Islam, M. A., Islam, M. R.: Mining trailers data from youtube for predicting gross income of movies. In 2017 IEEE Region 10 Humanitarian Technology Conference (R10-HTC) (pp. 551–554). IEEE (2017)
12. Zubiaga, A., Spina, D., Martínez, R., Fresno, V.: Real-time classification of twitter trends. J. Am. Soc. Inf. Sci. **66**(3), 462–473 (2015)
13. Satyavani, A. V., Raveena, M., Poojitha, B.: Analysis and prediction of television show popularity rating using incremental K-Means Algorithm IJMET, vol. 9, pp. 482–489 (2018)

14. Schmit W., Wubben S.: Predicting ratings for new movie releases from Twitter content. Workshop on Computational Approaches to Subjectivity, Sentiment and Social Media Analysis (WASSA 2015), pp. 122–126 (2015)
15. Wang, H., Zhang, H.: Movie genre preference prediction using machine learning for customer-based information. Int. J. Comput. Inform. Eng. **11**, 1329–1336 (2017)
16. Battu, V. et al.: Predicting theLsed on its Synopsis, 32nd Pacific Asia Conference on Language, Information and Computation Hong Kong, pp. 52–62 (2018)
17. Maloof, M. A. (Ed.): Machine learning and data mining for computer security: methods and applications. Springer Science & Business Media (2006)
18. Khan, R., Urolagin, S.: Airline sentiment visualization, consumer loyalty measurement and prediction using Twitter data. Int. J. Adv. Comput. Sci. Appl. **9**(6), 380–388 (2018)
19. Bhardwaj, P., Gautam, S., Pahwa, P.: A novel approach to analyze the sentiments of tweets related to TripAdvisor. J. Inf. Optim. Sci. **39**(2), 591–605 (2018)
20. Jindal, R., Taneja, S.: A lexical-semantics-based method for multi-label text categorization using word net. Int. J. Data Mining Model. Manage. **9**(4), 340–360. Publisher: Inderscience (2017)
21. Banik, R.: The Movies Dataset, (Version 7), [Metadata on over 45,000 movies. 26 million ratings fromver 270,000 users.]. Retrieved from https://www.kaggle.com/rounakbanik/the-movies-dataset/metadata [Last Accessed: 15 October 2019] (2017)

An Empirical Evaluation on Nomophobia: Mobile Phone Dependence Among Medical Students

Vijay Rana and Sunny Sharma

1 Introduction

A remarkable enhancement in the use of mobile phones among medical students has come forward through developing medical knowledge implementations, the improvement of mobile devices, and the constantly increasing specialty of medicine. Mobile phones are an appropriate way of communication that has provided enthusiastic and almost entirely accepted globally in equally developed and developing countries. India has the second highest mobile and Internet accessible stakeholder in the globe after China, with more than 87 connections per 100 peoples [1]. Mobile phones for medical students are enormously valuable for obtaining medical information, enhancing communications [2]. It can help to change the way of living and it is beginning to change the way of learning. Literature reviews [3–5] showed that previous work on influence of mobile applications in medical education mainly focused on effectiveness of mobile technology as an educational work and resource, infrastructure to maintain e-learning, advantages, issues, and proper utilization.

Unfortunately, mobile phone technology has some negative effects also that make several common health issues caused by utilizing them frequently and for a long period of time; such as headache followed by irritability from constant use of mobile applications, lack of concentration especially among the medical students because of the regular messages and phone calls that results in students' lack of focus would divert from finishing their tasks and will unenthusiastically affect their academic performance [6].

That leads the individuals addiction of devices, which is a key social issue caused by extreme use of mobile phones. It works on various forms such as mobile phone

V. Rana (✉)
GNA University, Phagwara, Punjab, India

S. Sharma
Kathua Campus, University of Jammu, Jammu & Kashmir, India

B. Agarwal et al. (eds.), *Proceedings of International Conference on Intelligent Cyber-Physical Systems*, Algorithms for Intelligent Systems,
https://doi.org/10.1007/978-981-16-7136-4_26

addiction and it has been diversely termed as mobile phone dependence, mobile phone problematic exercise, trouble in mobile phone use, mobile phone exploitation, and especially nomophobia. Nomophobia is a socio-psychological disease. The characteristics of nomophobia fluctuate and could comprise frequent utilization of mobile phone, nervousness, or edginess caused in the lack of access to mobile phone or regular inspection of the mobile display for notifications, slumbering with mobile device in bed, and predilection for mobile communication opposed to physical interaction.

The prevalence of mobile phone addiction among the medical students is found to be 76 (38%). The research group consisted of 66 (33%) postgraduate students and 134 (67%) undergraduates' students and average of the participants in this work 197 (98.5%) owned a smartphone and standard age of starting to utilize a mobile phone in this work group was 18 years. When summarized the whole period of time spent on utilizing mobile phone per day average 112 (56%) responded that they utilize mobile phone for duration of 01 to 3.5 h per day pursued by 46 (23%) who utilized it for 0.5–01 h, whereas 32 (16%) used it for 3–5 h and 10 (5%) used for more than 5 h in a day.

2 Review

Excessive use of digital devices can cause addiction, prompting researchers to coin the term "nomophobia" to describe and define smartphone addiction. This research [20] specifically identified smartphone dependency with its symptoms, as well as associated problems, and incorporates evidence from cognitive, behavioral, neurobiological, and anthropological fields to support its existence. Despite the fact that there is very little work on nomophobia and smartphone addiction, this paper argued that the excess use of smartphones and the possible behavioral addictions they pose should be recognized.

The majority of people devote a considerable time on their smartphones. Physical illness, unpleasant sentiments, pathological addiction, and melancholy, as well as fear anxiety, performance, and low accomplishment, are all linked to heavy use of smartphones. As a result, while dealing with the excessive and uncontrolled use of smartphones, security practices must be addressed. The authors in [21] established the prevalence of nomophobia among secondary and high school students aged 12–18, as well as the demographic and academic factors that influence these levels. The population of this study, which was designed using a relational model, consisted of 612 students from all levels of secondary school and high school. The study employed a personal information form as well as two different scales. Further, descriptive and linear multiple regression analysis was performed on the collected data. On the basis of the founded results, there is a strong link between smartphone addiction and nomophobia.

This research [22] assessed nomophobia at the institution on 104 students. The non-representative sample revealed that a tiny group (3%) of students had severe nomophobia, whereas nearly 40% were somewhat nomophobic. The rest individuals were classified as slightly nomophobic, with no students classified as not nomophobic. This poses a concern for any teacher-led opportunities to develop the use of smartphones, such as mobile learning. As a result, it is suggested that this condition be watched and that nomophobia be addressed in future digital literacy initiatives.

The main central aim of our work is to examine the pattern of mobile phone usage and frequency of addiction of mobile phone among medical students.

3 Methods

A descriptive cross-sectional study is conducted on 200 graduate and postgraduate students with both day scholars and hostelers, especially those students who are using mobile phones. These students are arbitrarily selected and included in the study after their consent. This study is conducted among the students of Medical College, SBBSU, Jalandhar, India from September 2019 to November 2019. Each student was required to complete a pre-tested self-administered survey that highlighted their psychographic and socio-demographic pattern of mobile phone usage. The socio-demographic feature contains attributes and variables such as sex, age, and residence and education level. The psychographic elements contain student's approach in the context of mobile phones addiction and some similar issues. The data is compiled and evaluated with SPSS version 25 and articulated as occurrence and percentages.

A questionnaire was proposed on the basis of specific ranges of mobile phone dependency among the learning subjects and executed on a set of students who were not a central component of the experimental study. The students who are addicted to phones are categorized as homophobes elements. All questionnaires contained queries to obtain the demographic and psychographic attributes of the participants. There were 08 obligatory queries to be responded to measure nomophobia (Table 1).

The Kaiser–Meyer–Olkin (KMO) test determines if the collected data is suitable for this study. The KMO test assesses the sampling appropriateness of each model variable as well as for the entire model. The statistic is a measurement of the amount of variation that is common across parameters. The lower the percentage, the better our data is suited for factor analysis.

KMO returns values ranging from 0 to 1. If the KMO value is less than 0.6, the sampling is insufficient.

$$\mathrm{KMO_j} = \frac{\sum_{i \neq j} {r^2}_{ij}}{\sum_{i \neq j} r_{ij}^2 + \sum_{i \neq j} a_{ij}^2} \tag{1}$$

Chi-square (x^2) test (given in Eq. 2) is used to discover the significance of proposed parameters on categorical scale among patterns of mobile utilization and selected socio-demographic and psychographic feature variables.

Table 1 Questionnaire sample

Sr. No	Questions	Responses	
		Number (N = 200)	Percentage (%)
1.1	Do you require using mobile for enhanced amounts of time in order to attain satisfaction/improvement?	157	78.5
2.2	Has mobile utilization has made you spend minimum time with friends or family?	61	30.5
3.3	Experienced stress due to network errors or weak networks?	71	35.5
4.4	Do you spontaneously reply to Calls, SMS, where it does not allow (classroom, meeting, and group participation)?	43	21.5
5.5	Do you impulsively answer calls, SMS, where it is treacherous to do so (during driving, crossing roads, and working with machines)?	43	21.5
6.6	Is the mobile phone effective for doing medical study?	103	51.5
7.7	Does utilizing mobile facilitate you to defeat the awful moods (feeling of inferiority, vulnerability, anxiety, depression, etc.)?	131	65.5
8.8	Do you believe that you are getting addicted to mobile use?	65	32.5

$$(X)^2 = \sum \frac{(O_i - E_i)^2}{E_i} \tag{2}$$

where X^2, O_i, and E_i are denoted as chi-squared, observed value, and expected value, respectively.

4 Results

The experimental outcomes contained 110 (55%) male and 90 (45%) female students; of these 94 (47%) were day scholars and 106 (53%) were residents of hostels. Majority of students were belonging to the age group 18–26 years. The experimental data contained 66 (33%) postgraduate students and 134 (67%) undergraduates.

Majority of the participants in this work 197 (98.5%) owned a smartphone presently and the standard age of beginning to utilize a mobile phone in this work group was 18 years. Table 1 highlighted the features of the sample data and experiments (Table 2).

When outlined the complete interval of time spent on utilizing the mobile phone per day, 112 (56%) participants responded that they utilize mobile phone for time of 01–3.5 h per day followed by 46 (23%) who utilized phone for 0.5–1 h, whereas

Table 2 Characteristics of the sample participants

Sr. No	Items		Number (N = 200)	Percentage (%)
1	Gender	Male	110	55
		Female	90	45
2	Parental working class	Single parent working	127	63.5
		Both parents working	73	36.5
3	Family category	Single parent	25	12.5
		Nuclear	135	67.5
		Joint	40	20
4	Mobile phone type	Smart phone	197	98.5
		Non-smart phone	3	1.5
5	Average time spent on mobile phone usage in per day (hours)		1–53.5 h a day	–
6	Phone call per day	1–6	121	60.5
		6–12	67	33.5
		>2	12	6

32 (16%) utilized phone for 3–5 h and 10 (5%) are the candidates whose utilization period exceeds 5 h per day.

The nomophobic scores (42 and above) were higher in participants who utilized mobile phones for more than 3 h evaluated to applicants using less than 3 h in a day. Majority 152 (76%) of the participants have spent money of Rs. 250 to 500 per month on mobile data and 38 (19%) applicants described they would update their mobile software promptly when a latest version comes and 62 (31%) applicants answered that they would update or change the mobile phone once in 2 years.

The regularity of checking phones for calls, SMS, email, and social media in an hour was evaluated, and 112 (56%) participated answered that they would verify the mobile 2–3 times in an hour. When verified about how long they receive to answer a phone call, 103 (51.5%) participated answered that they can answer subsequently 2–4 phone rings, while 53 (26.5%) answered they would promptly reply to the phone call. About 46 (23%) applicants felt that they drop their attentiveness and get anxiety when they do not have their mobile phone around or their mobile has run out of data or power. Approximately, 185 (92.5%) participants answered that they stay with their mobile phone in a regular manner when they leave to snooze, 171 (85.5%) applicants utilized mobile phone during study hours, and only 18 (9%) of the applicants utilized it when completely required, while 148 (74%) participated answered that they would rarely utilize the mobile phone while driving (Table 3).

As per the experimental results shown that 156 (77%) participants agreed that mobile phone is an essential device to assist students in academic work while 22 (11%) felt it is not essential (Table 4).

Approximately, 88 (44%) participants described that mobile phones have not had a severe major role in academic accomplishment while 112 (56%) did not agree with

Table 3 Reason for mobile phone usage

Sr. No.	Reason	Number (N = 200)
7	Call to family members	26 (13%)
8	Call to friends	16 (8%)
9	Using Internet for study purpose	10 (5%)
10	Social networking	112 (56%)
11	Texting or messaging	20 (10%)
12	Playing games	4 (2%)
13	Listening and watching video/music	9 (4.5%)
14	Taking photos (selfies)	5 (2.5)

Table 4 Health hazard of mobile phone among

Main symptoms	Frequency (%)
Accidents	3.84
Earache	7.4
Headache	33
Eyestrain	38
Lack of sleep	65

this statement. About 54 (27%) of participants had tried to minimize the use of their phones but were ineffective. While about 30 (15%) of participants agreed that life without a mobile phone is easier, 76% didn't have such a consideration.

5 Discussion

Mobile phone has become a necessary element of present human life and contains various functions that make them extremely significant to student's study. This work mainly focused on emerging positives as well as harmful psychological as well as physical effects of extreme utilization of phones on medical students and experimental results also showed that excessive utilization of mobile phones leading to progress of symptoms evocative of dependence syndrome. Maximum students were regularly using mobile phones for social media, taking pictures or videos, playing games, and watching/listening to video/music other than for calling and messaging.

Nomophobia has several unique features such as using mobile frequently and spending maximum time on it. It consists of several phobia elements: feeling worried and anxious at the consideration of losing a mobile phone or whenever the mobile cannot be used due to low balance, constantly taking a charger, network problem, or battery issue. It also makes it a habit to regularly check the phone screen to check whether calls and messages have come, or immense outflow from utilizing the mobile are also deemed as characteristics of mobile dependence and nomophobia. It is present pathologies that have been intuitive as an outcome of the regular use of mobile phones that convenient technologies have had on students and the dependence enabled between students, especially toward smartphones. This phobia appears itself and is increased by the loss of instant access to information, to the communications of parents or friends. All these impacts are working with the expansion of the students' daily life (physiological, physical, social interaction, among others). Although the current research is in an initial stage, the issue has not been studied with the impact of teaching process, so this work defines a pioneering result with teaching and students' side, the main vision being to evaluate the occurrence of nomophobia in future study and medical education, as well as to ensure the occurrence of rest time in the levels of nomophobia.

The experimental outcomes highlighted that the result of nomophobia issue among medical students disclosed that out of 200 participants only 76 (38%) were nomophobic. In [7], it was highlighted that 17.8% participants have some symptoms of nomophobia and as per the research report of [8] it has been described that 37.6% of respondent fear of being without mobile phones. Similarly, in a research on problematic mobile phone utilization by the medical student's research conducted in [9], problematic mobile phone utilization consisted to be low in the female student's in comparison to the male's students. Bianchi [10] described that there is no dissimilarity among the females and males with regard to the mobile phone addiction and experimental results highlighted that mobile phone utilization is common and also equally worked between male and female students in colleges.

However, during COVID-19 pandemic and lockdown restrictions on face-to-face clinical study and the problems faced by medical students in delivering patient care, alternative modern technologies like telemedicine and smartphones are playing a key role. This research work had few limitations, because of a single center study and can't be implemented at all centers but it described proper study investigation that will help in future experiments with multiple centers.

6 Conclusion

With the rapid growth of information technology, the digital world and anxiety are the sufferings of life and the new addition to the anxiety group is "nomophobia," the panic of being out of mobile phone contact. Nomophobia is a catchy abbreviation for mobile phone addiction and it is found to be associated with timely use on mobile, calls per day, and money expenditure on data recharge per month of possession of mobile phone. The experimental outcomes highlighted that mobile phone dependency among medical students is increasing rapidly and present work recommended making educational curriculum to teach the students to utilize mobile phones meaningfully. Emergence of mobile phone technologies has an equally positive and negative impact that enables us to execute various tasks quickly and efficiently. On the other side, the long-term usage leads to addictive behavior that is evolving as a public health issue in a huge section of medical students. As per the experimental results, we conclude that mobile phone usage is prevalent between medical students, so certain measures required to be taken care. Proper counseling sessions are required to be performed regarding this problem between medical students so that mutual advantage and disadvantage may be explained.

Acknowledgements We thank the individuals who volunteered to participate in the nomophobia survey. We believe that the findings will prompt them to consider how addicted they are to their phones and whether they are making the best use of them.

References

1. ICT Facts and Figures. [Internet]. Available from: https://data.worldbank.org/indicator/IT.CEL.SETS.P2?locations=IN&most_recent_year_desc=true. Last accessed on 2020 June 08
2. Subhash, T.S., Bapurao, T.S.: Perception of medical students for utility of mobile technology use in medical education. Int. J. Med. Public Health, 303–11 (2015)
3. Jafari, H., Aghaei, A., Kkhatony, A.: The relationship between addiction to mobile phone and sense of loneliness among students of medical sciences in Kermanshah. Iran, BMC Research Notes **2019**, 3–5 (2019)
4. Jamal, A., Sedie, R.: Khadijah Abdul Haleem, Najla Hafiz, Patterns of use of 'smart phones' among female medical students and self-reported effects. J. Taibah Univ. Med. Sci. **7**(1), 45–49 (2012)
5. Thapa, K., Lama, S., Pokharel, R., Sigdel, R.: Surya Prasad Rimal mobile phone dependence among undergraduate students of a medical college of Eastern Nepal: a descriptive cross-sectional study. J. Nepal Med. Assoc. **58**(224), 234–239 (2020)
6. Lin, Y.-H., Lin, Y.-C., Lee, Y.-H.: Time distortion associated with smartphone addiction: Identifying smartphone addiction via a mobile application (App). J. Psychiatr. Res. **14**, 1–7 (2015)
7. Jafari, H., Aghaei, A.: Alireza khatony, The relationship between addiction to mobile phone and sense of loneliness among students of medical sciences in Kermanshah, Iran, BMC Research. Notes **12**, 1–5 (2019)
8. Karki, S., Singh, J.P., Paudel, G., Khatiwada, S., Timilsina, S.: How addicted are newly admitted undergraduate medical students to smartphones? A cross-sectional study from Chitwan medical college. Nepal, BMC Psychiatry **20**, 1–7 (2020)
9. Sharma, N., Advani, U., Sharma, L., Jain, M., Sharma, K.: Anand Mohan Dixit. Pattern of mobile phone usage among medical students **5**(2), 118–123 (2019)
10. Bianchi, A., Phillips, J.G.: Psychological predictors of problem mobile phone use. J. Cyber Psycho Behavior **8**, 39–51 (2005)
11. D.R. K, Subhashini, V., Padmini, U.V., Moorthy, G.K., Thangappan, R.K.: Mobile phone dependence among medical students – a prospective study. Indian J. Basic Appl. Med. Res. **8**(1), 613–8 (2018)
12. Mei, S., Chai, J., Wang, S.-B., Ng, C., Ungvari, G., Xiang, Y.-T.: Mobile phone dependence, social support and impulsivity in Chinese University students. Int. J. Environ. Res. Public Health **15**(3), 504 (2018)
13. Jain, P., Gedam, S.R., Patil, P.S.: Study of smartphone addiction: prevalence, pattern of use, and personality dimensions among medical students from rural region of central India. Open J. Psychiatry Allied Sci. **10**(2), 132 (2019)
14. Farooqui, I.A., Pore, P., Gothankar, J.: Nomophobia: An emerging issue in medical institutions? J. Ment Health **27**, 438–441 (2018)

15. Sapacz, M., Rockman, G., Clark, J.: Are we addicted to our cell phones? Comput. Hum. Behav. **57**, 153–159 (2016)
16. Choudhury, S., Saha, I., Som, T.K., Ghose, G., Patra, M., Paul, B.: Mobile phone involvement and dependence among undergraduate medical students in a Medical College of West Bengal. India. J. Educ. Health Promot. **8**, 1 (2019)
17. Kuss, D.J., Kanjo, E., Crook-Rumsey, M., Kibowski, F., Wang, G.Y., Sumich, A.: Problematic mobile phone use and addiction across generations: the roles of psychopathological symptoms and smartphone use. J. Technol. Behav. Sci. **3**(3), 141–149 (2018)
18. Toda, M., Monden, K., Kubo, K., Morimoto, K.: Mobile phone dependence and health-related life-style of university students. Soc. Behav. Pers. **34**, 1277–1284 (2006)
19. Loredo, E.S.M.P., de Souza Matos, B.D., da Silva, E.O., Lucchetti, A.L.G., Lucchetti, G.: The use of smartphones in different phases of medical school and its relationship to internet addiction
20. Tran, D.: Classifying nomophobia as smart-phone addiction disorder. UC Merced Undergraduate Res. J. **9**(1) (2016)
21. Durak, H.Y.: Investigation of nomophobia and smartphone addiction predictors among adolescents in Turkey: demographic variables and academic performance. Soc. Sci. J. **56**(4), 492–517 (2019)
22. Davie, N., Hilber, T.: Nomophobia: is smartphone addiction a genuine risk for mobile learning? Int. Assoc. Dev. Inform. Soc. (2017)

The Facets of Machine Learning in Lane Change Prediction of Vehicular Traffic Flow

Shreya Upadhyaya and Deepti Mehrotra

1 Introduction

The complexity in the traffic flow is exponentially magnifying in terms of congestion, erratic flow patterns and advanced responsive behaviour which impinges upon lesser human interference in the driving decisions and need for improved prediction in the Intelligent Transportation Systems (ITSs).

The varied applications of machine learning cover significant aspects involving the recognition of driving style, categorization of risky driving behaviour and the prediction of Vehicular Traffic Flow (VTF) which have provided an increased accuracy, safety and timeliness in prediction that has caused the timely emergence of an Intelligent Transport System (ITS). The field of transport-based research is witnessing the evolution of novel, efficient algorithms and techniques with growing expertise and accuracy in traffic flow prediction along with several other aspects concerning safety on road and its correlation to driving style.

The extensive application of SVM (support vector model) in areas of text categorization, voice identification and pattern recognition has proved the effectiveness, suitability and efficacy of the method in traffic flow prediction, capability of predicting the change in lane and recognizing the erratic driving styles and driver action responses as well [1–3]. Further the lateral control concerning the vehicle fundamentally relies upon the mutually exclusive events of lane switching and lane keeping which are essential to characterize vehicular motion [4]. The switching of lane initiates a process of disturbance in the adjacent lanes with a corresponding increase in the vehicular speed and rests upon the inter-relation of decisions taken by the drivers in a definite hierarchical pattern [5] while lane keeping refers to the task of staying and continuing to drive within the current lane without intending to leave the lane during a particular period of time.

S. Upadhyaya (✉) · D. Mehrotra
Amity University Uttar Pradesh, Noida, India

B. Agarwal et al. (eds.), *Proceedings of International Conference on Intelligent Cyber-Physical Systems*, Algorithms for Intelligent Systems,
https://doi.org/10.1007/978-981-16-7136-4_27

The ability of predicting the change in lane is one of the dimensions of driving behaviour prediction which can be formulated as a regression or classification problem. The regression evaluates positional coordinates and speed values to map motion of vehicle while in classification the vehicle states are discretized to enable clear, faster discrimination between vehicle behaviour in real time. The paper solely focuses on the accurate and precise prediction of the lane-switching pattern of the vehicle in vehicular traffic flow. In this regard, the Support Vector Machine (SVM) shows improved and accurate results when trained on a dataset to classify lane prediction and lane changing behaviour [4].

The paper comprehensively proposes the working of various Machine Learning (ML) techniques based on the predictive—capacity of model to determine the change in lane with improved accuracy and precision. The methodological framework for achieving the objective has high modularity and comprises of six sections where Sect. 1 deals with introduction to the abilities of Intelligent Transport Systems (ITSs), Sect. 2 covers description of the flow of applied methodology and related research, Sect. 3 presents experimental setup, Sect. 4 highlights implementation of Machine Learning (ML) algorithms, Sect. 5 presents result and discussion and Sect. 6 presents conclusion.

2 Related Research and Methodology

Vehicle maneuvres describe a vehicle's movement on the road in terms of its position and speed (4). Lane maintaining and lane changing are mutually exclusive in terms of lateral vehicle control on highways (12). Depending on how the driver perceives conditions in the current and target lanes, as well as environmental factors that impact the choice to change lanes, lane changes can be optional or forced.

Besides this there are other machine learning techniques, with the exception of unsupervised machine learning techniques like as clustering, require labelled or partially labelled driving behaviour data. The method of driving style labelling for each driver in the sample is critical to the recognition model's reliability in the field of driving style recognition. Prior researches duly stressed on driver drowsiness or tired driving which was labelled in several studies using facial movement or driving time [6, 7]. To name drivers in each clustering group, unsupervised clustering approaches such as K-means [8] and fuzzy clustering [9] are utilized.

"Machine learning (ML) is an efficient and automatic method of prediction and learning [10], the algorithms are designed to learn from past datasets, we input a large amount of data, the Machine Learning (ML) model analyses that data, and we can make predictions about the future based on that train model [9, 11, 12]". For the lane change prediction, the framework applied on the trajectory data is shown in Fig. 1.

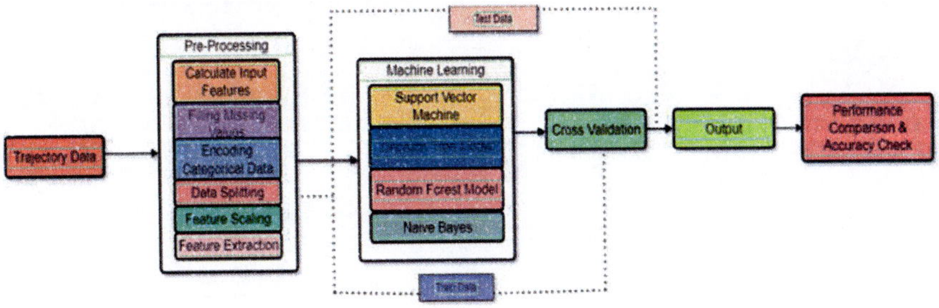

Fig. 1 Framework for lane prediction

2.1 *Overview*

The methodological approaches in main scope involve the application of four Machine Learning (ML)-based algorithms—Naïve Bayes classification, Support Vector Machine (SVM), decision tree-based classification and Random Forest (RF) classification that will predict the change in lane of the vehicles on the data experimentally obtained from the vehicles in the Peachstreet, Atlanta City (GA).

2.2 *Data Pre-processing*

The data were then pre-processed using "caret package" in R studio Integrated Development Environment (IDE). The features of the vehicular flow data were the training inputs of RF, SVM, Decision Tree (DT) and Naïve Bayes (NB) models. Next was the substitution of missing values and the lane id was encoded as a categorical factor for classification. The comparative significance of attributes is shown in Figs. 2 and 3. The localized x positional coordinate, space headway, time headway and vehicular velocity were found relatively more significant than other features in the trajectory dataset.

2.3 *Training the Model*

This section of the methodology holds drastic significance wherein each of the Random Forest (RF), Decision Tree (DT), SVM and Naïve Bayes model was trained. The distinguished models of each family were tested on training results. The process enables the model to cross validate the test set (D2) results with the training set (D1) results to make predictions of lane change with highest accuracy.

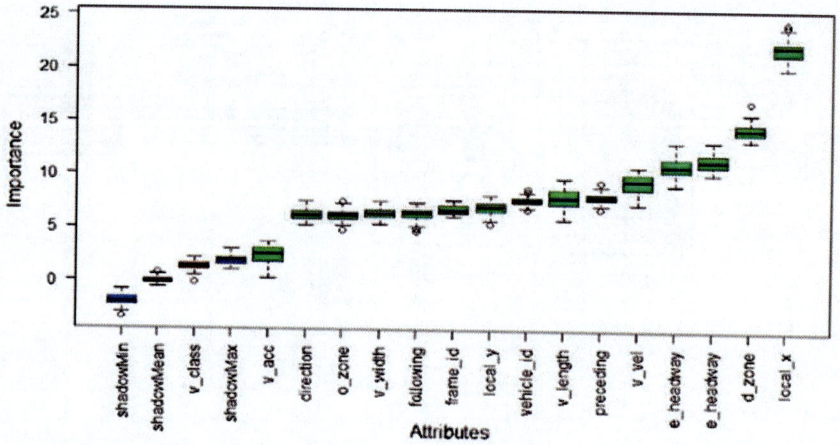

Fig. 2 Relative importance of the attributes in the traffic flow data

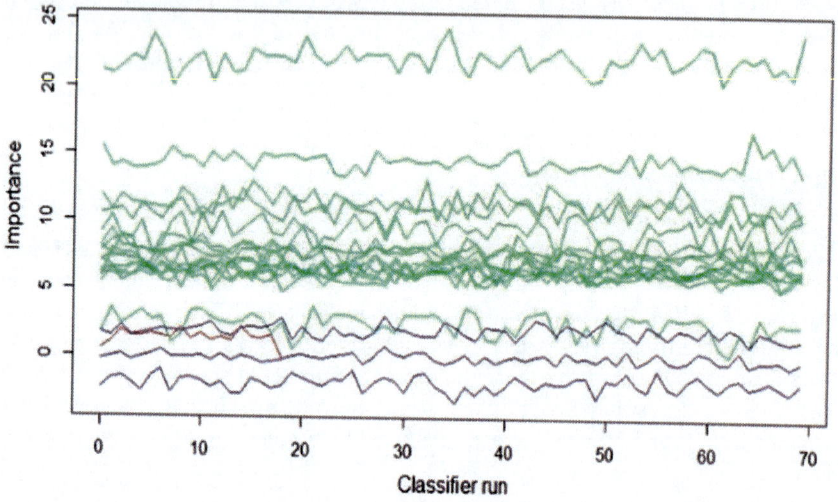

Fig. 3 The importance versus classifier run graph using Vehicular. Boruta package in R

3 Experimental Setup

3.1 Tool Description

R Studio is an Integrated Development Environment (IDE) used for the in-depth application and practical understanding of Machine Learning (ML) models to the vehicular flow data to predict the switching of lane by the vehicle.

3.2 Vehicular Traffic Flow Data

In 2005, the U.S. Federal Highway Administration (FHWA) had collected a high-fidelity vehicular flow dataset, by the aid of software called the "Next-Generation Simulation (NGSIM)". The transportation research has special focus in areas of traffic flow analysis and modelling, traffic-related estimation and prediction and working of Intelligent Transport Systems (ITSs) which makes this trajectory data widely useful even after years of its collection. The vehicle trajectory data provided the detailed lane positions and locations relative to other vehicles.

4 Implementation of the Machine Learning Algorithms

"Machine learning (ML) is an automatic learning method [11], and the algorithms are built for effective comprehension as machine learning (ML) models analyse data and train the model based on it, allowing us to generate predictions about future possibilities [12, 13]". For the lane change prediction, various techniques of Machine Learning (ML) effectively applied on the vehicular flow data are given below.

4.1 Naïve Bayes Classification (NB)

In the event of a big training dataset, this model is frequently used to make assumptions. This methodology is essentially based on the concept of conditional probability (P(A|B)), which uses the Bayesian approach to compute the probability [8]. It has the highest accuracy in estimating probabilities for noisy data, which is commonly used as an input [9]. But in this case, it underperformed w.r.t other models.

4.2 Decision Tree Classification (DT)

Decision tree (DT) [14] is a significant Classification and Regression (C&R) model. It is highly useful in cases of large, complicated data. The study data was split into two datasets, namely, training dataset (D1) and validation dataset (D2) such that training dataset (D1) is utilized to build the decision tree model and the dataset for validation is utilized to decide on the appropriateness of tree size to build the optimal final model. While building the model, the significant and core input variables are identified and categorized into bins based on status of the input variables. Decision tree (DT) method utilizes CART [15], C4.5 [6] and C5.0 [7]. The generated Decision Tree (DT) is shown in Fig. 4. The decisive factors found significant on algorithm's application are destination zone, time headway, vehicle velocity, destination zone, origin zone and space headway.

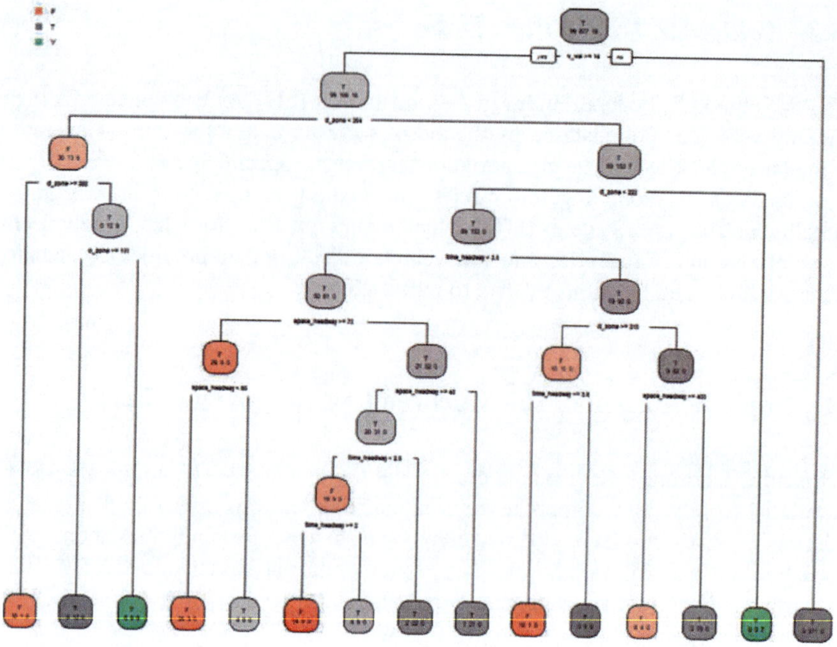

Fig. 4 The Decision Tree (DT) was generated using five major factors affecting the decision of lane change significantly, namely, *d_zone,space_headway,time_headway,v_velocity,o_zone* and the levels of classification are denoted by T, F, Y indicating the *lane_category*

4.3 Random Forest Classification (RF)

Based on supervised learning [10], the Random Forest algorithm [16] solves both Classification and Regression (C&R) issues. It is a fundamental part of Machine Learning (ML) for predicting new data, varied possible relations that can be possibly derived from previous dataset [17]. It is an ensemble technique that performs classification tasks by forming multiple decision trees where a bootstrap sample is randomly chosen and each node of RF- trees comprises d randomly selected features [18]. Figures 5 and 6 show the details obtained on implementation of Random Forest (RF) technique with two dimensions taken into account, namely, localized x and y positional coordinates of vehicle.

4.4 Support Vector Machine (SVM)

It is a supervision-based learning technique that can be used to solve both Classification and Regression (C&R) problems [17]. It is made up of both theoretical and numerical functions that are used to solve the regression problem. The approach

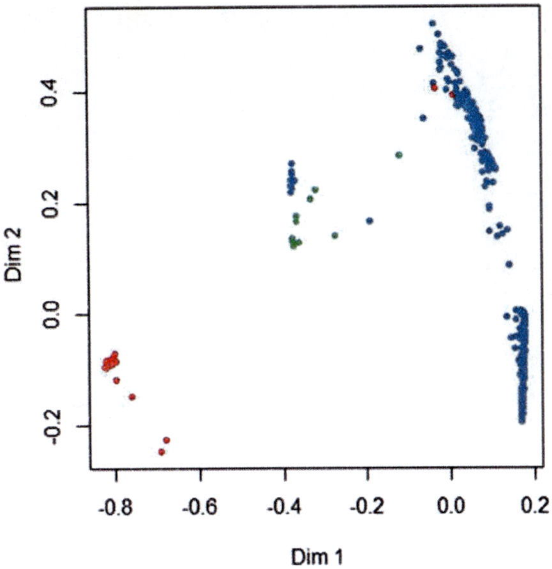

Fig. 5 The plot shows the proximity matrix for the Random Forest (RF) model with two dimensions as *Dim1 (local_x), Dim2 (local_y)*

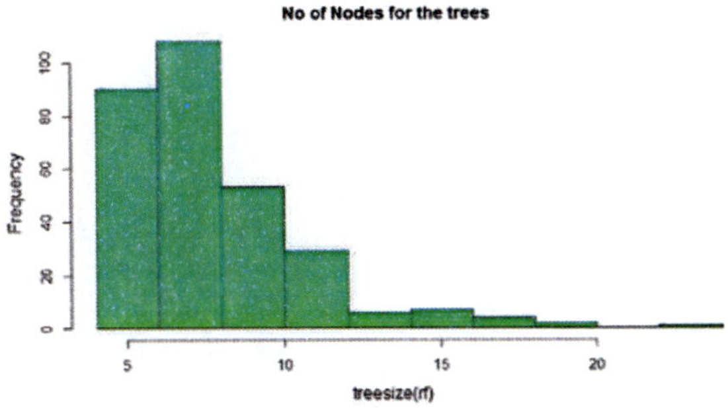

Fig. 6 The graph shows *treesize* versus *frequency* for the Random Forest (RF) model. This denotes the number of nodes for the trees

offers a high prediction accuracy when dealing with massive datasets [19]. It's a powerful method that uses 3D and 2D modelling [20, 21]. The SVM involves the use of the Maximal Margin Classifier (MMC) and epsilon (ε) which denotes the loss sensitive function determining the number of support vectors, the cost parameter (C) which controls the tolerance of deviations larger than the real values and gamma parameter (γ) that determines the radius of influence (RoI) of support vectors. Figures 7 and 8 are obtained using linear SVM model while Figs. 9 and 10 show plots of kernel SVM which outperformed the linear SVM results.

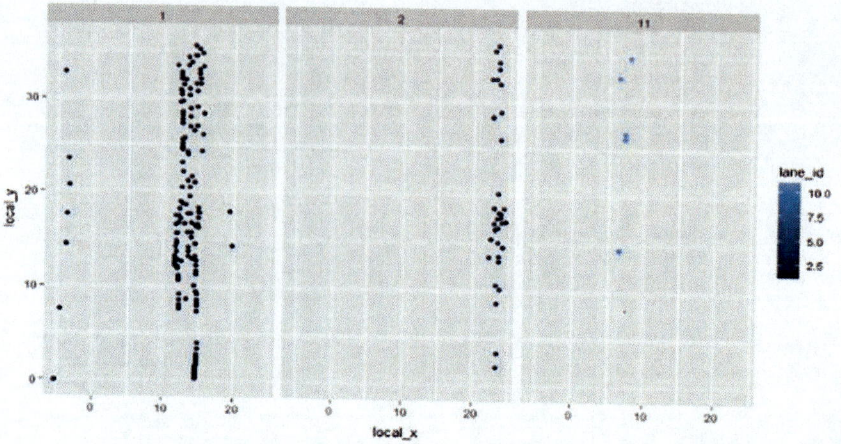

Fig. 7 The plot denotes the position of the vehicles on the basis of the lane (1,2,11) with *local_x (local x-coordinate), local_y (local y coordinate)* in case of linear SVM

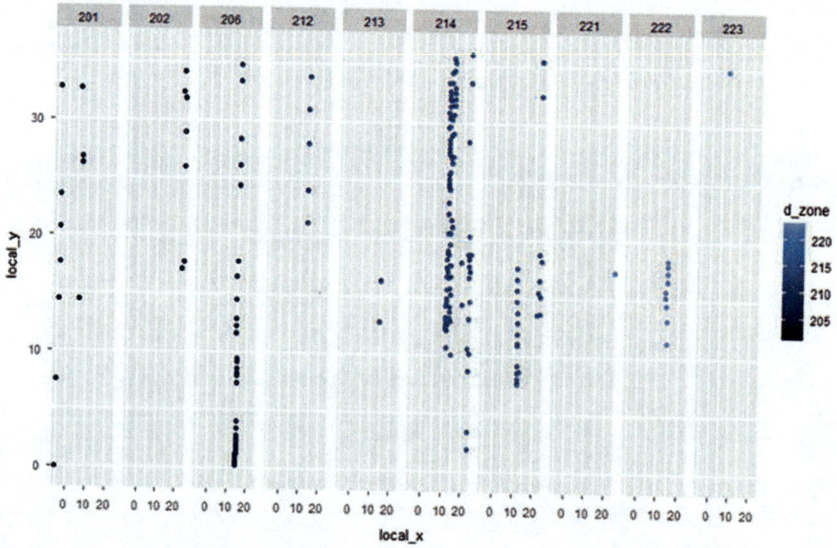

Fig. 8 The plot denotes the position of the vehicles on the basis of *d-zone (destination zone)* using linear SVM

5 Results and Discussion

In priorly conducted studies, manoeuver classification, recognition of swift and rapid driving styles have been the areas of chief focus. Contrastively, our work

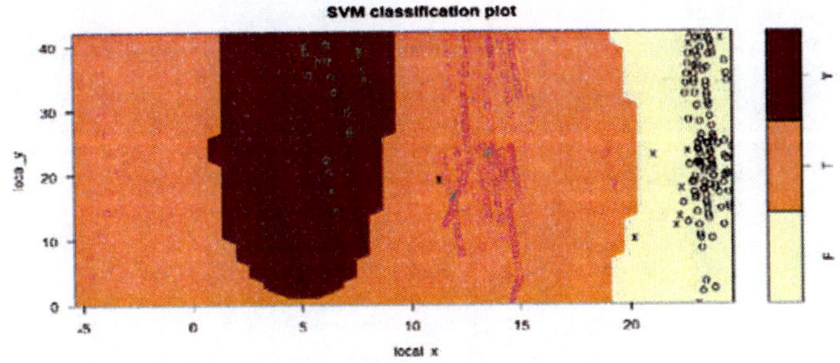

Fig. 9 The SVM classification plot obtained using kernel SVM where kernel is *Radial basis function*. The *lane id* is used as the factor for classification. With *local_x, local_y (local x- and y-coordinates)*

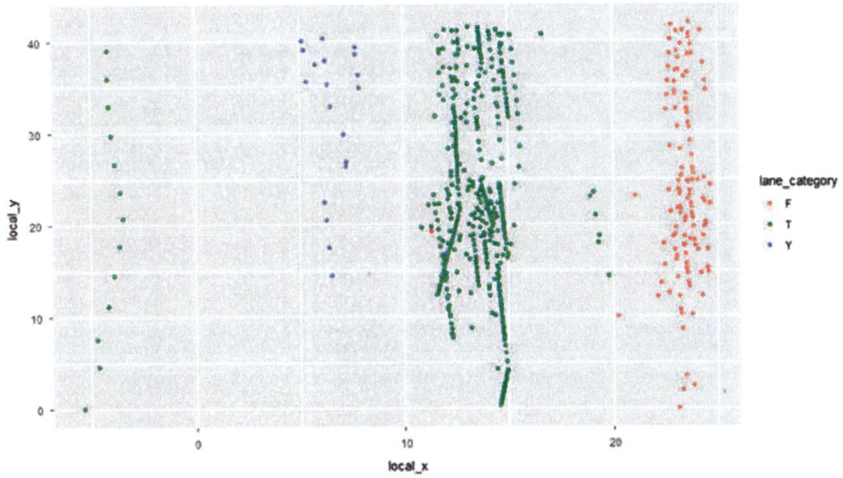

Fig. 10 The plot denotes the position of the vehicles on the basis of lane (1,2,11) using *lane id* as the factor for classification in case of *Radial basis function* kernel having *local_x, local_y (local x- and y-coordinates)*

involves the kernel SVM with a radial basis function that has not been put to application for detecting lane change. Furthermore, it also shows accuracy-based effective comparison among salient machine learning models highlighting our study (Fig. 11).

5.1 Evaluation

Table 1 illustrates the impact of variations in cost (C) parameter on the percentage of accuracy and Kappa values obtained on the application of Linear Support Vector

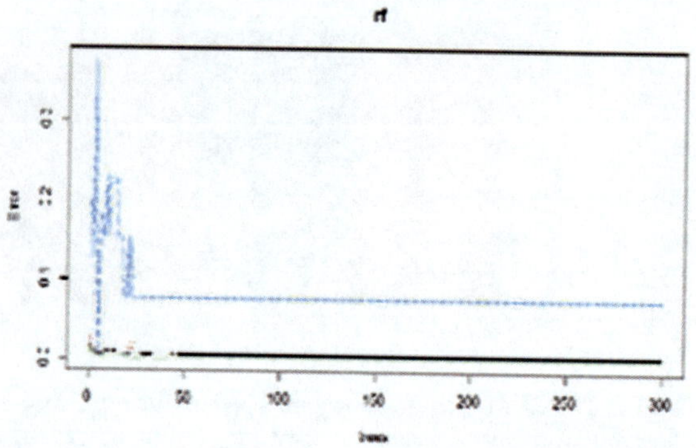

Fig. 11 Error versus trees curve for Random Forest (RF) model

Table 1 Table for cost, accuracy and Kappa value of linear SVM

C	Accuracy	Kappa
0.01	0.9747	0.8918
0.05	0.9857	0.9403
0.10	0.9906	0.9615
0.25	0.9928	0.9706
0.50	0.9924	0.9685
0.75	0.9916	0.9658
1.00	0.9914	0.9642
1.25	0.9919	0.9662
1.50	0.9914	0.9640
1.75	0.9916	0.9646
2.00	0.9914	0.9639
5.00	0.9909	0.9909

Machine (SVM) technique. The accuracy was selected as the parameter to obtain the optimal model having the largest value. The Kappa parameter determines the accuracy in classification and the magnitude of Extent of Reliability (EoR) preserved in varying scenarios to evaluate whether used data in study is concrete representation of measure attribute or not.

The final value utilized for the model was C = 0.25 as it obtains value which is the largest in terms of accuracy and one of the finest Kappa measured (Tables2 and 3).

In Table 4, the accuracy (%) values obtained for the applied Machine Learning (ML) models are illustrated. The segment of test data was presented to varied, priorly trained Machine Learning (ML) models mentioned in the methodological framework and the accuracies were obtained to analyse performance for comparative, effective evaluation. We observed that the kernel SVM model outperforms other techniques applied on the dataset.

Table 2 Parameters of the Random Forest (RF) classification

Parameters	Lane class		
	Class T	Class F	Class Y
Sensitivity	1.0000	1.0000	1.0000
Specificity	0.9963	1.0000	1.0000
Positive predicted value	0.9629	1.0000	1.0000
Negative predicted value	1.0000	0.9714	1.0000
Prevalence	0.08814	0.8847	0.02712
Detection rate	0.08814	0.8814	0.02712
Detection prevalence	0.09153	0.8814	0.02712
Balanced accuracy	0.99814	0.9981	1.0000

Table 3 Parameters of the Naïve Bayes (NB) classification

Parameters	Values
Sensitivity	1.0000
Specificity	0.4475
Positive predicted value	0.2406
Negative predicted Vvlue	1.0000
Prevalence	0.1490
Detection rate	0.1490
Detection prevalence	0.6192
Balanced accuracy	0.7237

Table 4 Accuracy-based comparison for the Machine Learning (ML) models

Machine learning (ML) algorithm	Accuracy (%)
Decision tree	92.8
Random Forest	99.66
Linear SVM	99.28
Kernel SVM	99.7
Naive Bayes	52.98

The linear SVM on application upon the vehicular flow dataset gave high accuracy as shown in Fig. 12. While for the Random Forest (RF) model, the error versus trees curve in Fig. 11 indicates the misclassification error which was minimized by training the model on the dataset.

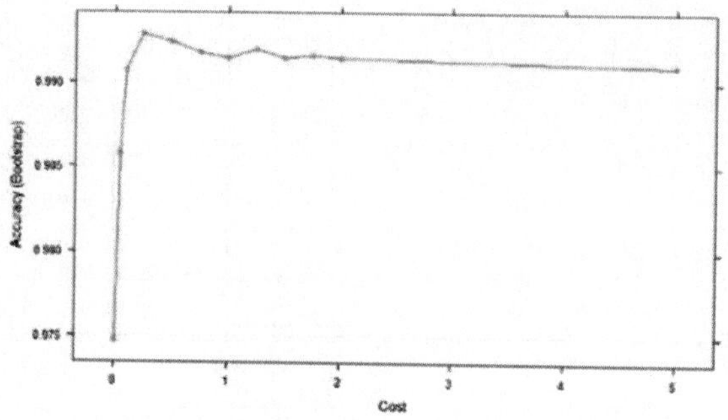

Fig. 12 Accuracy versus cost curve of linear SVM

6 Conclusion

We conclude through this insightful study that among the Machine Learning (ML) models executed on the trajectory data, kernel SVM outperforms the other distinct Machine Learning (ML) models. The parameters extracted and chosen for the evaluation were, namely, destination zone, lane id and lane category. Furthermore, the accuracy of the kernel SVM and Random Forest (RF) models was shown to be higher than the accuracy of the other models evaluated in the investigation. The study results can be used as a baseline for additional Machine Learning (ML) models that are applied to the ad hoc data set. In the future, we will be able to establish surrogates for forecasting unpredictable driving styles in order to assure road safety.

References

1. Bengler, K., Dietmayer, K., Farber, B., Maurer, M., Stiller, C., Winner, H.: Three decades of driver assistance systems: review and future perspectives. IEEE Intell. Transp. Syst. Mag. **6**(4), 6–22 (2014)
2. Zheng, Y., Li, S.E., Wang, J., Cao, D., Li, K.: Stability and scalability of homogeneous vehicular platoon: study on the influence of information flow topologies. IEEE Trans. Intell. Transp. Syst. **17**(1), 14–26 (2015)
3. Park, K., Han, S. H., Kwahk, J.: Toward trustworthy and comfortable lane keeping assistance system: an empirical study of the level of haptic authority. Int. J. Hum. Comput. Interact. 1–17 (2021)
4. Makaba, T., Doorsamy, W., Paul, B.S.: Exploratory framework for analysing road traffic accident data with validation on Gauteng province data. Cogent Eng. **7**(1), 1834659 (2020)
5. de Zepeda, M. V. N., Meng, F., Su, J., Zeng, X. J., Wang, Q.: Dynamic clustering analysis for driving styles identification. Eng. Appl. Artif. Intell. **97**, 104096 (2021)
6. Budiman, E., Kridalaksana, A. H., Wati, M.: Performance of decision tree C4. 5 algorithm in student academic evaluation. International Conference on Computational Science and Technology (pp. 380–389). Springer, Singapore (2017)
7. Pandya, R., Pandya, J.: C5. 0 algorithm to improved decision tree with feature selection and reduced error pruning. Int. J. Comput. Appl. **117**(16), 18–21 (2015)

8. Ibrahim, A. A., Hashad, A. I., Shawky, N. E. M.: A comparison of open source data mining tools for breast cancer classification. In Handbook of Research on Machine Learning Innovations and Trends (pp. 636–651). IGI Global (2017)
9. Mahmoodzadeh, A., Mohammadi, M., Ali, H. F. H., Abdulhamid, S. N., Ibrahim, H. H., & Noori, K. M. G. (2021). Dynamic prediction models of rock quality designation in tunneling projects. *Transportation Geotechnics*, *27*, 100497.
10. Tuggener, L., Amirian, M., Rombach, K., Lörwald, S., Varlet, A., Westermann, C., Stadelmann, T.: Automated machine learning in practice: state of the art and recent results. In 2019 6th Swiss Conference on Data Science (SDS) (pp. 31–36). IEEE (2019)
11. Dey, A.: Machine learning algorithms: a review. Int. J. Comput. Sci. Inform. Technol. **7**(3), 1174–1179 (2016)
12. Xue, Q., Wang, K., Lu, J. J., Liu, Y.: Rapid driving style recognition in car-following using machine learning and vehicle trajectory data. J. Adv. Transp. (2019)
13. Mahajan, V., Katrakazas, C., Antoniou, C.: Prediction of lane-changing maneuvers with automatic labeling and deep learning. Transp. Res. Rec. **2674**(7), 336–347 (2020)
14. Sharma, H., Kumar, S.: A survey on decision tree algorithms of classification in data mining. Int. J. Sci. Res. (IJSR) **5**(4), 2094–2097 (2016)
15. Mahmood, A. M., Imran, M., Satuluri, N., Kuppa, M. R., Rajesh, V.: An improved CART decision tree for datasets with irrelevant feature. In: International Conference on Swarm, Evolutionary, and Memetic Computing (pp. 539–549). Springer, Berlin, Heidelberg (2011)
16. Sipper, M., Moore, J.H.: Conservation machine learning: a case study of random forests. Sci. Rep. **11**(1), 1–6 (2021)
17. Tran, H.: A survey of machine learning and data mining techniques used in multimedia system. 113, 13–21 (2019)
18. Yu, B., Wang, H., Shan, W., Yao, B.: Prediction of bus travel time using random forests based on near neighbors. Comput. Aided Civil Infrastruct. Eng. **33**(4), 333–350 (2018)
19. Kumar, T.S.: Data Mining Based Marketing Decision Support System Using Hybrid Machine Learning Algorithm. J. Artif. Intell. **2**(03), 185–193 (2020)
20. Yao, J., Xia, H., Zhang, N., Yu, B.: Prediction on building vibration induced by moving train based on support vector machine and wavelet analysis. J. Mech. Sci. Technol. **28**(6), 2065–2074 (2014)
21. Abou Elassad, Z. E., Mousannif, H., Al Moatassime, H., Karkouch, A.: The application of machine learning techniques for driving behavior analysis: A conceptual framework and a systematic literature review. Eng. Appl. Artif. Intell. **87**, 103312 (2020)

A Review on Face Recognition Methods Using Infrared Images

Mohit Pandey and Abhishek Gupta

1 Introduction

The working of visible imaging-based face recognition methods highly depends on illumination conditions. The change in illumination intensity of the light source and change in position of the camera degrade the performance significantly. The visible light face recognition systems may not work efficiently in outdoor settings. In outdoor settings, it is difficult to control the intensity of light. The visible light face recognition systems do not work in dark environments and night vision applications. To overcome above-stated limitations, infrared imaging-based recognition systems were found in the literature. Infrared imaging-based recognition systems do not depend on the illumination conditions unlike visible light systems. For the night vision application, passive thermal sensors are proposed recently. These sensors take the radiation having 3–14 μm of wavelengths emitted by objects. In critical security applications, the use of infrared imaging for face recognition has been expanding. Various setups have been proposed for face recognition using infrared imaging, including some methods that are used in the visible light-based face recognition system. These methods work using local features and take out typical local structural information. Some methods are fusion based, in which visible and thermal images or features are fused to get the final output. In cross modality-based methods, distinctive features of visible and thermal images are matched. In deep learning-based techniques, observe mapping between thermal images and corresponding visible images of features.

M. Pandey (✉) · A. Gupta
School of Computer Science and Engineering, Shri Mata Vaishno Devi University, Jammu and Kashmir 182320, India
e-mail: 19DCS005@smvdu.ac.in

A. Gupta
e-mail: abhishek.gupta@smvdu.ac.in

B. Agarwal et al. (eds.), *Proceedings of International Conference on Intelligent Cyber-Physical Systems*, Algorithms for Intelligent Systems,
https://doi.org/10.1007/978-981-16-7136-4_28

The infrared (IR) portion of the electromagnetic spectrum is further split into four bandwidths named as LWIR (Long-Wave-Infrared), MWIR (Medium-Wave-Infrared), SWIR (Short-Wave-Infrared), and NIR (Near-Infrared). The Long-wave-IR is also known as Thermal-IR, and it has drawn the most notice because of its strength as compared to others. Thermal sensors are based on the heat amount produced from the object; unlike visible light sensors, they don't work on reflected light rays. Thermal sensors can work under different light conditions even in dark environments. The heat produced from the object has less dominance to dispersion and adsorption by smoke and dust. Thermal sensors also disclose human face anatomical information that makes them capable to detect disguised faces [1]. The objective of this study is to convey novel research in human face recognition using the thermal imaging field.

The rest of this paper is organized as follows. In Sect. 2, the literature review of work related to infrared-based face recognition systems is presented. In Sect. 3, the challenges faced by researchers in the thermal face recognition system are discussed. In Sect. 4, the two datasets used by various researchers in their work are presented. The conclusion of this study is given in Sect. 5.

2 Literature Review

The system that has face recognition/identification aims to find out and learn the unique features of the face. Along with the learning unique features, it is also crucial to maximize the similarity between different images of the same person. Different images of the same person mean images taken in different conditions like the distance between sensor and face, lighting conditions, mood, pose, etc.

Various researchers proposed well-performing human face detection and recognition methods based on visible imaging as shown in Table 1. However, the various poses of the face and illumination conditions are limitations.

Local as well as global features of human face images are required in visible light-based human face identification system. The same features are extracted under illumination controlled environment. Some methods were focused on the pose of a person in which images or features are transformed into a subspace where the intra-class lection is minimized and inter-class lection is maximized for finding better taxonomy of the human facial images.

Recently, researchers have been showing their interest in deep learning for face recognition, mainly in the convolutional filters. Many neural networks have been proposed to overcome the limitation of face recognition systems like the pose of a person, low resolution, distance between face and sensor, variation in lighting conditions, etc. The major challenge in neural network methods is acquiring the huge dataset required for training purposes and due to the large dataset, computational cost also increases.

Visible light face recognition methods are performing well but they can fail in dark environments or even in improper light conditions. Whereas infrared sensors work

Table 1 Different techniques used by researchers for face recognition systems in thermal images

References	Techniques	Results
Socolinsky and Socolinsky [2]	Local features: PCA, ICA, LDA	Worst-to-mean performance for LWIR attained 0.936
Mendez et al. [3]	Local features: PCA, ICA, LDA, LFA, LBP	They achieved identification rate average of 97.3%
Ahonenet et al. [4]	Local features: LBP, LDA	They achieved 0.81 mean recognition rates of the LBP
Bebis et al. [5]	Fusion, genetic algorithm	Facial Expression and with eyeglass mean accuracy on fused wavelet domain is 91% and 93%, respectively
Singh et al. [6]	Fusion using 2-granular SVM	94.8% (LBP, Notre Dame dataset) and 95.85% (2D log polar Gabor, Notre Dame dataset)
Wu et al. [7]	CNN	Recognition rates in different experimental setups, 98% in head rotations, 99.4% in variant expressions, and 100% in different illumination conditions
Sarfraz et al. [8]	CNN	Achieved rank-1 with 83.73% identification accuracy

well even in a completely dark environment, and they do not require any external light source.

2.1 *Infrared Face Recognition Techniques*

For human identification in security systems, the biometric feature of a human face can be used. For face recognition, firstly input face image through a camera and then try to accurately match with face image already stored in the database of the same person. In this method, major challenges are illumination conditions at the time of capturing images for input purposes, finding an accurate match of the same person with the large database, and dealing with the disguised faces. Unlike the visible light face recognition system, infrared imaging techniques use facial thermograms.

The infrared sensors computing heat radiations from the objects and capturing thermal images are independent of the effect of illumination conditions. Analogous to visible light human face recognition system, thermal imaging-based system also take unique features of the image and learns them for recognition as shown in Fig. 1 [9]. There are certain methods that infuse thermal images as well as visible images before the actual feature extraction as shown in Fig. 2 [9].

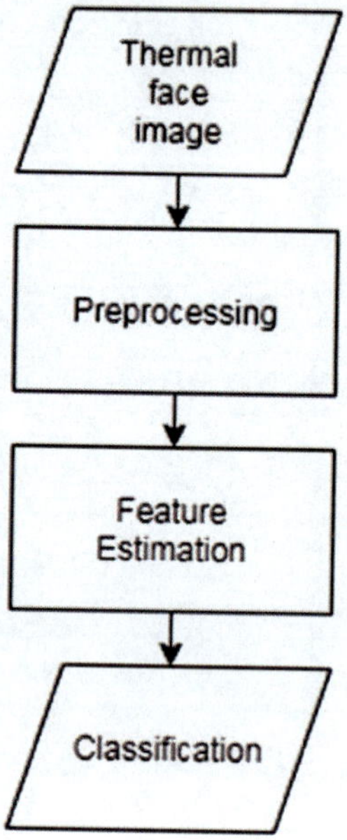

Fig. 1 One infrared face recognition

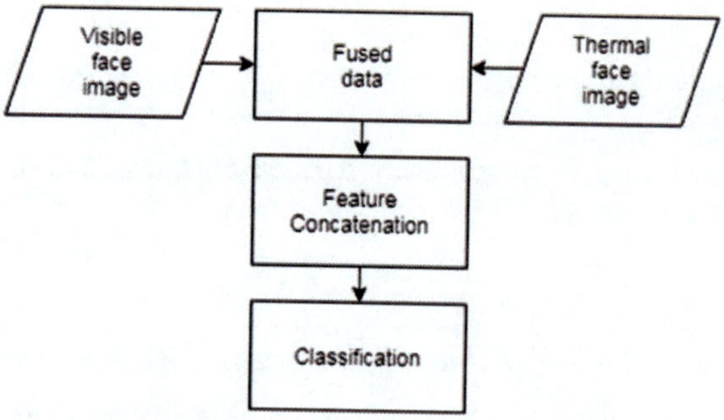

Fig. 2 Cross modality in the face recognition system

This fused image has the best features of both modalities, and for testing, it can use as per requirements. For a 24 × 7 surveillance system, cross modality-based methods are most suitable. For improving efficiency, researchers also concatenated features of visible and thermal images as shown in Fig. 3 [9].

Figure 4 [9] shows another infrared face recognition system, in which features of both modalities images are projected into a common subspace using Canonical Correlation Analysis (CCA) and like algorithms.

Some researchers also try for extracting common features using a deep neural network, in which the network learns a mapping between visible light image and thermal image features as shown in Fig. 5 [9]. After network training and testing with mapping of the thermal image, the network is able to map it to corresponding visible

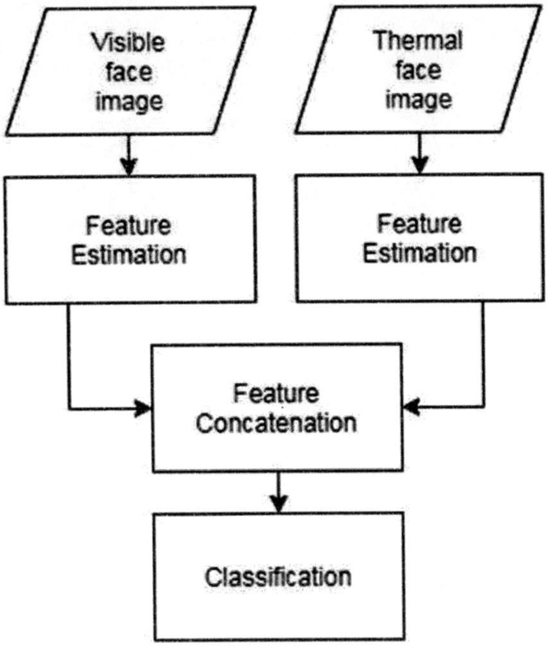

Fig. 3 Heterogeneous face recognition using concatenated features

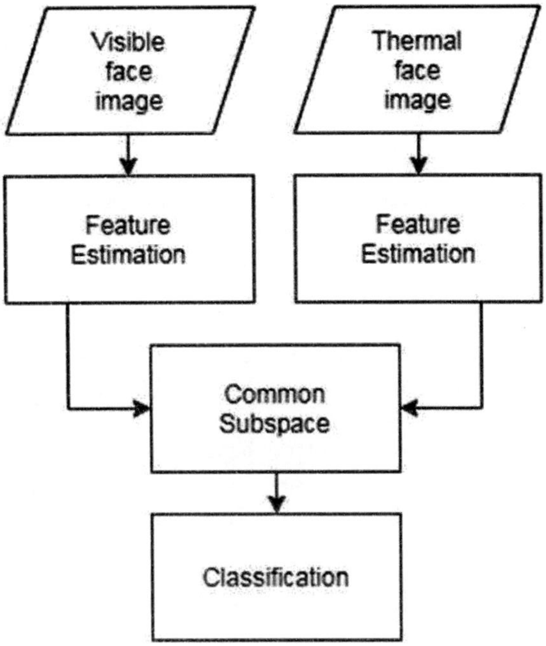

Fig. 4 Heterogeneous recognition using common subspace for learning

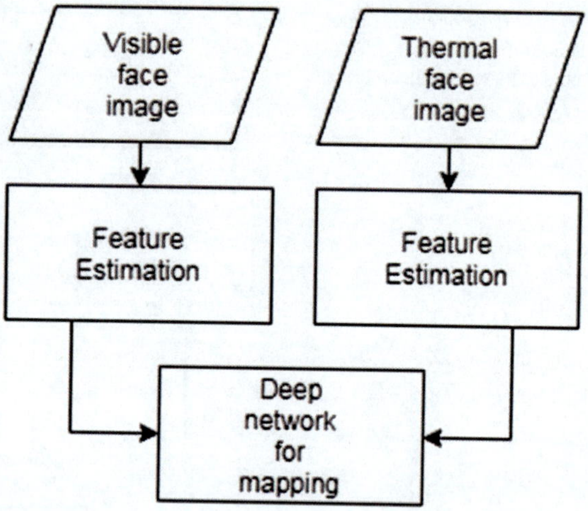

Fig. 5 Features mapping using DNN

mode images. Nowadays, deep learning can be used for many different purposes in face recognition systems.

2.2 *Preprocessing in Infrared Face Recognition Techniques*

For pull-out quality features from facial images, pre-processing is an important step. The face part of the image is cropped and normalized after face detection in the whole image and removed noise by using a low pass filter. Low pass filter is also helpful for removing illumination variations in the images. To remove illumination variations, the Difference of Gaussian (DOG) filter is frequently used for visible face recognition [10]. After applying convolution on the original image using a two-dimensional DOG filter, a DOG filtered image is constructed. DOG filter is used to reduce illumination variations in the visible facial imagery, and in the thermal images, it is used for reducing local variations due to the temperature distribution of the face. By reducing local variations in the thermal image using DOG filter, make image closer to visible light image and lift edge information. For balancing positive and negative components for face recognition, the σ of the Gaussian filter is opt precisely. The DOG filter is beneficial for the methods using local features, as the DOG filter removes noise and enhances edges. DOG filter can be used in both visible and thermal images by correctly tuning σ values.

2.3 Local Features in Infrared Face Recognition Techniques

In beginning, researchers consider visible light face recognition methods for developing thermal face recognition system. Diego A. Socolinsky et al. found visible and IR imaging for human face recognition algorithms based on the appearance of features [2]. They considered many analysis tools for measuring performance like PCA (principle component analysis), ICA (Independent component analysis), and LDA (linear discriminant analysis).

Mendez et al. achieved better performance in infrared face recognition than visible light imaging [3]. The authors use one hundred dimensions for PCA, ICA, and LFA (Local feature analysis), and the LDA for different algorithms; they discovered recognition performance was worst when illumination and facial expression of training and testing sets are not the same.

Ahonenet al. introduced LBP (local binary pattern)-based face recognition system [4], and afterward, various extensions of the original operator have been proposed [11]. LBP is an efficient texture operator, producing features with high recognition ability [12]. LBP is resistant to lighting effects because it are invariant to gray level transformations. In uniform facial regions, investigation is highly recommended for improving features robustness.

A dataset of LWIR face images contains 40 frames of 91 persons, with different settings, and this includes front pose, speaking action face, with or without glass, and various illumination conditions [13]. Xiaoyang Tan et al. have conducted a study using the above-stated dataset with focusing on images with and without glasses. The researchers perform the comparison between LBA and LDA and they found LBP and LDA are performing similarly by taking the average of both in the case of images without eyeglass, whereas in the case of images with glasses, performance degrades significantly. In the computed feature space for computing similarity, several tests are performed including Chi square test, log-likelihood statistic, and histogram intersection in face recognition with nearest neighbor classifier [10].

Several researchers [14–16] have done comparison-based studies for comparing many thermal face recognition techniques with UCH thermal face and Equinox datasets. They found Weber Local Descriptor (WLD) as the best performing method over Equinox including the case of images having glasses and simple images. They also found appearance-based techniques are not performing well especially with disguised faces or faces with different expressions. On UCH thermal face dataset, Speeded up Robust Features (SURF) and Scale-invariant Feature Transform (SIFT) show better results with rotations and facial expressions.

Recently, researchers show their interest in wide baseline matching approaches and achieved significant improvement. In these methods, local interest points take it out separately from the test and corresponding image and designate by invariant descriptors, and then recursively descriptors are matched to get the original transformation of two images. D. Lowe shows SIFT descriptors are a better method for object recognition systems including a probabilistic hypothesis rejection approach with real-time operating, recognition capabilities [17]. Additionally, SIFT features can also be used for registering thermal and visible light images [18].

2.4 Fusion Techniques

Image fusion is possible in many ways like at image level, feature level, match score level, and decision level in both thermal and visible light images. Concatenation of feature vectors of fused images (visible and thermal) is the simplest way of fusion along with Eigenspace fusion of images also proposed. Fusion is also done by transformation into wavelet domain [19] before training on 2 V-GSVM whereas Singh et al. [6, 20] fused images at image level and domain level.

Bebis et al. [5] studied the effects of the facial clog by eyeglasses on thermal face recognition and they found in their experiment recognition efficiency let down when images of human faces having spectacles are available in the gallery image but not in the screening image and conversely in thermal imaging. To solve this hard problem, they take advantage of the fusion technique, by fusing the thermal image with the visible light image. In multi-resolution level fusion, features with different spatial extend to be fused at the resolution at which they are most salient. Compounded multi-scale representation is created by using some specific fusion rules from multi-scale transformation of thermal and visible light images [21]. The final fused image is acquired by performing an inverse multi-scale transform.

2.5 Deep Learning in Face Recognition Techniques

Recently, researchers' interest in deep learning increases and applied in various fields like computer vision, artificial intelligence, and pattern recognition. Many works show its benefits by using it in various fields. Deep learning exhibits a very strong learning ability. The main advantage in using deep learning in face recognition includes automatic feature design study results reduced manual overhead, and classifier fixation and feature extraction/selection are done in a single step unlike traditional methods.

A convolutional neural network (CNN) is a type of deep neural network in deep learning which is wildly used for analyzing visual images. Wu et al. [7] used CNN for face recognition in thermal imaging. In this work, the RGB-D-T dataset of thermal images is used to train CNN and learn efficient features of thermal images. CNN is more efficient in terms of recognition rate than modern methods like Histograms of Oriented Gradients (HOGs), Local Binary Patterns (LBPs), and moments invariant. They achieved significant recognition rates in different experimental setups, 98% in head rotations, 99.4% in variant expressions, and 100% in different illumination conditions.

In thermal face recognition, screening image is thermal and need to be mapped with the stored visible image in the database which is a very challenging task because thermal and visible images are of completely different modes. Deep Neural Networks (DNN) were used and Deep Perceptual Mapping (DPM) records non-linear relation between both modes [8]. CNN is capable of holding person identity information when it learns the non-linear mapping from infrared to visible light images. All visible light image mapped descriptors are affixed to form a single vector of the feature. After normalization of the created long vector from visible light images, it is matched with the thermal image vector. Thermal is also constructed in a similar way as for visible images. The dataset contains 4584 images of 82 persons, having images of both visible and thermal mode, achieved rank-1 with 83.73% identification accuracy.

3 Challenges in Thermal Face Recognition

The thermal face recognition systems measure the temperature of the face, and that temperature may vary with several conditions like environment temperature, the person with fever or other health issues, drunk person, etc. along with the person with eyeglass.

Eyeglass is a serious challenge in thermal imaging because thermal sensors cannot measure temperature beyond the glass; additionally, eyeglass is an obstacle between face and camera, which results in the sensor not being able to record important information.

The surroundings temperature, person mood, and health status also affect the object temperature which may lead to performance degradation especially in the case of MWIR and LWIR images. Alcohol dunked person's facial temperature can change results recognition efficiency losses significantly.

4 Various Datasets

4.1 ND Collection X1 [22]

The dataset collection was conducted by the University of Notre Dame (UND) to help research work of human recognition algorithms. This database contains a total of 2292 pairs of visible and IR facial frontal images captured from 82 subjects from 2002 to 2004. Merlin uncooled LWIR sensor is used for capturing thermal images and for visible images, advanced resolution sensors are used.

4.2 *Equinox Dataset [13]*

This LWIR images dataset contains 40 frames from 91persons with three sequences of each frame and is collected by Equinox Corporation NY. Both left and right lateral and frontal settings are used for external light. Images were captured when the person is speaking, smiling, frowning, and giving a surprise expression, and additionally, taken extra shots of the person with eyeglasses.

5 Conclusion

In this study, we reviewed various related thermal face recognition works from the literature. We found in literature both local and global features are used and methods based on local features give better recognition efficiency than global features-based methods. If both visible light and infrared images are available, then methods based on fusion and DNN are used for taking advantages of both imaging techniques. Medium wave infrared face recognition performs well in a dark environment. Texture- and appearance-based methods also give a significant performance in Medium wave infrared face recognition. Low wave infrared imagery is strong in the change in illumination settings and has less intra-class variations. When thermal images are available for training, cross spectral matching can be used for recognition in dark and outdoor settings. A recognition system gives better performance if both visible and thermal images are used for learning. For the future, a few interesting approaches are Common Representation Learning, Canonical Correlation Analysis, and Deep Neural Networks for face recognition systems.

References

1. Ghiass, R.S., et al.: Infrared face recognition: a literature review. In: IJCNN. IEEE (2013)
2. Socolinsky, D.A., Selinger, A.: comparative analysis of face reco.performance with visible and thermal infrared imagery. In: Object Recognition Supported by User Interaction for Service Robots, vol. 4. IEEE (2002)
3. Méndez, H.: Face recognition with LWIR imagery using local binary patterns. In: International Conference on Biometrics. Heidelberg (2009)
4. Ahonen, T., Hadid, A., Pietikäinen, M.: Face recognition with local binary patterns. In: European conference on Computer Vision. Heidelberg (2004)
5. Bebis, G. et al.: Face recognition by fusing thermal infrared and visible imagery. Image aVis. Comput. **24**(7), 727–742 (2006)
6. Singh, R., Vatsa, M., Noore, A.: Integrated multilevel image fusion and match score fusion of visible and infrared face images for robust face recognition. Pattern Recognit. **41**(3), 880–893 (2008)
7. Wu, Z., Peng, M., Chen, T.: Thermal face recognition using convolutional neural network.In: 2016-ICOIP. IEEE (2016)
8. Sarfraz, et al.: Deep perceptual mapping for thermal to visible face recognition. arXiv:1507.02879 (2015)

9. Kakkirala, K.R., Chalamala, S.R., Jami, S.K.: Thermal infrared face recognition: a review. In: 2017 UKSim-AMSS 19th International Conference on Computer Modelling & Simulation (UKSim). IEEE (2017)
10. Tan, X., Triggs, B.: Enhanced local texture feature sets for face recognition under difficult lighting conditions. IEEE Trans. Image Process. **19**(6), 1635–1650 (2010)
11. Marcel, S., Rodriguez, Y., Heusch, G.: On the recent use of local binary patterns for face authentication (No. ARTICLE) (2007)
12. Ghatge, S.S., Dixit, V.V.: Robust face recognition under difficult lighting conditions. IJTEL **1**(1) (2012)
13. Selinger, A., Socolinsky, D.A.: Appearance-based facial recognition using visible and thermal imagery: a comparative study. EQUINOX NY (2006)
14. Hermosilla, G. et al.: A comparative study of thermal face recognition methods in unconstrained environments. Pattern Recognit. **45**(7), 2445–2459 (2012)
15. Klare, B.F., Jain, A.K.: Heterogeneous face recognition using kernel prototype similarities. IEEE Trans. Pattern Anal. Mach. Intell. **35**(6), 1410–1422 (2012)
16. Bourlai, T. et al.: A study on using mid-wave infrared images for face recognition. In: Sensing Technologies for Global Health, Military Medicine, Disaster Response, and Environmental Monitoring II; and Biometric Technology for Human Identification IX, vol. 8371. International Society for Optics and Photonics (2012)
17. Lowe, D.G.: Distinctive image features from scale-invariant keypoints. Int. J. Comput. Vis. **60**(2), 91–110 (2004)
18. Kalamkar, K.D., Mohod, P.S.: Feature extraction of heterogeneous face images using SIFT and MLBP algorithm. In: ICCSP-2015. IEEE (2015)
19. Bi, Y. et al.: Multi-feature fusion for thermal face recognition. Infrared Phys. Technol. **77**, 366–374 (2016)
20. Singh, R., Vatsa, M., Noore, A.: Hierarchical fusion of multi-spectral face images for improved recognition performance. Inf. Fusion **9**(2), 200–210 (2008)
21. Chellappa, R. et al.: A feature based approach to face recognition. In: Conference on Computer Vision and Pattern Recognition. IEEE (1992)
22. https://cvrl.nd.edu/projects/data/#nd-collection-x1. Last accessed 28 Oct 2020

Multi-Objective Sparrow Search Algorithm-Based Clustering and Routing in Wireless Sensor Networks

Panimalar Kathiroli and S. Kanmani

1 Introduction

In recent days, Wireless Sensor Network (WSN) has become popular due to its extensive applications in real-time scenarios [1]. WSN comprises a collection of nodes that are useful because of its easy deployment, and self-identification with other sensors creating a dynamic system. Since the nodes are energy-constrained, the clustering and routing process are well thought out as an effective way for achieving energy efficiency and maximizing network lifetime. A WSN contains a powerful sink for deploying routes with specific transmission configurations [2]. Clustering groups nodes into a cluster and from each cluster, Cluster Head (CH) is elected with an intention of maximizing energy and network lifetime. CH accumulates data from Cluster Members (CM), eliminates the duplicate data, computes it, and limits the energy degradation of a system. CH consumes maximum energy while data reception, data aggregation, and data forwarding to BS result in an additional energy drain. The CH uses data fusion to discriminate data redundancy in the individual cluster. The capability of data processing in CH is relational to the energy. That is, if CH fails due to energy shortage, the devices dependent on CH also get affected. The data gets transferred from one CH to another either directly or through BS. In WSN, Routing is one of the major challenging issues. Sample architecture of cluster-based routing is shown in Fig. 1. Multi-hop routing [4] finds the shortest path for inter-cluster data transmission. When a path miscarries, data can be sent by another path. This boosts better data transmission, minimizes excess load for path discovery, and increases the

P. Kathiroli (✉)
Department of Computer Science and Engineering, Pondicherry Engineering College, Puducherry, India
e-mail: panimalar2688@pec.edu

S. Kanmani
Department of Information Technology, Pondicherry Engineering College, Puducherry, India

B. Agarwal et al. (eds.), *Proceedings of International Conference on Intelligent Cyber-Physical Systems*, Algorithms for Intelligent Systems,
https://doi.org/10.1007/978-981-16-7136-4_29

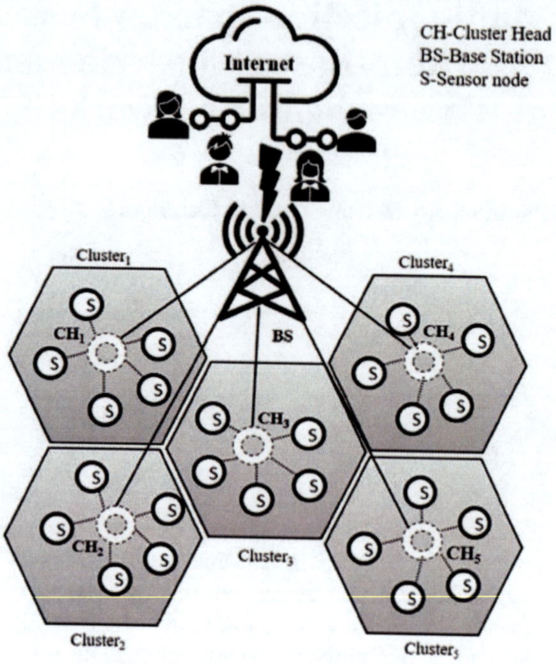

Fig. 1 Cluster-based routing process in WSN

throughput of the network. To instigate better communication among BS and nodes, a route discovery model must be initialized for choosing the next hop in the network which depends upon the topological infrastructure and QoS metrics of neighbour nodes. In case of any error, the source node sends the data to BS over different best-case multipaths within a limited time under the round-robin path selection technique that shares the traffic load. Hence, the probable routing protocol might offer the best throughput at the time of reducing packet loss with limited delay and improving the network lifetime.

Several clustering and routing algorithms have been reputed that are dissimilar to wired and wireless networks. Low-Energy Adaptive Clustering Hierarchy (LEACH) [5] is a notable clustering model where one of the nodes among the others in the network is certain as a CH for the cluster. The CHs distant from the sink exhaust their energy faster than nearer ones. Presuming if a node with minimum energy is chosen as a CH, then it may die hastily. Also, this algorithm progressively changes the CHs for the purpose of good load balancing, since it does not use any data relaying methods. Inaptly, LEACH is not appropriate for large area networks as a lot of shortcomings exist. For solving the major challenge of the death of low-energy CHs, various LEACH algorithms have been offered for selecting optimal CHs and cluster formation. Energy-efficient CH is selected using the energy and distance parameters. But the algorithm agonizes from the connectivity issue of the selected CHs. Many relevant heuristics works have also been suggested for clustering-based routing in WSNs. Prominently, these traditional routing protocols are not suitable in a direct manner as sensor networks varied from ad hoc networks, weightless routing protocols as well as adaptive communication patterns [3]. Clustering and routing are considered as NP-hard problems, and metaheuristic algorithms are found useful to determine optimal solutions. The vital aim of this model is curtailing the load of nodes by allocating massive concern to BS and to present the weightless clustering with multipath routing protocol by assuming the multiple QoS measures. Particle Swarm Optimization-Based Selection (PSOBS) [10] is applicable to identify closer optimal rendezvous points for effective management of network resources. Also, a weight measure is determined for all nodes depending on the packets received from other sensors. Shuffled Complex Evolution with Particle Swarm Optimization (SCE-PSO) [11] clustering model is an evolutionary algorithm that assumes cluster distance. Each particle is assessed by a fitness function. Likewise, in Karthick and Palanisamy [12], CH election is done with the help of GA and KH models. GA is aided for CH selection using residual energy, inter- and intra-cluster distance. KHA enhances the network lifetime, routing to BS. The nodes from the cluster assist in enhancing the energy distribution and exploring the WSN protocol performance. In this paper to resolve these problems, a multi-objective sparrow search algorithm for clustering and routing is proposed. The Sparrow Search Algorithm (SSA) [16] is based on the interesting foraging and anti-predation behaviours of sparrows. The nature of sparrows helps to effectively select the CHs and optimal routing paths to BS. The proposed algorithm derives a fitness function for the clustering and routing process based on several measures. SSA has the capability to select CHs and an optimal set of paths to perform inter-cluster communication. An exhaustive simulation analysis is carried out for ensuring the goodness of SSA under varying node counts. The remaining portions of the paper are organized as follows: Sect. 2 discusses the work related to the proposed method. Section 3 discusses the energy and network model; Sect. 4 introduces the presented model and Sect. 5 validates the proposed model. Lastly, Sect. 6 concludes the work.

2 Sparrow Search Algorithm (SSA)

The sparrows are friendly omnivorous birds with strong creativity and memory power spread in every part of the world [16]. Moreover, sparrows utilize behavioural strategy flexibly switching between producer and scrounger. The producers vigorously look for the food, whereas the scroungers acquire food from them. The energy reserves of the folks might have an essential task when the bird selects the various foraging strategies, and the birds with small energy assets scrounge more. The birds located on the periphery of the population are more possible to be attacked by predators and repetitively try to get an improved position. The sparrow placed in the middle might travel towards their neighbours to reduce the risk of danger. It is applied for discovering the region of enriched food. Once the sparrow detects a predator, it begins to chirp and invokes an alert signal. If the alarm value is higher than the required threshold, then producers must move to safer zones for saving the life of other sparrows. Every sparrow is implied as a producer when it explores the best food, however, the ratio of producers and scroungers are the same. Sparrows with maximum power are producers. Scroungers obey producers which offer optimal food. Simultaneously, a few scroungers observe producers and contend for food to improve the predation value. Sparrows present in the border of a group fly immediately to the safer zone and protect themselves from danger, whereas the sparrows in an intermediate position of a group move in a random manner and reach the edge. Sparrows are utilized for identifying the better food in the direction of variables d that must be optimized. The producers with maximum fitness measures accomplish the best food in the searching process. Also, the producers are in charge to explore food and assist the action of the whole population. Thus, the producers can find food in wider ranges when compared to scroungers. Based on Eq. (1) for existing iteration t, $j = 1, 2, \ldots, d$; $P^t_{i,j}$ is the rate of ith sparrow, I_{max} is maximum iterations, arbitrary $\alpha \in (0, 1]$. An alarm value $AV \in [0, 1]$ and safety threshold $ST \in [0.5, 1.0]$. Q is a random value and L is a matrix of $1 \times d$ for all elements within 1. The process involved in SSA is shown in Fig. 4. When$AV < ST$, no predators exist, and the producer gets into wider search mode. When$AV \geq ST$, then a few sparrows have found the predator, and it is essential to safeguard themselves by flying to safer regions.

$$P^{(t+1)}_{i,j} = \{P^t_{i,j} \cdot \exp\left(\frac{-i}{\alpha \cdot I_{\max}}\right) \; if\ AV < ST \; P^t_{i,j} + Q \cdot L \; if\ AV \geq ST \tag{1}$$

Only some scroungers follow the producers as in Eq. (2). If the producers find the best food, then they leave the place to compete for food. If the competition is successful, then they can obtain the food of a producer, else Eq. (3) is implemented. The location upgrading expression for a scrounger is given using P_{Best} which refers to the best place applied by a producer, P_{Worst} implies the recent global worst position, A showcases a matrix of $1 \times d$ for an element within –1 or 1, and $A^+ = A^T(AA^T)^{-1}$. If $i > n/2$, it recommends that ith scrounger with ineffective fitness is more starving. As a result, the sparrows which are away from risk would have additional lifetime.

$$P^{(t+1)}_{i,j} = \left\{Q \cdot \exp\left(\frac{P^t_{Worst} - P^{t_i}_{i,j}}{J^2}\right)\right. \; if\ i > n/2 \; P^{(t+1)}_{Best} + \left|P^t_{j,j} - P^{(t+1)}_{Best}\right| \cdot A^+ \cdot L \; otherwise \tag{2}$$

The positions of sparrows are produced randomly where β is step size control parameter of random value (mean value=0 and variance = 1), $K \in [-1, 1]$ is a random measure, f_i is fitness value, f_{gb} and f_w are global best and worst fitness, and ε is a minimum constant which avoid zero-division-error. If $f_i > f_{gb}$ validates that the sparrow is at the border of a group. Here, $f_j = f_g$ denotes that the sparrows are in the middle of a population that is aware of a risk and migrate closer to the edge. K is a direction where a sparrow moves and a step size control coefficient (Fig. 2).

Fig. 2 Flowchart of sparrow search algorithm

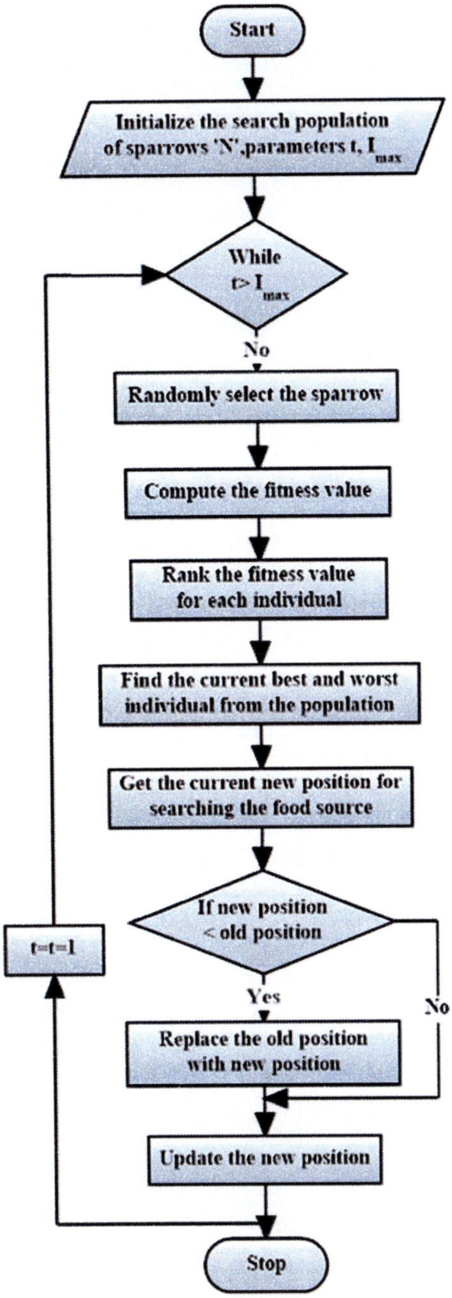

$$P_{i,j}^{(t+1)} = \{P_{GBest}^t + \beta \cdot |P_{i,j}^t - P_{GBest}^t| \ if \ f_i > f_{gb} P_{i,j}^t + K \cdot \left(\frac{|P_{i,j}^t - P_{Worst}^t|}{(f_j - f_w) + \varepsilon} \right) \ if \ f_i = f_{gb} \quad (3)$$

3 Network Model and Energy Model

Assume a WSN with arbitrarily placed nodes each with CHs, and it becomes static once the deployment is completed. A node could be allocated to any type of CH only when it is inside the transmission radius of a node. Hence, it is composed of a few pre-defined CHs where specific sensor nodes might be declared. Similar to LEACH, the data collection task is classified into rounds. For all rounds, nodes should accumulate the local data and transmit it to concerned CH. Once the data is received, CHs collect the data and remove the duplicate information as well as unwanted data. Finally, the error-free data is forwarded to BS through CH that is a next hop relay node. Among neighbouring nodes, they switch off the radios for power consumption. The communication is carried out using a wireless link. Here, the wireless link has been deployed among 2 nodes when it is present inside the transmission radius. Recent developing models are highly intended in TDMA to offer MAC layer communication. CHs apply the CSMA/CA MAC protocol to interact with BS. There are various network duration descriptions provided, such as the time until first node expires, time until last node expires and the time until a target percentage of nodes die. Additionally, network lifetime is assumed to be the period until the whole area is concealed. The radio method for energy applied in this work is as in the literature [16].

An energy model represented schematically in Fig. 3 is measured in the physical and MAC layers of WSN based on the model of energy depletion which is directly proportional to the distance. '*d*' is the distance between sender and receiver. The consumption of energy by the CH and sensor node impacts the network lifetime. Here, the free space as well as multipath fading channels was employed based on distance from a sender and receiver. If distance '*d*' is minimum than a threshold

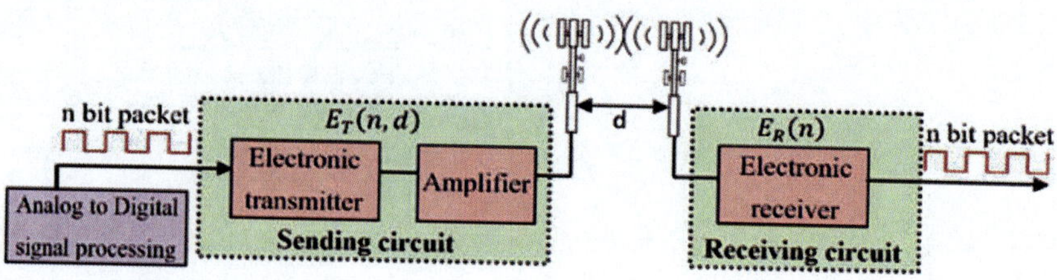

Fig. 3 Energy model

distance value d_0, then it uses the free space (fs) model else it uses the multipath (mp) model. If the distance amid the sender and receiver is less than d_0, then free space is used. The energy consumption by an amplifier in free space is $E_{fs}*d^2$. If the distance amid the sender and receiver is larger than d_0, then multipath space communication is used, and energy consumption in multipath channel $E_{mp} * d^4$ is based on the distance among a sender, receiver, and the reasonable bit-error rate. For clear understanding as mentioned in the equation E_{elec}, E_{fs} and E_{mp} are the amount of energy needed by electronic circuits, energy required by an amplifier in free space and energy required by multipath correspondingly. The energy consumed by electronic circuit E_{elec} is based on various aspects like digital coding, modulation, filtering, and signal distribution. If the sender sends an n-bit message, then the total energy essential is given as in Eq. (4). Here, $E_{T-elec}(n)$ is the energy needed to run the sender circuit and to send the n bit message. It can also be perceived that the energy consumption of the sender circuit depends on message length and not on the distance. $E_{T-amp}(n, d)$ is the energy to increase the signal to reach the receiver circuit and it includes both n, d.

$$E_T(n, d) = E_{T-elec}(n) + E_{T-amp}(n, d) \tag{4}$$

Besides, the energy required by a radio for sending n-bit message across a distance d meters is given in Eq. (5). The threshold $d_0 = \sqrt{\frac{E_{fs}}{E_{mp}}}$ is the crossover distance.

$$E_T(n, d) = \{n * E_{elec} + n * E_{fs} * d^2, d < d_0 n * E_{elec} + n * E_{mp} * d^4, \ d \geq d_0 \tag{5}$$

Since the receiving circuit devours more energy, the energy required by the receiver to effectively receive the message is expressed in Eq. (6), where $E_{R-elec}(n)$ is energy vital to receive n-bits at the receiver side.

$$E_R(n) = E_{R-elec}(n) \tag{6}$$

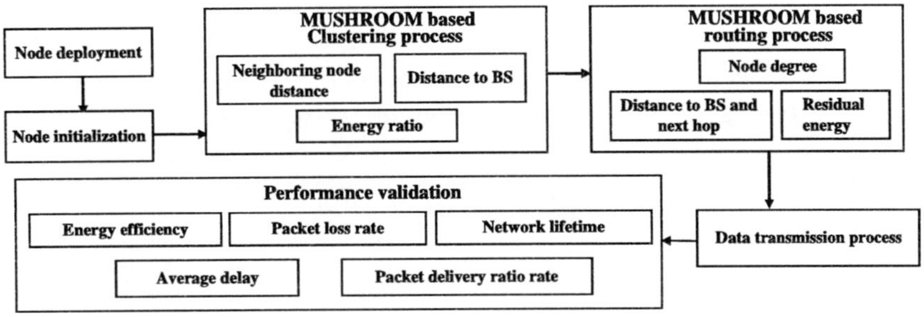

Fig. 4 Block diagram of the MUSHROOM model

Specifically, it is pointed that it is a simple method. Generally, radio wave propagation is extremely differentiable and complex for development.

4 The Proposed Multi-objective Sparrow SearcH Algorithm-Based Clustering and ROuting Model (MUSHROOM)

The proposed MUSHROOM model selects the CHs and finds optimal paths based on the fitness function derived which is based on the fascinating behaviour of sparrows to perform the clustering and routing process.

4.1 *MUSHROOM-Based Clustering Process*

The primary intention of the clustering problem is to improve the life of the network and reduce the power utilization of sensor nodes. Hence, network duration could be improvised while reducing the energy utilization of CHs. Power application of sensor nodes could be limited by diminishing the distance from nodes and concern CHs.

Maximum neighbour node distance: It is applicable to elect CHs with minimum distance between adjacent nodes. In the intra-cluster communication procedure, sensor nodes intake power for CH communication. When the neighbour node distance is reduced, then the power of intra-cluster communication is also minimized.

$$f_1 = \sum_{j=1}^{m} \frac{1}{l_j} \left(\sum_{i=1}^{l_j} dis(CH_j, s_i) \right) \tag{7}$$

Average distance to BS: It is referred to as the distance among a Cluster Head CH_j and sink, where $dis(CH_j, BS)$ is as in Eq. (8). Sink distance is highly important in selecting CHs closer to BS. Massive CHs adjacent to BS are applicable in removing the improper energy application to develop unequal clusters.

$$f_2 = \sum_{j=1}^{m} \frac{1}{m}(dis(CH_j, BS)) \tag{8}$$

Energy ratio: It is a proportion of power utilized by CHs to RE of CHs as in Eq. (9). When a CH intakes low energy on events like sensing, processing, and communicating then its RE becomes high. As the energy ratio is low, the CH election becomes more possible.

$$f_3 = \sum_{j=1}^{m} \frac{E_c(CH_j)}{E_R(CH_j)} \tag{9}$$

In the MUSHROOM method, it is required to reduce the linear integration of the objective functions.

$$Potential\,energy = \alpha_1 f_1 + \alpha_2 f_2 + \alpha_3 f_3$$

where $0 < f_1, f_2, f_3 < 1; \alpha_1 + \alpha_2 + \alpha_3 = 1, \alpha_2 \geq (\alpha_1 + \alpha_3)$

The main aim is intended to reduce the potential energy purpose. When the potential energy purpose is minimum, the proposed model improves its efficiency and refers that it has shown superior performance in the CH election. Once the optimal CH has been elected, non-CH sensors are allocated to CHs according to the obtained cost function.

4.2 MUSHROOM-Based Routing Process

In the multi-hop routing method, the fitness score of sparrows implies a complete path from CHs to BS. If $M_{1i} = \left[Y_{i,1}(t), Y_{i,2}(t), \ldots, Y_{i,D}(t)\right]$ is ith sparrow from a *Pop,* the sparrow count is the same as the number of CHs, in which $Y_{i,d}(t)$ chooses upcoming hop CH_ID as communication of a CH, $1 \leq i \leq N_M, 1 \leq d \leq D$. The producer is used to elect the upcoming hop CH inside a communication range. The deployed model is autonomous for the initial generation of sparrows.

Residual energy of Next Hop node: CH to upcoming hop node is based on RE of the next hop node. A CH node can select the future hop CH over feasible hop nodes where it is composed of maximum RE.

$$g_1 = \sum_{j=l}^{m} NextHop(E_R(CHj)) \tag{10}$$

Next Hop node and Sink distance: For all CHs to next hop, it is relied on the distance to the next hop node and distance from the CH to sink. A CH node can select the CH over feasible next hop nodes where it is embedded with minimum distance between alternate feasible nodes and from the corresponding node to BS distance.

$$g_2 = \sum_{j=l}^{m} dis(CH_j, NextHop(CH_j) + dis(Next\ hop(CH_j), BS) \tag{11}$$

Node degree: The primary intention of each CH to the next hop node is based on a node degree. The CH is declared to the upcoming hop node with a lower node degree.

$$g_3 = \sum_{j=l}^{m} Node\, degree\, of\, NextHop(CHj) \tag{12}$$

The total sum for optimization process depends on the objectives which are reliable to one another $Potential\, energy = \beta_1 g_1 + \beta_2 g_2 + \beta_3 g_3,$ where $0 < \beta_1, \beta_2, \beta_3 < 1\, and\, 0 < g_1, g_2, g_3 < 1$. Therefore, the MUSHROOM model will choose the next hop node with maximum energy. Then, the CHs forward the aggregated data from its members to BS via optimal path.

5 Experimental Results and Performance Validation

The performance evaluation has been simulated using MATLAB R2018a on an Intel (R) Core (TM) i5 processor, 2.80 GHz CPU, and 16 GB RAM running on a Windows 10 platform. The experiments are done by presuming a WSN setup with nodes with initial energy 1 J and CH positioned in 100 × 100 m^2 area with sensing range 10 m. The MUSHROOM model is equated with state-of-the-art techniques and is estimated in terms of network lifetime, energy consumption, the number of nodes that die, and packet loss (Table 1).

Figure 5 illustrates the network lifetime analysis of the MUSHROOM model over the compared methods under varying numbers of rounds. Figure 5 states that the number of alive nodes gets significantly reduced in LEACH protocol at the earlier operating rounds. Simultaneously, the number of alive nodes is gradually reduced in the KHA and Cuckoo Search with Harmony Search (CS-H). Likewise, the Optimized QoS-based Clustering with Multipath Routing Protocol (OQoS-CMRP) model is a slightly better outcome than the compared methods. At the same time, the SCE-PSO model has delayed the death of the sensor nodes and attained effective network lifetime. But the MUSHROOM model has depicted maximum network stability and increased the network lifetime to a significant extent. Better network stability means

Table 1 Simulation parameters

Parameters	Value
Number of nodes	1000
Packet size	4000 bits
E_{elec}	50 nJ/bit
E_{fs}	10 pJ/bit/m^2
E_{mp}	0.0013 pJ/bit/m^4
d_0	85.55

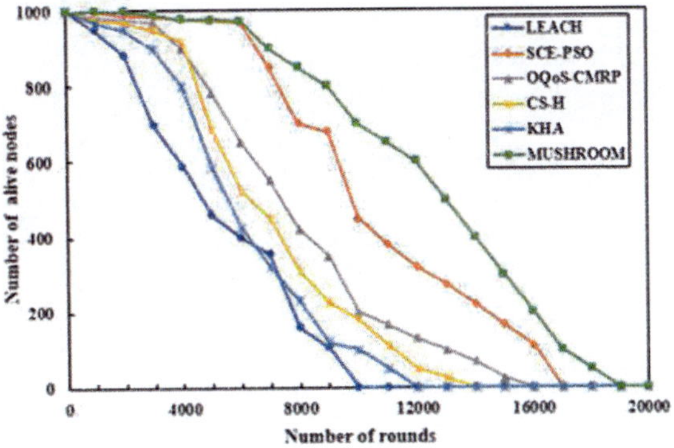

Fig. 5 Number of alive node analysis of MUSHROOM algorithm

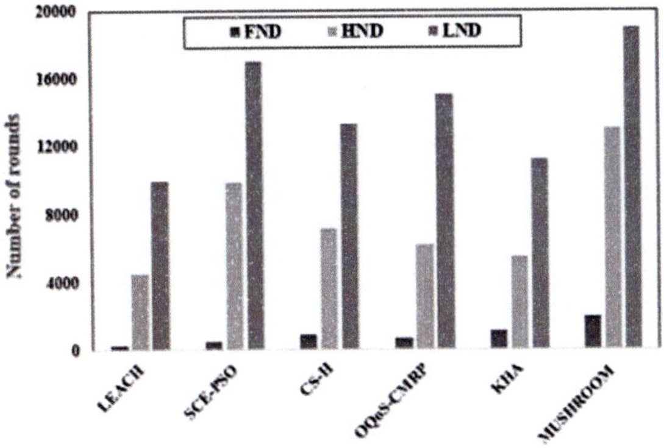

Fig. 6 Network lifetime analysis of MUSHROOM algorithm

the better network lifetime defined using Fig. 6 in terms of FND, LEACH, SCE-PSO, OQoS-CMRP, CS-H, and KHA algorithms resulted in the FND of 254, 485, 685, 898, and 1114 rounds. However, MUSHROOM has delayed the FND to 1985 rounds. Also, determining the network lifetime in terms of HND, LEACH, SCE-PSO, OQoS-CMRP, CS-H, and KHA resulted in the HND of 4554, 9840, 6253, 7115, and 5496 rounds. However, MUSHROOM has delayed the HND to 13,049 rounds. Likewise, for LND, the LEACH, SCE-PSO, OQoS-CMRP, CS-H, and KHA algorithms resulted in the LND of 9955, 17,021, 15,049, 13,296, and 11,192 rounds. However, the MUSHROOM algorithm has delayed the LND to 18,974 rounds.

Figure 7 illustrates the energy efficiency investigation of the MUSHROOM for RE under varying iteration. The LEACH is the worst performer exhibiting the maximum amount of energy consumption. The KHA and CS-H have surpassed LEACH and attained slightly higher and closer average RE. Simultaneously, the OQoS-CMRP model has a better average RE. Besides, the SCE-PSO has shown competitive RE compared to all other methods except the MUSHROOM model. At last, the MUSHROOM model has demonstrated effective performance by the attainment of

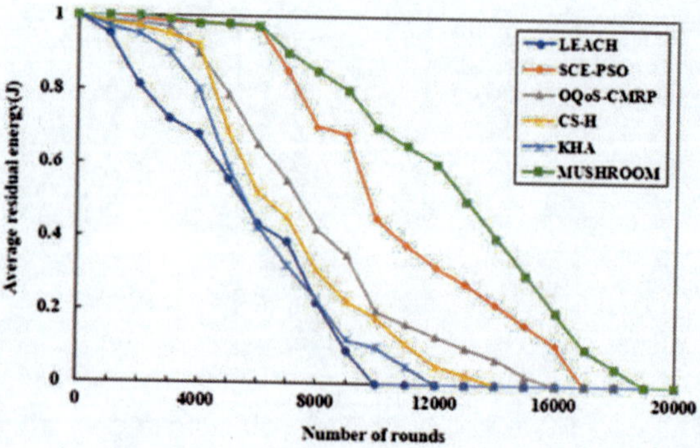

Fig. 7 Average residual energy analysis of MUSHROOM algorithm

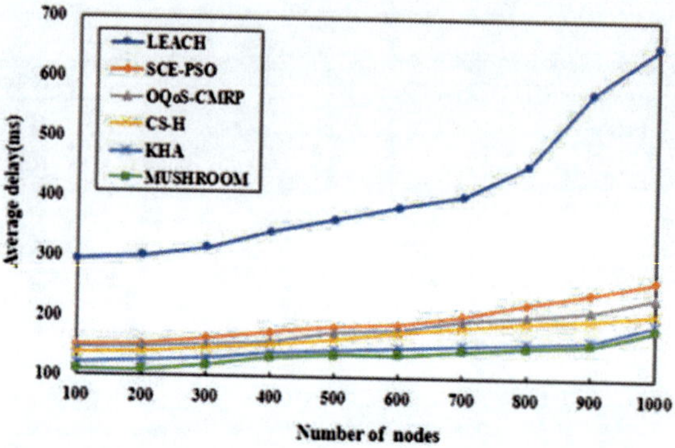

Fig. 8 Average delay analysis of MUSHROOM algorithm

maximum RE. It indicates that the MUSHROOM model has consumed less amount of energy and it gets reduced with an enhancement in a number of rounds.

Figure 8 depicts the average delay analysis of the MUSHROOM model under varying node counts. The LEACH protocol incurs high delay due to single-hop communication and arbitrary selection of CHs. The SCE-PSO algorithm has shown significantly lower average delay compared to LEACH, but not than other algorithms. In line with, the OQoS-CMRP model has demonstrated lower average delay over the compared algorithms. But, the CS-H and KHA algorithms have required low and near-identical delay under varying nodes. However, the MUSHROOM model has exhibited superior performance by minimum delay under varying node counts.

Figure 9 depicts the packet loss rate of the MUSHROOM model under varying node counts. This portrays that the LEACH protocol attains high packet loss over compared models. The SCE-PSO algorithm has revealed considerably low packet loss compared to LEACH. OQoS-CMRP, CS-H, and KHA algorithms had lower packet loss compared to LEACH and SCE-PSO algorithms. But the MUSHROOM model has proved an effective outcome by attaining a lower packet loss rate beneath all the varying node counts. Figure 10 investigates the PDR analysis under varying rounds. Here, LEACH is the ineffective technique that has achieved minimum PDR. Besides, OQoS-CMRP and SCE-PSO have outpaced LEACH and attained better

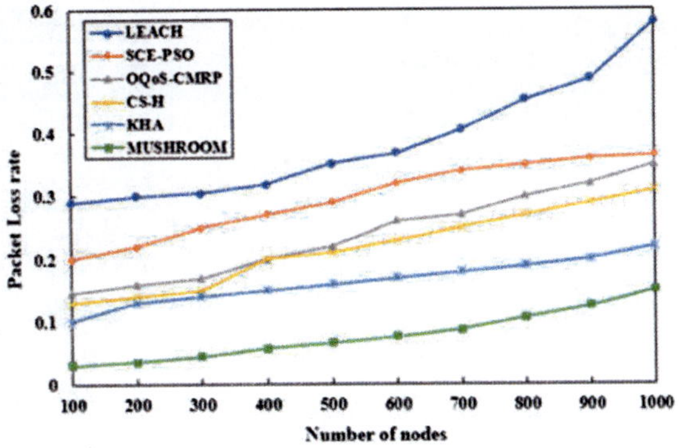

Fig. 9 Packet loss rate analysis of MUSHROOM model

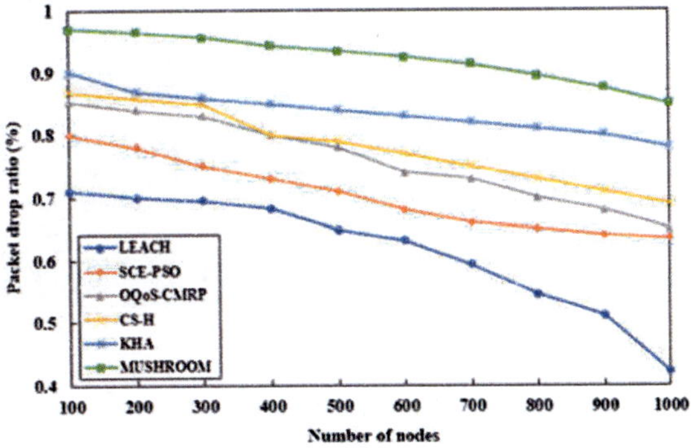

Fig. 10 Packet drop ratio analysis of MUSHROOM algorithm

PDR. Concurrently, CS-H attained high PDR over previous methods. Finally, the MUSHROOM model has obtained superior results by the attainment of maximum PDR. It indicates that the MUSHROOM model has successfully received a maximum number of packets compared to other algorithms.

To further verify the effective performance of the MUSHROOM model, the number of packets reached at the CHs and BS is shown in Figs. 11 and 12. The MUSHROOM model has achieved a maximum number of packets attained at the CHs and BS over the other algorithms. Particularly, the MUSHROOM model has attained a maximum of 2050 packets at CHs and 18,454 packets at BS.

6 Conclusion

This paper has developed a new clustering and routing model called MUSHROOM based on the fascinating behaviour of sparrows. The cluster head is chosen, and

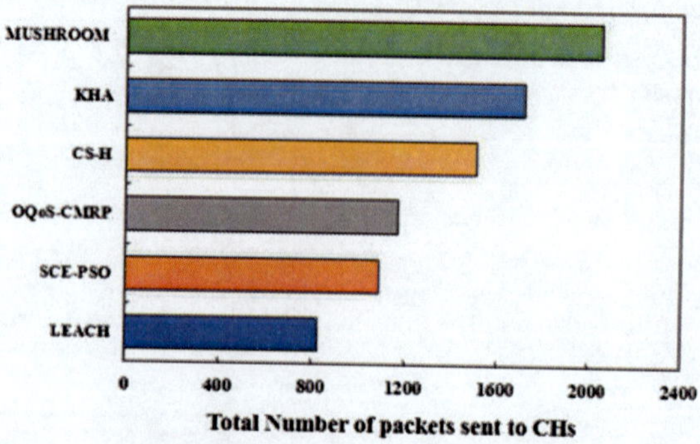

Fig. 11 Number of packets sent to CHs by MUSHROOM algorithm

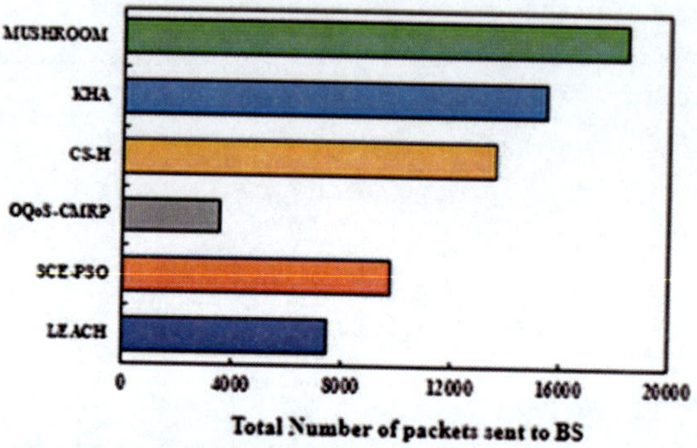

Fig. 12 Number of packets sent to BS by MUSHROOM algorithm

optimal paths are discovered based on the derived fitness function. A detailed simulation study ensures the proficient performance of MUSHROOM and the experimental outcome outpaced the compared approaches in relation to energy efficiency, network lifetime, delay, packet loss rate, and packet delivery ratio. In future, the MUSHROOM model performance can be further improved using data aggregation methods at CHs.

References

1. Akyildiz, I.F., Su, W., Sankarasubramaniam, Y., Cayirci, E.: Wireless sensor networks—a survey. IEEE Commun. Mag. **40**(8), 102–114 (2002)
2. Yetgin, H., Cheung, K.T.K., El-Hajjar, M., Hanzo, L.H.: A survey of network lifetime maximization techniques in wireless sensor networks. IEEE Commun. Surv. Tutor. **19**(2), 828–854 (2017)
3. Xiang, W., Wang, N., Zhou, Y.: An energy-efficient routing algorithm for software-defined wireless sensor networks. IEEE Sens. J. **16**(20), 7393–7400 (2016)
4. Toldan, P., Kumar, A.A.: Design issues and various routing protocols for wireless sensor networks. In: Proceedings of the National Conference on New Horizons in IT, pp. 65–67 (2013). ISBN: 978-93-82338-79-6

5. Heinelman.W.R., Chandrakasan A., Balakrishnan.H.: Energy-efficient communication protocol for wireless microsensor networks. In: Proceedings of the 33rd Annual Hawaii International Conference on System Sciences, USA, pp. 4–7 (2002)
6. Tripathi, M., Gaur, M.S., Laxmi, V., Battula, R.B.: Energy efficient LEACH-C protocol for wireless sensor network. In: 3rd International Conference on Computational Intelligence and Information Technology (CIIT 2013). IEEE, pp. 402–405 (2013)
7. Ihsan, A., Saghar, K., Fatima, T.: Analysis of LEACH protocol(s) using formal verification. In: 2015 12th International Bhurban Conference on Applied Sciences and Technology (IBCAST). IEEE, pp. 254–262 (2015)
8. Hong, J., Kook, J., Lee, S., Kwon, D., Yi, S.: T-LEACH: The method of threshold-based cluster head replacement for wireless sensor networks. Inf. Syst. Front. **11**, 513 (2008)
9. Amirthalingam, K.A.: Improved LEACH: a modified LEACH for wireless sensor network. In: 2016 IEEE International Conference on Advances in Computer Applications (ICACA). IEEE, pp. 255–258 (2016)
10. Tabibi, S., Ghaffari, A.: Energy-efficient routing mechanism for mobile sink in wireless sensor networks using particle swarm optimization algorithm. Wirel. Pers. Commun. **104**, 199–216 (2019)
11. Edla, D.R., Kongara, M.C., Cheruku, R.: SCE-PSO based clustering approach for load balancing of gateways in wireless sensor networks. Wireless Netw. **25**(3), 1067–1081 (2019)
12. Karthick, P.T., Palanisamy, C.: Optimized cluster head selection using krill herd algorithm for wireless sensor network. Automatika **60**(3), 340–348 (2019)
13. Dattatraya, K.N., Rao, K.R.: Hybrid based cluster head selection for maximizing network lifetime and energy efficiency in WSN. J. King Saud Univ. Comput. Inf. Sci. (2019)
14. Jadhav, A.R., Shankar, T.: Whale optimization-based energy-efficient cluster head selection algorithm for wireless sensor networks. arXiv:1711.09389 (2017)
15. Chandirasekaran, D., Jayabarathi, T.: Cat swarm algorithm in wireless sensor networks for optimized cluster head selection: a real time approach. Springer J. Clust. Comput. **22**, 11351–11361 (2019)
16. Xue, J., Shen, B.: A novel swarm intelligence optimization approach: sparrow search algorithm. Syst. Sci. Control. Eng. **8**(1), 22–34 (2020)
17. Deepa, O., Suguna, J.: An optimized QoS-based clustering with multipath routing protocol for wireless sensor networks. J. King Saud Univ. Comput. Inf. Sci. (2017)
18. Wang, G.G., Gandomi, A.H., Zhao, X., et al.: Hybridizing harmony search algorithm with cuckoo search for global numerical optimization. Soft Comput. **20**, 273–285 (2016)

Author Index

B. Agarwal et al. (eds.), *Proceedings of International Conference on Intelligent Cyber-Physical Systems*, Algorithms for Intelligent Systems,
https://doi.org/10.1007/978-981-16-7136-4